Metabolism of Alimentary Compounds by the Intestinal Microbiota and Health

François Blachier

Metabolism of Alimentary Compounds by the Intestinal Microbiota and Health

 Springer

François Blachier
UMR PNCA, Nutrition Physiology
and Alimentary Behavior
Université Paris-Saclay,
AgroParisTech, INRAE
Paris, France

ISBN 978-3-031-26324-8   ISBN 978-3-031-26322-4   (eBook)
https://doi.org/10.1007/978-3-031-26322-4

This Springer imprint is published by the registered company Springer Nature Switzerland AG
The registered company address is: Gewerbestrasse 11, 6330 Cham, Switzerland

Preface

Gut microbiota is one of the recent topics that has greatly stimulated the scientific and medical community interest and activity in the last decade. Indeed, by entering the term Gut Microbiota in the PubMed database (that gives the articles dedicated to this subject in the medical and biological areas), more than 60 000 publications are listed. While in 2010, 467 articles were dedicated to that subject, in 2021, more than 12 000 articles related to gut microbiota were written.

In an effort of integration, what can be learned from this impressive amount of data gathering experimental, clinical, and epidemiological works, and more particularly, what do the studies revealed in terms of the ways the compounds present in our alimentation have an impact on our microbiota metabolic activity? What consequences of such metabolic activity for communication between microbes and for lodging host metabolism and physiological functions? How this bacterial activity modifies the risk of specific pathologies, either positively or negatively, notably in the large intestine, but also in the peripheral organs? The answers to these questions are clearly of major importance if we want to understand better how utilization of alimentary compounds by microbes influences our health and well-being.

The aim of this book is then to give a state of the art regarding these aspects and to indicate some emerging practical implications deduced from these data. Given the intrinsic complexity of the subject, that implicates numerous fields of research including nutrition, microbiology, metabolism, physiology, toxicology, and pharmacology, the choice has been taken to give numerous typical examples that are intended to illustrate how alimentary compounds influence the complex metabolic relationships between microbes and between microbes and lodging host tissues. More precisely, the present book, after a recapitulation of the different elements of the gut ecosystem, focuses on the effects of bacterial metabolites derived from components present in the food on the host's tissues. Firstly, the effects of the bacterial metabolites on intestinal mucosa and epithelium metabolism and functions are examined. Secondly, the effects of bacterial metabolites on peripheral tissues, after absorption in blood, and for some of these metabolites, after modification by the host cells, are reviewed. The ways by which this diet-dependent metabolic crosstalk between our microbiota and ourselves have an impact on the functioning of our body in different situations are then developed, together with new perspectives for future research and applications.

Paris, France   François Blachier

Acknowledgments

I wish to express all my gratitude to my mentors in France, Canada, Belgium, and in the United Kingdom; to my talented colleagues and students of my laboratory that give me the honor to work with me, discuss Science, and sometimes contradict me. I wish to also thank my highly beloved wife Anne for her help in the drawing of figures and for her constant patience; my 4 precious children Martin, Louise, Lucie, and Clémence; and my adored grandchildren for their love and encouragement.

Contents

About the Author

François Blachier worked as Research Director at the National Institute for Agriculture, Alimentation and Environment (INRAe). He was deputy director of the Nutrition Physiology and Alimentary Behavior laboratory, which is one of the research units belonging to the Université Paris-Saclay/AgroParisTech/INRAe consortium. After working as research assistant at McGill University in Montréal (Canada), at Brussels Free University (Belgium), and at the Institute of Animal Physiology and Genetics in Cambridge (UK), he received his PhD in Cellular and Molecular Pharmacology from Université Pierre and Marie Curie (Paris, France), and then joined INRAe. Presently, his main research interest is related to the metabolic crosstalk between the bacteria living in our gut and ourselves and to the consequences of such bacterial metabolic activity towards dietary compounds on the intestinal metabolism and physiology in different situations including healthy state and mucosal healing after an inflammatory episode. He has published over 170 articles (source PubMed) that have been cited over 9 000 times. He has worked as an expert for numerous international scientific organizations and received numerous invitations to give lectures in Europe, USA, Canada, China, Japan, Vietnam, Korea, Brazil, Chile, Egypt, and Togo. François Blachier worked as associate editor for the journals Amino Acids; as member of the editorial board of the journals Microorganisms, Nutrition Research and Practice, Animal Nutrition and Journal of Animal Science and Biotechnology; and as academic editor for PLoS ONE.

Physiological and Metabolic Functions of the Intestinal Epithelium: From the Small to the Large Intestine

Abstract

The intestinal epithelium in both the small and the large intestine represents a selective barrier between the luminal fluid and the bloodstream. This structure is rapidly renewed from the division of stem cells in crypts followed by their migration and differentiation in specialized cells with specific functions including absorption of alimentary compounds, secretion of mucus and hormones, and immune homeostasis in a context of microbial loads. Fully mature epithelial cells are finally exfoliated in the luminal fluid allowing the maintenance of the epithelial structure. Absorptive intestinal epithelial cells are characterized by an intense energy metabolism that allows macromolecule synthesis and movement of nutrients in the small intestine, and water and electrolyte absorption in the large intestine. These cells which are polarized can receive their fuels from the luminal content and from the bloodstream. In the case of the large intestine, the colonocytes can absorb large amounts of numerous metabolites produced by the intestinal microbiota and metabolize a part of them during their transfer from the luminal fluid to the bloodstream. Metabolism in enterocytes of the small intestine and in the colonocytes is important not only for energy production but also for intracellular signaling and inter-organ relationships.

In this chapter, the morphology and organization of the intestinal mucosa in relationship with the diverse physiological and metabolic functions of the intestinal epithelium in the small and large intestines are presented. These parameters are important to be considered, as it helps in the understanding of the interplay between the microbial and host metabolic activities.

When comparing the proximal parts of the small intestine, and the distal parts of the large intestine, the structure and functions of the intestinal epithelium have some homology, but also marked differences. The metabolism of the substrates provided from the intestinal content and blood in polarized epithelial cells is described.

F. Blachier, *Metabolism of Alimentary Compounds by the Intestinal Microbiota and Health*, https://doi.org/10.1007/978-3-031-26322-4_1

Finally, the roles played by the binding of microbial components on epithelial cells are briefly presented.

1.1   The Intestine Displays Multiple Physiological Functions in Addition to Absorption of Components Originating from Food

The intestine is made up of different parts from the small to the large intestine. The small intestine comprises the duodenum, the jejunum, and the ileum, while the large intestine is composed of the caecum, colon, and finally rectum (Figs. 1.1 and 1.2). Each part of the intestine is characterized by specific physiological functions. Briefly, the small intestine is mainly responsible for nutrient, vitamin, mineral, and

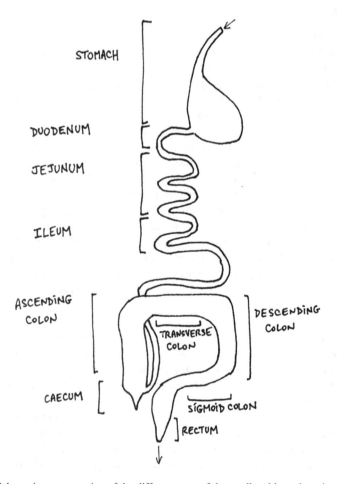

Fig. 1.1 Schematic representation of the different parts of the small and large intestine

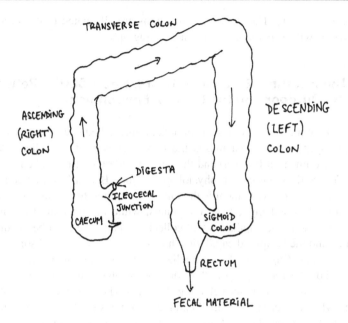

TRANSVERSE COLON

ASCENDING
(RIGHT)
COLON

DESCENDING
(LEFT)
COLON

DIGESTA

ILEOCECAL
JUNCTION

CAECUM

SIGMOID
COLON

RECTUM

FECAL MATERIAL

Fig. 1.2 The human large intestine

plant micronutrient absorption. In the healthy state, most of the food components are digested and absorbed in the duodenum and jejunum, while the ileum absorbs notably some specific dietary compounds, like vitamin B12, magnesium, and endogenous compounds like bile salts [1–3]. The large intestine is mainly responsible for water and electrolyte absorption. In fact, the human large intestine under basal conditions absorbs approximately 2 liters per day but displays a maximal absorptive capacity of around 6 liters per day [4].

The process of absorption of the food components is accomplished by specialized cells of the small intestine epithelium (called enterocytes), while in the large intestine, the colonocytes, in addition to water and electrolyte absorption, display capacities for specific bacterial metabolite absorption.

However, the intestine is not only responsible for the absorption of dietary compounds, but also plays a major role in the overall metabolism of the body, and as an organ with important endocrine and immune functions. Indeed, specialized minor subpopulation of intestinal epithelial cells secretes a panel of enteroendocrine hormones that play multiple physiological roles including pancreatic endocrine secretion, gastrointestinal motility, and food intake [5–8]. Regarding the intestinal immune system, this system is in a situation of constant exposure to microbial loads, while playing a central role in the maintenance of immune homeostasis and barrier function [9]. As will be developed in Chap. 3, several bacterial metabolites derived from dietary compounds are active on the intestinal enteroendocrine and intestinal immune cells [10, 11]. These different functions are all related to the structure of the intestine as will be explained in the next paragraph.

1.2   The Intestinal Epithelium Represents a Border Between the Outdoor and the Internal Environment

In all parts of the intestine, there is an epithelial lining called epithelium that is made of a single layer of cells, that makes the border between the "milieu extérieur" (outdoor environment in English) and the "milieu intérieur" (internal environment) as defined in 1850 by the French physiologist Claude Bernard. In that conception, the intestinal outdoor environment is the luminal intestinal fluid which contains the intestinal microbes and myriad of compounds, while the internal environment is represented by the blood capillaries that collect and carry the absorbed hydrophilic compounds, and the lymphoid capillaries that collect and carry the absorbed hydrophobic compounds (Fig. 1.3). The epithelial layer is situated on a collagen matrix, itself situated on the *lamina propria* that contains the blood and lymphoid capillaries together with numerous immune cells (Fig. 1.3) [12]. The structure of the epithelium is different when comparing the small and the large intestine. In the small intestine, morphological analysis shows a structure of protruding villi separated by crypts which contain intestinal stem cells that are at the origin of the epithelium renewal [13]. In the large intestine, there is no villi, but instead, crypts separated by the surface epithelium (Fig. 1.4). The stem cells are situated at the bottom of the colonic crypts. Thus, the individual functional unity in the small intestine is the crypt-villus

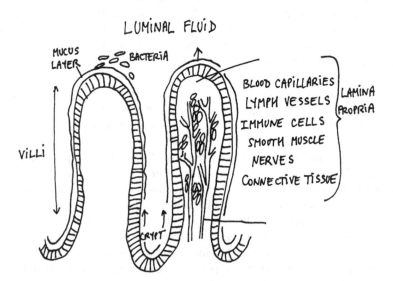

Fig. 1.3 Schematic representation of the epithelium of the small intestine, resting on the lamina propria. Lamina propria contains blood and lymph capillaries together with immune cells, nerves, smooth muscles, and connective tissue. The small intestine epithelium is covered by a mucus layer

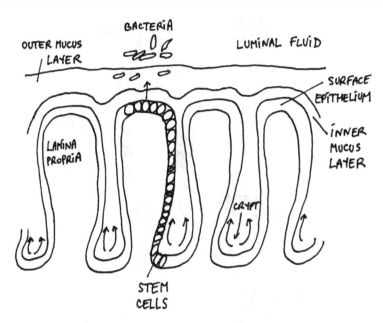

Fig. 1.4 Schematic representation of the epithelium of the large intestine. The renewal of the epithelium is made by stem cell division in the colonic crypts, followed by cell migration and differentiation, and finally exfoliation of fully mature cells in the luminal fluid. The large intestine epithelium is covered by two layers of mucus

axis, while in the large intestine, such functional unity is constituted by crypts and surface epithelium.

The Intestinal Epithelium Represents a Selective Frontier Between the Luminal Content and the Host Internal Medium
The epithelial layer in both the small and large intestine can be viewed as a selective barrier between the outdoor and internal environment, allowing the passages of numerous dietary and endogenous compounds, together with some of their metabolites, including notably the metabolites produced by the intestinal microbiota from the available substrates. This selective barrier also avoids, or at least drastically limits, the passages of microbes present in the intestinal content. This barrier function is depending notably on complex molecular structures, including the so-called tight junctions, that link neighboring cells of the epithelium and play a major role in the so-called intestinal barrier function [14, 15].

If we consider the intestinal epithelium in greater detail, the first important notion to be considered regarding this structure facing the intestinal content is related to the fact that the renewal of the intestinal epithelium is one of the fastest among the different cell systems in the organism. Indeed, in mammals in general, and in humans in particular, the intestinal epithelium in both the small and the large intestine is entirely renewed within 4–5 days [13, 16, 17]. This fast renewal is made possible by the presence of a population of stem cells which, during the

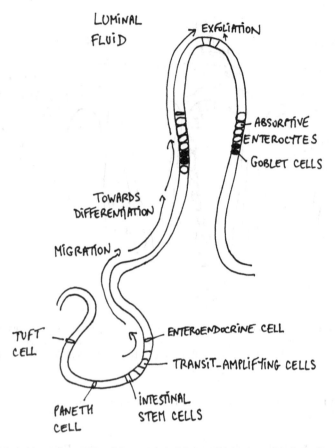

Fig. 1.5 Schematic representation of the renewal of the small intestine epithelium. The renewal of the epithelium is made by stem cell division followed by cell migration and differentiation, and finally exfoliation of fully mature cells in the luminal fluid

whole life, give rise in graded steps to several differentiated cell phenotypes that carry the physiological functions of the epithelium [18]. Fully differentiated mature epithelial cells are finally exfoliated in the intestinal content [19] (Figs. 1.4 and 1.5). This process of epithelium renewal is highly regulated by numerous external and internal factors, thus allowing the maintenance of intestinal epithelial homeostasis [17], as defined by internal stability while adapting to changing external conditions as originally defined by Claude Bernard. The understanding of the mechanisms that allow to maintain such epithelial homeostasis, as well as the events that are at the origin of a loss of such homeostasis, as well as the different consequences of such a loss in different situations, have been the object of intense research with main discoveries being presented in the next chapters of this book.

1.3   Structure and Functions of the Small Intestine Epithelium: Metabolism of Nutrients in Epithelial Cells

As said above, the small intestine epithelium is renewed within few days in mammals [20, 21]. This renewal is made possible by the division of the intestinal stem cells situated at the bottom of the crypts. The cells at the boundary of the stem cell niche undergo then differentiation into progenitor cells [22, 23]. At this step, cells undergo multiple rounds of cell division, while migrating out of the crypt toward the villus compartment. Then the cells undergo lineage into either cells of absorptive type (enterocytes and microfold cells), or into cells of the secretory type (enteroendocrine cells, goblet cells, tuft cells, and Paneth cells) (Fig. 1.5). The fully mature epithelial cells are finally exfoliated in the intestinal lumen allowing mainte-nance of epithelium homeostasis [19]. Adult intestinal epithelial stem cells have been shown to play a critical role as drivers of epithelial homeostasis and regeneration [24].

A Minor Part of Nutrients Is Metabolized in the Enterocytes of the Small Intestine During Absorption from the Intestinal Content to the Blood
The metabolism of nutrients in epithelial cells of the small intestine has been nearly exclusively studied in the absorptive enterocytes, but very little in the other epithelial cell types. The absorptive enterocytes are the most abundant cells among the different epithelial cell types [25]. Absorptive enterocytes are polarized cells with one side (the luminal side) facing the intestinal content, while the other side (the baso-lateral side) facing the blood/lymph capillaries (Fig. 1.6). At the luminal side, cells are characterized by the presence of brush-border membranes equipped with numerous transporters responsible for the entry of the different dietary compounds [26]. At the baso-lateral side, cells are also equipped with a battery of transporters allowing the export of nutrients in the circulation in the post-prandial phase and the import of nutrients from blood in the inter-prandial phase [27]. Some of the transporters are specifically present on the baso-lateral side, while others are present at both the luminal and baso-lateral sides. As indicated in Fig. 1.6, neighboring enterocytes are equipped with macromolecular structures like tight junction, adherens junction, and desmosomes that are central elements for establishing the intestinal barrier function.

Some dietary compounds like vitamins, minerals, and some plant micronutrients can directly enter the enterocytes, while macronutrients like proteins, carbohydrates, and lipids need to be digested before nutrient absorption. During absorption, enterocytes of the small intestine metabolize a minor part of the transported nutrients for their own metabolism and functions, and in an inter-organ metabolic interplay. The vast majority of the dietary and recycled endogenous compounds that have not been metabolized within the absorbing enterocytes are released in the portal vein that drain the blood from the intestine (and other visceral tissues) to the liver (Fig. 1.7). In the proximal parts of the small intestine, since the concentrations of bacteria are low, and the transit time short, the utilization of nutrients by the microbiota and the production of bacterial metabolites from the numerous available substrates are

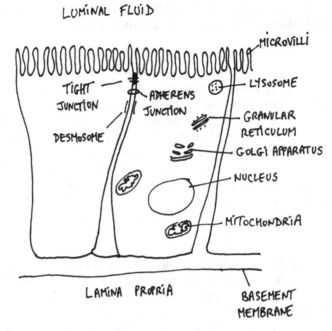

Fig. 1.6 Schematic representation of absorptive enterocytes. The enterocytes are presented with some of their intracellular organelles. Between neighboring cells, several macromolecular structures that are playing a central role for the intestinal barrier function are indicated

Fig. 1.7 Schematic representation of the transfer of compounds absorbed by the intestine to the liver

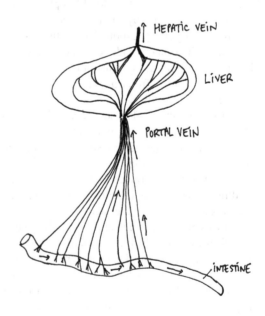

limited. The situation is different in the distal part of the small intestine (that is ileum) where the concentration of bacteria is much more important, representing in humans 10^7–10^8 bacteria/g of content [28].

As we will see in Chaps. 1 and 3, intestinal epithelial cells can on some occasions enter in competition with the bacterial cells of the intestinal microbiota for the utilization of some nutrients [29], notably in case of short supply.

The Enterocytes Use a Minor Part of Nutrients During Their Absorption for Their Metabolic and Physiological Needs, and for Inter-Organ Metabolism
A minor part of the nutrients released by digestion of the dietary compounds is thus utilized by the intestinal mucosa for their own metabolism. For instance, the rate of protein synthesis from available amino acids in the rodent small intestine represents around 15–20% of the whole-body protein synthesis [30, 31], and in humans, the intestinal mucosa contributes to a substantial utilization of amino acids for protein synthesis when compared to the whole-body protein synthesis [32]. This active protein synthesis in the intestinal mucosa is a process that requires a high amount of energy in the form of ATP [33]. Within the intestinal mucosa, the epithelium itself requires a high amount of ATP due to its rapid renewal, and to transport of electrolytes across the epithelium [34, 35]. This partly corresponds to the fact that although the gastrointestinal tract represents approximately 5% of the body weight, it is responsible for around 20% of the whole-body oxygen consumption [36, 37].

The available amino acids for protein synthesis in the small intestine originate mostly from the digestion of dietary and endogenous proteins. These endogenous proteins, which can be considered as recycled proteins, include the proteins contained in the intestinal epithelial cells exfoliated in the intestinal content in the process of the epithelial renewal, and the proteins present in the exocrine secretion, like the digestive enzymes secreted by the pancreas and the mucus secreted by the epithelial goblet cells [38–40] (Fig. 1.8) as will be explained in the next paragraphs.

Enterocytes Utilizes a Part of Available Nutrients for Energy Production Dedicated to Macromolecule Synthesis and Transport of Electrolytes
Importantly, the nutrients are not exclusively available to the absorptive enterocytes from the luminal content but can also be provided to enterocytes from blood through the baso-lateral side, notably in the fasting state [41]. In the intestinal content, the amino acids and oligopeptides released from proteins by the action of proteases and peptidases present in the exocrine pancreatic fluid enter the enterocytes. Then, the oligo-peptidases present in the enterocytes release amino acids in a step referred as terminal digestion [42]. Major part of these amino acids is then transferred to the blood capillaries, while a minor part is used for macromolecule synthesis in the enterocytes. These macromolecules are mainly proteins which are synthesized from the 20 amino acids, and purines and pyrimidines which involve the amino acids glutamine, aspartate, and glycine for their synthesis. The purines and pyrimidines are then used for nucleotide synthesis that are incorporated in DNA and RNA in enterocytes. The synthesis of these macromolecules requires for numerous metabolic steps chemical energy in the form of ATP.

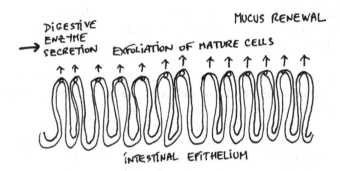

Fig. 1.8 Schematic representation of some major endogenous proteins present in the luminal fluid. Major endogenous proteins include enzymes secreted by the exocrine pancreas, proteins recovered from exfoliated mature epithelial cells, and mucus secreted by goblet cells

Regarding energy production in enterocytes, several specific amino acids, like glutamine, glutamate, and to a minor extent aspartate, are used by absorptive enterocytes which oxidize them in their mitochondria [34, 43, 44]. Indeed, after protein digestion, amino acids like glutamine and glutamate, which are present in large amounts in most dietary proteins [45], are avidly metabolized in the intestinal mucosa when being transferred from the luminal fluid to the bloodstream [46, 47]. In enterocytes, the first step of the utilization of glutamine in mitochondria releases ammonia and glutamate, glutamate being then converted to alpha-keto-glutarate, one intermediary of the tricarboxylic acid cycle that allow the synthesis of reduced cofactors used then for energy production in the mitochondria of the absorptive cells (Fig. 1.9). Glutamate is thus used as an oxidative fuel in enterocytes, but at difference with glutamine which produces large amount of ammonia, produces few ammonia [48].

Incidentally, although the concentration of glutamate is low in the peripheral blood plasma obtained from healthy volunteers after an overnight fast, glutamine concentration is the highest among the 20 amino acids in this compartment [49] This can be explained by the fact that glutamine can be synthesized from glutamate in numerous tissues, including notably perivenous liver and muscles [50, 51], and this production is followed by glutamine release in the bloodstream. Glutamine, glutamate, and aspartate are not the exclusive fuels in enterocytes, since these cells can also use, although to a minor extent, glucose, and fatty acids for energy production through mitochondrial oxidation of both substrates [35, 52]. In enterocytes, glucose is also used in the cytosolic glycolytic pathway, but in that case with a much lower yield of ATP production [53]. Glucose metabolism in the pentose phosphate pathway allows in enterocytes the synthesis of ribose-5 phosphate used for nucleotide synthesis, that are used as substrates for DNA and RNA synthesis; as well as the

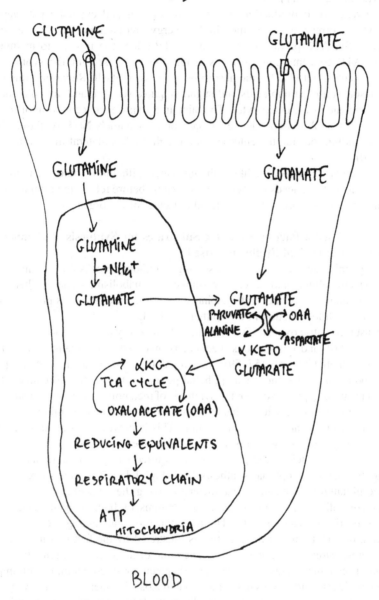

Fig. 1.9 Metabolism of glutamine and glutamate by the absorptive enterocytes for energy production. In this scheme, glutamine and glutamate are provided to the enterocytes from the luminal fluid, and oxidation of these amino acids in mitochondria allows energy supply in the form of ATP

reduced equivalent NADPH used in numerous anabolic pathways (synthesis pathways) in these cells [54].

The overall ATP production in the mitochondrial and cytosolic pathways in enterocytes corresponds to the high energy requirement of these cells. Schematically, ATP production must meet ATP utilization in cells to maintain a constant ATP concentration. ATP, as stated above, is used for the synthesis of macromolecules, but also for the absorptive process by itself. Regarding this latter point, absorption of amino acids and sugars requires sodium, and this mineral can then accumulate inside enterocytes. To maintain a constant concentration of sodium in enterocytes, an ATP-dependent enzymatic process, namely Na/K ATPase, allows to export sodium outside the enterocytes through the baso-lateral membranes [55] (Fig. 1.10).

Then, things are happening like if the intestinal epithelium would use in the first place the substrates it needs for its own functioning, before releasing vast majority of nutrients that are then available for the other tissues of the organism.

Enterocytes Used a Part of Available Substrates for Synthesis of Compounds Active on the Intestinal Epithelium Itself
Regarding amino acids, in addition to serving as building blocks for protein synthesis in the epithelium, and as energy source for anabolism, some of them, like arginine, glutamate, glycine, and cysteine, are used in enterocytes as precursors for metabolites with local action in the epithelium itself.

For instance, nitric oxide (NO) is produced from a very tiny amount of arginine in enterocytes [56], and this gaseous mediator is involved in the maintenance of the intestinal epithelial integrity [57–59], in the regulation of intestinal motility [60], and in the modulation of the intestinal epithelial permeability [61, 62]. The amounts of amino acids used as precursors for the synthesis of metabolites with biological action in cells are in general very low when compared to the amounts of amino acids used for protein synthesis and energy production. This is however not always the case. For instance, glutathione which is synthesized in enterocytes from the three amino acids glutamate, cysteine, and glycine [63], required substantial amounts of these three amino acids for optimal synthesis [64] (Fig. 1.11). The ratio of reduced to oxidized glutathione is a central parameter for fixing the intracellular redox status, and for controlling the intracellular concentrations of both oxygen-reactive and nitrogen-reactive species [65, 66]. The importance of maintaining a constant concentration of glutathione in the intestine is illustrated by the fact that inhibition of mucosal glutathione synthesis provokes alteration of intestinal functions [67].

In addition, amino acids like glutamate are precursor of compounds with important metabolic regulatory roles in the intestinal mucosa. In enterocytes, a tiny amount of glutamate is converted to N-acetylglutamate [68], and this compound then activates the enzyme carbamoyl phosphate synthetase 1, resulting in increased synthesis of citrulline mainly from the amino acid precursor arginine, and to a much lower extent from glutamine [41, 69]. Glutamine increases the synthesis of citrulline from arginine mainly by providing ammonia during conversion to glutamate [70] (Fig. 1.12). Citrulline is an amino acid that is not present in proteins but is

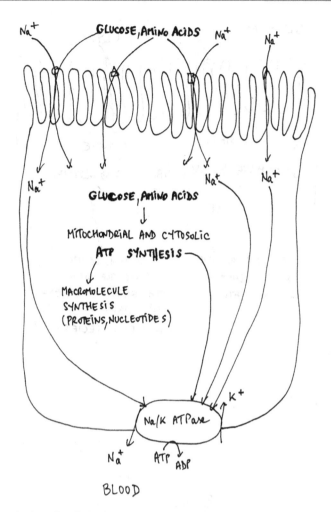

Fig. 1.10 Utilization of ATP in absorptive enterocytes for macromolecule synthesis and Na/K ATPase enzymatic activity. In this scheme, nutrients are provided from the luminal side to the enterocytes, and their absorption from the luminal fluid to the blood implies sodium accumulation in the cells. Sodium is then exported to the bloodstream by the Na/K ATPase activity. Protein and nucleotide synthesis in enterocytes requires notably amino acids and glucose as well as energy in the form of ATP synthesized in mitochondria, and to a much lower extent in the cytosol of enterocytes

a precursor of arginine in kidney [71–74], thus allowing increased synthesis of this amino acid in specific situations of increased need, and/or short supply (Fig. 1.12), thus representing an illustration of the inter-organ metabolic crosstalk (thus here between intestine and kidney) as detailed in the next paragraph.

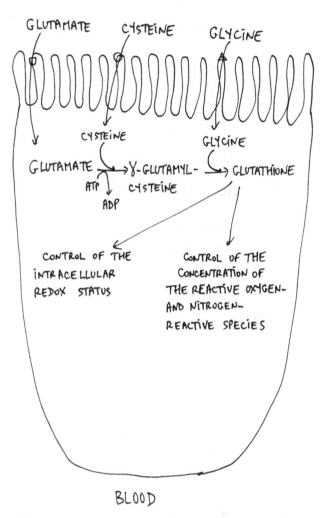

Fig. 1.11 Synthesis of glutathione in absorptive enterocytes and major associated functions. In this scheme, glutathione is synthesized in enterocytes from luminal glutamate, cysteine, and glycine. Adequate concentration of reduced glutathione in enterocytes is required for adequate intracellular redox status and control of the concentrations of reactive oxygen- and nitrogen-reactive species

Some Metabolites Produced from Nutrients in the Enterocytes Are Used for Inter-Organ Crosstalk

Other metabolites produced in the epithelium from amino acids are thus used at distance in other tissues, in the so-called inter-organ metabolism. For instance, ornithine, like citrulline, is not present in dietary proteins but can be produced together with urea in enterocytes mostly from arginine [75, 76]. This ornithine can be released in the portal vein and used in liver hepatocytes in the urea cycle [77, 78]. This cycle controls the concentration of ammonium in the blood,

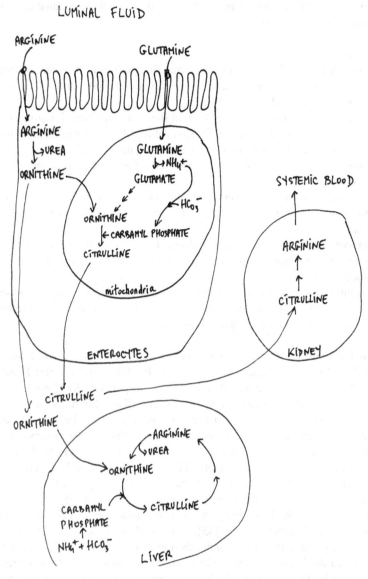

Fig. 1.12 Production of ornithine from arginine, and production of citrulline from arginine and glutamine in enterocytes. The utilization of ornithine for ammonia disposal in the liver and the utilization of citrulline for the synthesis of arginine in the kidney are presented as illustrative examples of inter-organ amino acid metabolism

originating from either the ammonia-producing pathways not only in the host tissues, including the intestine, but also from the intestinal microbiota metabolic activity, thus avoiding accumulation, and thus deleterious effects of this compound, notably on the central nervous system [79] (Fig. 1.12), as will be detailed in Sect. 5.6 devoted to the gut–brain axis.

Key Points
- The intestinal epithelium is a selective and efficient barrier between the luminal fluid, which contains bacteria and complex mixture of compounds of dietary and endogenous origin, and the bloodstream.
- The renewal of the intestinal epithelium in the small and large intestine is made every few days through stem cell division followed by cell migration along the villi, and exfoliation of fully mature cells in the luminal fluid.
- This process is highly coordinated allowing the maintenance of epithelial homeostasis.
- Stem cell division in the intestinal epithelium is followed by differentiation into the different cell phenotypes responsible for the physiological and metabolic functions.
- Absorptive enterocytes, the most abundant cells in the small intestine epithelium, metabolize a minor part of nutrients during their transfer from the luminal fluid to the bloodstream.
- The metabolism in enterocytes of the small intestine allows energy production, used for macromolecule synthesis and electrolyte transport, and the production of metabolites with local and peripheral effects.

The Mucin-Secreting Goblet Cells Are Important Players for Protecting the Intestinal Epithelium and for Maintaining the Intestinal Barrier Function
Goblet cells are polarized epithelial cells that are responsible for the synthesis of specialized glycosylated proteins called mucins, as well as other minor proteins with different functions [80, 81]. These proteins are excreted at the luminal side of the goblet cells, thus representing a process of exocrine secretion. The mucins cover the epithelium, thus facilitating the progression of the luminal content in the gut and allowing protection of the epithelium. The proportion of goblet cells among differentiated epithelial cells in the small intestine increases from the proximal to the distal parts of the small intestine, representing 4% in the duodenum, 6% in the jejunum, and 12% in the ileum [82]. Mucus is one main element implicated in the intestinal epithelium barrier function [83]. In the small intestine, the mucus layer that is discontinuous and non-attached limits the number of bacteria that can reach the epithelium and the Peyer's patches. These latter structures are made of aggregated lymphoid vesicles that belong to the intestinal immune system [84]. In addition to the roles described above, the mucus is involved in the presentation of luminal antigens to the immune system [85]. Mucins in the small intestine and in the colon represent sites of fixation for specific bacteria called the adhering bacteria [86, 87]. These adhering bacteria, which include notably *Lactobacillus* species, adhere to mucus glycoproteins [88].

Intestinal mucins in the intestinal mucosa are particularly rich in threonine among the other amino acids present in these glycoproteins [89, 90]. Interestingly, a

moderate threonine deficiency is responsible for an alteration of intestine functionality in terms of paracellular permeability [91].

**The Enteroendocrine Cells Are Playing Important Roles in the Secretion
of Various Hormones Involved in Numerous Physiological Functions**
Enteroendocrine cells are epithelial cells that are present all along the gastrointestinal
tract and secrete various hormones [92]. They represent not more than 1–2% of the
whole epithelial cells. These cells secrete in blood peptides that are involved in very
different functions including appetite regulation, gastric emptying, intestinal transit,
exocrine secretion, and insulin release and thus glycemia regulation [93, 94]. These
cells are polarized with a luminal side characterized by the presence of brush-border
membranes, and a baso-lateral side in contact with the blood capillaries.
Enteroendocrine cells express a high number of receptors and channels that enable
them to secrete hormones in response to different stimuli [95]. Different
enteroendocrine subtypes differ by their hormonal production, but also according
to the receptors they express, enabling a fine control for the secretion of the different
hormones [96]. A wide range of nutrients have been identified as implied in the
expression and secretion of the various hormones, some of these nutrients acting on
the luminal side, while others acting at the baso-lateral side [97]. These nutrients
include simple carbohydrates, several amino acids, as well as medium- and long-
chain fatty acids [98].

The Paneth Cells Are Involved in the Host-Microbes Homeostasis
Paneth cells are located within the crypts of the small intestine. These cells secrete
antimicrobial peptides which play important roles in the maintenance of host-
microbes homeostasis [99]. In addition, Paneth cells display regulatory roles on
intestinal stem cells [100].

The Tuft Cells Are Playing a Role in Intestinal Immunity
Tuft cells are a minor type of secretory epithelial cells that produce compounds
linked to intestinal immunity [101]. These cells produce a mixture of active
compounds implicated in processes like allergy and neurotransmission. These
compounds are known to initiate both beneficial and adverse effects, depending on
the context, through population of immune and neuronal cells [101].

The Microfold (M) Cells Play a Role in the Immune Response
Microfold cells are located within the intestinal epithelium covering mucosa-
associated lymphoid tissues and participate in the induction of immune responses
to some mucosal antigens [102, 103].

Key Points
• Goblet cells in the epithelium of the small intestine secrete mucus that protects the
  intestinal epithelium, plays a role in the intestinal barrier function, facilitates
  transit, and plays a role in the presentation of luminal antigens to the intestinal
  immune system.

- Mucus secreted by goblet cells represents a site of fixation for adhering bacteria.
- Enteroendocrine cells in the small intestine epithelium secrete numerous hormones with action notably on food intake, gastrointestinal transit, and exocrine/endocrine secretion.
- Paneth cells in the small intestine are implicated in the equilibrium between host cells and microbes.
- Tuft cells are involved in intestinal immunity.

1.4   Structure and Functions of the Large Intestine Epithelium: Metabolism of Nutrients in the Colonocytes

The colonic epithelium is made of different specialized cells with some homology with the epithelial cells found in the small intestine epithelium.

The Colonocytes Use Substrates in Blood and Bacterial Metabolites Present in the Luminal Fluid for Energy Production and Synthesis of Compounds with Biological Activity
The absorptive colonocytes are polarized cells that are present mainly in the surface epithelium. These cells, which ensure water absorption, both absorb and secrete electrolytes, with an overall net movement toward electrolyte absorption in the large intestine [104]. This function allows the progressive dehydration of the luminal content along the large intestine, with the luminal content being rather liquid in the proximal colon, while becoming more solid in the rectal part of the large intestine. The brush-border membrane situated at the luminal side of the colonocytes, which are much shorter than the one situated at the luminal side of enterocytes, are equipped with a battery of transporters for different minerals, including sodium, potassium, and chloride [105]. Absorptive colonocytes, like absorptive enterocytes, display high energy demand, corresponding to the rapid turnover of epithelial cells and to Na/K ATPase activity which regulates sodium concentration in colonocytes [106]. This ATP production can be obtained through the metabolism of compounds that can be supplied to colonocytes from both the apical and the baso-lateral sides (Fig. 1.13).

Luminal compounds that are supplied to colonocytes through the apical side of the cells for energy production are mainly produced by the bacteria as will be developed in Chap. 2 of this book. Indeed, since in usual dietary and physiological conditions, glucose is absorbed in the small intestine [27], almost no dietary glucose is available from the luminal fluid for colonic absorption. Regarding amino acids, they also do not represent luminal substrates available from the luminal side for colonocytes since the colonocytes do not absorb these compounds to any significant extent [107]. Colonic absorptive cells can however use fuels provided through the baso-lateral side by the arterial capillaries (Fig. 1.13). Glutamine is a major fuel from blood origin [108]. This amino acid is oxidized in the mitochondria of the colonocytes allowing ATP synthesis. Of note, it has been proposed that colonocyte ATP synthesis through mitochondrial oxidative phosphorylation, and thus oxygen

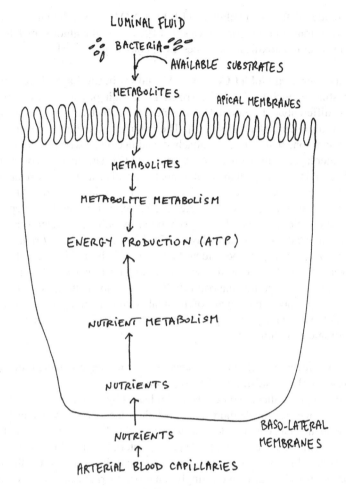

Fig. 1.13 Energy production in absorptive colonocytes from metabolites produced by bacteria and from nutrients present in the arterial blood capillaries. In this scheme, metabolites produced by the intestinal bacteria from the available substrates are absorbed through the apical membranes while the nutrients present in the arterial capillaries are absorbed through the baso-lateral membranes

consumption, participates in hypoxic condition that prevails nearby colonocytes. Such hypoxic condition participates in the maintenance of microbial community dominated by obligate anaerobic bacteria [109]. Anaerobic bacteria require reduced oxygen tension for growth [110]. Blood glucose is also metabolized in the colonocytes mainly in the glycolytic pathway, thus with a low yield of ATP production [108]. In colonocytes, glucose metabolism in the pentose phosphate pathway allows the synthesis of ribose-5 phosphate used for nucleotide synthesis, as well as the reduced equivalent NADPH used in numerous biosynthesis pathways [111]. In addition to energy production, several amino acids are used by the colonocytes as precursors for the synthesis of molecules with biological activity like nitric oxide (NO) and glutathione [112, 113]. NO has been shown to be involved in electrolyte transport in colonocytes [114], while glutathione metabolism is involved in several detoxification processes in colonocytes [115].

The Mucus-Secreting Goblet Cells Are Abundant in the Large Intestine
Large intestine epithelium is characterized by a high proportion of goblet cells among the differentiated epithelial cells since the goblet cells represent approximately 20% of all epithelial cells [116]. The colonic epithelium is covered by two layers of mucus, one that is firm and attached, and in contact with the epithelial cells, while the outer layer is unattached and less dense [117]. The thickness of the mucous layer is bigger in the large than in the small intestine, and this is related to the presence of high concentration of bacteria in the luminal content. In the large intestine, the inner mucus layer separates the bacteria from the host's epithelium, while the outer colonic mucus layer is inhabited by adhering bacteria that bind to different components of the mucus layer [87] (Fig. 1.4). Invasive *Escherichia coli* and *Clostridium difficile* have been identified as specific bacteria which adhere to the mucus layer [118, 119]. Increased mucin-degrading bacteria leads to thinner mucus layer and aggravates experimental colitis [120]. In addition to the process of bacterial adhesion, encapsulation by mucus of luminal fluid-containing bacteria has been identified in the mice model as a process by which bacteria remain in close contact with the colonic epithelium [121].

The Enteroendocrine Cells in the Large Intestine Release Hormones Mainly in Response to Microbial-Derived Compounds
The different peptides that are produced and released by the enteroendocrine cells of the large intestine are much more restricted than the ones released in the small intestine, with the main enteroendocrine peptides being PYY and GLP-1. In the large intestine, as will be developed in Chap. 3, bacterial metabolites and microbial components are able to stimulate the expression and the secretion of different gut entero-hormones [122, 123]. Interestingly, enteroendocrine cells in the colon have been shown to be involved in the regulation of metabolic functions despite not being directly stimulated by nutrients [124, 125], but by microbial-derived compounds [126–128].

Regarding Tuft cells, they have also been identified in the large intestine epithelium [129]. In the large intestine, Paneth cells are not detected, but antimicrobial peptides secretion is performed by colonocytes [130].

Key Points
- Absorptive colonocytes are provided by fuels from blood and by fuels from the luminal content, these luminal fuels originating from microbial metabolic activity.
- Fuel oxidation allows ATP synthesis required for colonic epithelium renewal and dehydration of the luminal fluid.
- Mucous-producing cells are more abundant in the large than in the small intestine. Specific bacteria can bind to the outer layer of the mucus.

- Enteroendocrine cells in the colon secrete hormones which are stimulated by microbial-derived compounds.

References

1. O'Learhy F, Samman S. Vitamin B12 in health and disease. Nutrients. 2010;2(3):299–316.
2. Schuchardt JP, Hahn A. Intestinal absorption and factors influencing bioavailability of magnesium. An update. Curr Nutr Food Sci. 2017;13(4):260–78.
3. Ticho AL, Malhotra P, Dudeja PK, Gill RK, Alrefai WA. Intestinal absorption of bile acids in health and disease. Compr Physiol. 2019;10(1):21–56.
4. Debongnie JC, Phillips SF. Capacity of the human colon to absorb fluid. Gastroenterology. 1978;74(4):698–703.
5. Hermansen K. Effects of cholecystokinin (CCK)-4, nonsulfated CCK-8, and sulfated CCK-8 on pancreatic somatostatin, insulin, and glucagon secretion in the dog: studies in vitro. Endocrinology. 1984;114(5):1770–5.
6. Liddle RA, Morita ET, Conrad CK, Williams JA. Regulation of gastric emptying in humans by cholecystokinin. J Clin Invest. 1986;77(3):992–6.
7. Lo CC, Davidson WS, Hibbard SK, Georgievsky M, Lee A, Tso P, Woods SC. Intraperitoneal CCK and fourth-intraventricular Apo AIV require both peripheral and NTS CCK 1R to reduce food intake in male rats. Endocrinology. 2014;155(5):1700–7.
8. Meyer BM, Werth BA, Beglinger C, Hildebrand P, Jansen JB, Zach D, Rovati LC, Stalder GA. Role of cholecystokinin in regulation of gastrointestinal motor functions. Lancet. 1989;2 (8653):12–5.
9. Honda K, Littman DR. The microbiota in adaptive immune homeostasis and disease. Nature. 2016;535(7610):75–84.
10. Beaumont M, Blachier F. Amino acids in intestinal physiology and health. Adv Exp Med Biol. 2020;1265:1–20.
11. Wang G, Huang S, Wang Y, Cai S, Yu H, Liu H, Zeng X, Zhang G, Qiao S. Bridging intestinal immunity and gut microbiota by metabolites. Cell Mol Life Sci. 2019;76(20):3917–37.
12. Roulis M, Flavell RA. Fibroblasts and myofibroblasts of the intestinal lamina propria in physiology and disease. Differentiation. 2016;92(3):116–31.
13. van der Flier LG, Clevers H. Stem cells, self-renewal, and differentiation in the intestinal epithelium. Annu Rev Physiol. 2009;71:241–60.
14. Camilleri M, Madsen K, Spiller R, Greenwood-Van Meerveld B, Verne GN. Intestinal barrier function in health and gastrointestinal disease. Neurogastroenterol Motil. 2012;24(6):503–12.
15. Shen L, Weber CR, Raleigh DR, Yu D, Turner JR. Tight junction pore and leak pathways: a dynamic duo. Annu Rev Physiol. 2011;73:283–309.
16. Blander JM. On cell death in the intestinal epithelium and its impact on gut homeostasis. Curr Opin Gastroenterol. 2018;34(6):413–9.
17. Santos AJM, Lo YH, Mah AT, Kuo CJ. The intestinal stem cell niche: homeostasis and adaptations. Trends Cell Biol. 2018;28(12):1062–78.
18. Beumer J, Clevers H. Cell fate specification and differentiation in the adult mammalian intestine. Nat Rev Mol Cell Biol. 2021;22(1):39–53.
19. Vereecke L, Beyaert R, van Loo G. Enterocyte death and intestinal barrier maintenance in homeostasis and disease. Trends Mol Med. 2011;17(10):584–93.
20. Darwich AS, Aslam U, Ashcroft DM, Rostami-Hodjegan A. Meta-analysis of the turnover of intestinal epithelia in preclinical animal species and humans. Drug Metab Dispos. 2014;42 (12):2016–22.
21. Potten CS. Epithelial cell growth and differentiation. II. Intestinal apoptosis. Am J Physiol. 1997;273:G253–7.

22. Gehart H, Clevers H. Tales from the crypt: new insights into stem cells. Nat Rev Gastroenterol Hepatol. 2019;16(1):19–34.
23. Hageman JH, Heinz MC, Kretzschmar K, van der Vaert J, Clevers H, Snippert HJG. Intestinal regeneration: regulation by the microenvironment. Dev Cell. 2020;54(4):435–46.
24. Barker N. Adult intestinal stem cells: critical drivers of epithelial homeostasis and regeneration. Nat Rev Mol Cell Biol. 2014;15(1):19–33.
25. Dauça M, Bouziges F, Colin S, Kedinger M, Keller MK, Schilt J, Simon-Assmann P, Haffen K. Development of the vertebrate small intestine and mechanisms of cell differentiation. Int J Dev Biol. 1990;34(1):205–18.
26. Bröer S. Amino acid transport across mammalian intestinal and renal epithelium. Physiol Rev. 2008;88(1):249–86.
27. Chen C, Yin Y, Tu Q, Yang H. Glucose and amino acid in enterocyte: absorption, metabolism and maturation. Front Biosci (Landmark Ed). 2008;23(9):1721–39.
28. Nishiyama K, Sugiyama M, Mukai T. Adhesion properties of lactic acid bacteria on intestinal mucin. Microorganisms. 2016;4(3):34.
29. Wasielewski H, Alcock J, Aktipis A. Resource conflict and cooperation between human host and gut microbiota: implications for nutrition and health. Ann N Y Acad Sci. 2016;1372(1): 20–8.
30. McNurlan MA, Garlick PJ. Contribution of rat liver and gastrointestinal tract to whole-body protein synthesis in the rat. Biochem J. 1980;186(1):381–3.
31. Preedy VR, Peters T. Protein metabolism in the small intestine of the ethanol-fed rat. Cell Biochem Funct. 1989;7(4):235–42.
32. Nakshabendi IM, Obeidat W, Russell RI, Downie S, Smith K, Rennie MJ. Gut mucosal protein synthesis measured using intravenous and intragastric delivery of stable tracer amino acids. Am J Phys. 1995;269:E996–9.
33. Pontes MH, Sevostyanova A, Groisman EA. When too much ATP is bad for protein synthesis. J Mol Biol. 2015;427(16):2586–94.
34. Blachier F, Boutry C, Bos C, Tomé D. Metabolism and functions of L-glutamate in the epithelium of the small and large intestines. Am J Clin Nutr. 2009;90(3):814S–21S.
35. Duée PH, Darcy-Vrillon B, Blachier F, Morel MT. Fuel selection in intestinal cells. Proc Nutr Soc. 1995;54(1):83–94.
36. Vaugelade P, Posho L, Darcy-Vrillon B, Bernard F, Morel MT, Duée PH. Intestinal oxygen uptake and glucose metabolism during nutrient absorption in the pig. Proc Soc Exp Biol Med. 1994;207(3):309–16.
37. Yen JT, Nienaber JA, Hill DA, Pond WG. Oxygen consumption by portal vein-drained organs and by whole animal in conscious growing swine. Proc Soc Exp Biol Med. 1989;190(4): 393–8.
38. Fouillet H, Mariotti F, Gaudichon C, Bos C, Tomé D. Peripheral and splanchnic metabolism of dietary nitrogen are differently affected by the protein source in humans as assessed by compartmental modeling. J Nutr. 2002;132(1):125–33.
39. Hoskins LC, Boulding ET. Mucin degradation in human colon ecosystems. Evidence for the existence and role of bacterial subpopulations producing glycosidases as extracellular enzymes. J Clin Invest. 1981;67(1):163–72.
40. Layer P, Gröger G. Fate of pancreatic enzymes in the human intestinal lumen in health and pancreatic insufficiency. Digestion. 1993;54:10–4.
41. Windmueller HG, Spaeth AE. Intestinal metabolism of glutamine and glutamate from the lumen as compared to glutamine from blood. Arch Biochem Biophys. 1975;171(2):662–72.
42. Kenny AJ, Maroux S. Topology of microvillar membrane hydrolazes of kidney and intestine. Physiol Rev. 1982;62(1):91–128.
43. Darcy-Vrillon B, Posho L, Morel MT, Bernard F, Blachier F, Meslin JC, Duée PH. Glucose, galactose, and glutamine metabolism in pig isolated enterocytes during development. Pediatr Res. 1994;36(2):175–81.

44. Windmueller HG, Spaeth AE. Metabolism of aspartate, asparagine, and arginine by rat small intestine in vivo. Arch Biochem Biophys. 1976;175(2):670–6.
45. Beyreuther K, Biesalski HK, Fernstrom JD, Grimm P, Hammes WP, Heinemann U, Kempski O, Stehle P, Steinhart H, Walker R. Consensus meeting: monosodium glutamate. An update. Eur J Clin Nutr. 2007;61(3):304–13.
46. Reeds PJ, Burrin DG, Jahoor F, Wykes L, Henry J, Frazer EM. Enteral glutamate is almost completely metabolized in first pass by the gastrointestinal tract of infant pigs. Am J Phys. 1996;270:E413–8.
47. Reeds PJ, Burrin DG. Glutamine and the bowel. J Nutr. 2001;131:2505S–8S.
48. Madej M, Lundh T, Lindberg JE. Activity of enzymes involved in energy production in the small intestine during suckling-weaning transition of pigs. Biol Neonates. 2002;82(1):53–60.
49. Cynober LA. Plasma amino acid levels with a note on membrane transport: characteristics, regulation, and metabolic significance. Nutrition. 2002;18(9):761–6.
50. Häussinger D. Liver glutamine metabolism. J Parenter Enteral Nutr. 1990;14:56S–62S.
51. Wu GY, Thompson JR, Baracos VE. Glutamine metabolism in skeletal muscles from the broiler chick (Gallus domesticus) and the laboratory rat (Rattus norvegicus). Biochem J. 1991;274:769–74.
52. Cherbuy C, Guesnet P, Morel MT, Kohl C, Thomas M, Duée PH, Prip-Buus C. Oleate metabolism in pig enterocytes is characterized by an increased oxidation rate in the presence of a high esterification rate within two days after birth. J Nutr. 2012;142(2):221–6.
53. Newsholme EA, Carrié AL. Quantitative aspects of glucose and glutamine metabolism by intestinal cells. Gut. 1994;35:S13–7.
54. Blachier F, M'Rabet-Touil H, Darcy-Vrillon B, Posho L, Duée PH. Stimulation by D-glucose of the direct conversion of arginine to citrulline in enterocytes isolated from pig jejunum. Biochem Biophys Res Commun. 1991;177(3):1171–7.
55. Palaniappan B, Arthur S, Sundaram VL, Butts M, Sundaram S, Mani K, Singh S, Nepal N, Sundaram U. Inhibition of intestinal villus cell Na/K-ATPase mediates altered glucose and NaCl absorption in obesity-associated diabetes and hypertension. FASEB J. 2019;33(8):9323–33.
56. M'Rabet-Touil H, Blachier F, Morel MT, Darcy-Vrillon B, Duée PH. Characterization and ontogenesis of nitric oxide synthase activity in pig enterocytes. FEBS Lett. 1993;331(3):243–7.
57. MacKendrick W, Caplan M, Hsueh W. Endogenous nitric oxide protects against platelet-activating factor-induced bowel injury in the rat. Pediatr Res. 1993;34(2):222–8.
58. Miller MJ, Zhang XJ, Sadowska-Krowicka H, Chotinaruemol S, McIntyre JA, Clark JA, Bustamante SA. Nitric oxide release in response to gut injury. Scand J Gastroenterol. 1993;28(2):149–54.
59. Stark ME, Szurszewski JH. Role of nitric oxide in gastrointestinal and hepatic function and disease. Gastroenterology. 1992;103(6):1928–49.
60. Calignano A, Whittle BJ, Di Rosa M, Moncada S. Involvement of endogenous nitric oxide in the regulation of rat intestinal motility in vivo. Eur J Pharmacol. 1992;229(2–3):273–6.
61. Kubes P. Nitric oxide modulates epithelial permeability in the feline small intestine. Am J Phys. 1992;262:G1138–42.
62. Kubes P. Ischemia-reperfusion in feline small intestine: a role for nitric oxide. Am J Phys. 1993;264:G143–9.
63. Coloso RM, Stipanuk MH. Metabolism of cyst(e)ine in rat enterocytes. J Nutr. 1989;119(12):1914–24.
64. Shoveller AK, Stoll B, O'Ball R, Burrin DG. Nutritional and functional importance of intestinal sulfur amino acid metabolism. J Nutr. 2005;135(7):1609–12.
65. Chakravarthi S, Jessop CE, Bulleid NJ. The role of glutathione in disulphide bond formation and endoplasmic-reticulum-generated oxidative stress. EMBO Rep. 2006;7(3):271–5.
66. Kemp M, Go YM, Jones DP. Nonequilibrium thermodynamics of thiol/disulfide redox systems: a perspective on redox system biology. Free Radic Biol Med. 2008;44(6):921–37.

67. Martensson J, Jain A, Meister A. Glutathione is required for intestinal function. Proc Natl Acad Sci U S A. 1990;87(5):1715–9.
68. Uchiyama C, Mori M, Tatibana M. Subcellular localization and properties of N-acetylglutamate synthase in rat small intestinal mucosa. J Biochem. 1981;89(6):1777–86.
69. Blachier F, M'Rabet-Touil H, Posho L, Darcy-Vrillon B, Duée PH. Intestinal arginine metabolism during development. Evidence for de novo synthesis of L-arginine in newborn pig enterocytes. Eur J Biochem. 1993;216(1):109–17.
70. Guihot G, Blachier F, Colomb V, Morel MT, Raynal P, Corriol P, Ricour P, Duée PH. Effect of an elemental vs complex diet on citrulline production from L-arginine in rat isolated enterocytes. J Parenter Enter Nutr. 1997;21(6):316–23.
71. Cynober L. Can arginine and ornithine support gut functions? Gut. 1994;35:S42–5.
72. Dhanakoti SN, Brosnan JT, Herzberg GR, Brosnan ME. Renal arginine synthesis: studies in vitro and in vivo. Am J Phys. 1990;259:E437–42.
73. Marini JC, Agarwal U, Robinson JL, Yuan Y, Didelija IC, Stoll B, Burrin DG. The intestinal-renal axis for arginine synthesis is present and functional in the neonatal pig. Am J Phys. 2017;313(2):E233–42.
74. van de Poll MC, Siroen MP, van Leeuwen PA, Soeters PB, Melis GC, Boelens PG, Deutz NE, Dejong CH. Interorgan amino acid exchange in humans: consequences for arginine and citrulline metabolism. Am J Clin Nutr. 2007;85(1):167–72.
75. Blachier F, Darcy-Vrillon B, Sener A, Duée PH, Malaisse WJ. Arginine metabolism in rat enterocytes. Biochim Biophys Acta. 1991;1092(3):304–10.
76. Wu G. Urea synthesis in enterocytes of developing pigs. Biochem J. 1995;312:717–23.
77. Lund P, Wiggins D. The ornithine requirement of urea synthesis. Formation of ornithine from glutamine in hepatocytes. Biochem J. 1986;329:773–6.
78. O'Sullivan D, Brosnan JT, Brosnan ME. Catabolism of arginine and ornithine in the perfused rat liver. Am J Phys. 2000;278(3):E516–21.
79. Sepehrinezhad A, Zarifkar A, Namvar G, Shahbazi A, Williams R. Astrocyte swelling in hepatic encephalopathy: molecular perspective of cytotoxic edema. Metab Brain Dis. 2020;35 (4):559–78.
80. Johansson ME, Thomsson KA, Hansson GC. Proteomic analyses of the two mucus layers of the colon barrier reveal that their main component, the Muc2 mucin, is strongly bound to the Fcgbp protein. J Proteome Res. 2009;8(7):3549–57.
81. Rodriguez-Pineiro AM, Bergström JH, Ermund A, Gustafsson JK, Schütte A, Johansson ME, Hansson GC. Studies of mucus in mouse stomach, small intestine, and colon. II. Gastrointestinal mucus proteome reveals Muc2 and Muc5ac accompanied by a set of core proteins. Am J Phys. 2013;305(5):G348–56.
82. König J, Wells J, Cani PD, Garicia-Rodenas CL, MacDonald T, Mercenier A, Whyte J, Troost F, Brummer RJ. Human intestinal barrier function in health and disease. Clin Transl Gastroenterol. 2016;7(10):e196.
83. Peterson LW, Artis D. Intestinal epithelial cells: regulators of barrier function and immune homeostasis. Nat Rev Immunol. 2014;14(3):141–53.
84. Pelaseyed T, Bergström JH, Gustafsson JK, Ermund A, Birchenough GMH, Schütte A, van der Post S, Svensson F, Rodriguez-Pineiro AM, Nyström EEL, Wising C, Johansson MEV, Hansson GC. The mucus and mucins of the goblet cells and enterocytes provide the first defense line of the gastrointestinal tract and interact with the immune system. Immunol Rev. 2014;260(1):8–20.
85. McDole JR, Wheeler LW, McDonald KG, Wang B, Konjufca V, Knoop KA, Newberry RD, Miller MJ. Goblet cells deliver luminal antigen to CD103+ dendritic cells in the small intestine. Nature. 2012;483(7389):345–9.
86. Johansson MEV, Holmen-Larsson JM, Hansson GC. The two mucus layers of colon are organized by the MUC2 mucin, whereas the outer layer is a legislator of host-microbial interactions. Proc Natl Acad Sci U S A. 2011;108:4659–65.

87. Kim HS, Ho SB. Intestinal goblet cells and mucins in health and disease: recent insights and progress. Curr Gastroenterol Rep. 2010;12(5):319–30.
88. Van Tassell ML, Miller MJ. Lactobacillus adhesion to mucus. Nutrients. 2011;3(5):613–36.
89. Fogg FJ, Hutton DA, Kumel K, Pearson JP, Harding SE, Allen A. Characterization of pig colonic mucins. Biochem J. 1996;316:937–42.
90. Schaart MW, Schierbeek H, van der Schoor SR, Stoll B, Burrin DG, Reeds PJ, van Goudoever JB. Threonine utilization is high in the intestine of piglets. J Nutr. 2005;135(4):765–70.
91. Hamard A, Mazurais D, Boudry G, Le Huërou-Luron I, Sève B, Le Floc'h N. A moderate threonine deficiency affects gene expression profile, paracellular permeability and glucose absorption capacity in the ileum of piglets. J Nutr Biochem. 2010;21(10):914–21.
92. Janssen S, Depoortere I. Nutrient sensing in the gut: new roads to therapeutics? Nutrients. 2013;24(2):92–100.
93. Gribble FM, Reimann F. Enteroendocrine cells: Chemosensors in the intestinal epithelium. Annu Rev Physiol. 2016;78:277–99.
94. Latorre R, Sternini C, De Giorgio R, Meerveld G-V. Enteroendocrine cells: a review of their role in brain-gut communication. Neurogastroenterol Motil. 2016;28(5):620–30.
95. Gribble FM, Reimann F. Function and mechanisms of enteroendocrine cells and gut hormones in metabolism. Nat Rev Endocrinol. 2019;15(4):226–37.
96. Billing LJ, Larraufie P, Lewis J, Leiter A, Li J, Lam B, Yeo GS, Goldspink DA, Kay RG, Gribble FM, Reimann F. Single cell transcriptomic profiling of large intestinal enteroendocrine cells in mice. Identification of selective stimuli for insulin-like peptide-5 and glucagon-like peptide-1 co-expressing cells. Mol Metab. 2019;29:158–69.
97. Furness JB, Rivera LR, Cho HJ, Bravo DM, Callaghan B. The gut as a sensory organ. Nat Rev Gastroenterol Hepatol. 2013;10(12):729–40.
98. Martin AM, Sun EW, Keating DJ. Mechanisms controlling hormone secretion in human gut and its relevance to metabolism. J Endocrinol. 2019;244(1):R1–R15.
99. Bevins CL, Salzmann NH. Paneth cells, antimicrobial peptides and maintenance of intestinal homeostasis. Nat Rev Microbiol. 2011;9(5):356–68.
100. Mei X, Gu M, Li M. Plasticity of Paneth cells and their ability to regulate intestinal stem cells. Stem Cell Res Ther. 2020;11(1):349.
101. Schneider C, O'Leary CE, Locksley RM. Regulation of immune responses by tuft cells. Nat Rev Immunol. 2019;19(9):584–93.
102. Kobayashi N, Takahashi D, Takano S, Kimura S, Hase K. The roles of Peyer's patches and microfold cells in the gut immune system: relevance to autoimmune diseases. Front Immunol. 2019;10:2345.
103. Mabott NA, Donaldson DS, Ohno H, Williams IR, Mahajan A. Microfold (M) cells: important immunosurveillance posts in the intestinal epithelium. Mucosal Immunol. 2013;6(4):666–77.
104. Rao MC. Physiology of electrolyte transport in the gut: implications for disease. Compr Physiol. 2019;9(3):947–1023.
105. Phillips SF. Functions of the large bowel: an overview. Scand J Gastroenterol. 1984;93:1–12.
106. Turnamian SG, Binder HJ. Electrolyte transport in distal colon of sodium-depleted rats: effect of sodium repletion. Am J Phys. 1988;255(3):G329–38.
107. van der Wielen N, Moughan PJ, Mensink M. Amino acid absorption in the large intestine of human and porcine models. J Nutr. 2017;147(8):1493–8.
108. Cherbuy C, Darcy-Vrillon B, Morel MT, Pégorier JP, Duée PH. Effect of germfree state on the capacities of isolated rat colonocytes to metabolize n-butyrate, glucose and glutamine. Gastroenterology. 1995;109(6):1890–9.
109. Litvak Y, Byndloss MX, Bäumler AJ. Colonocyte metabolism shapes the gut microbiota. Science. 2018;362(6148):eaat9076.
110. Nagy E, Boyanova L, Justesen US, ESCMID Study Group of Anaerobic Infections. How to isolate, identify and determine antimicrobial susceptibility of anaerobic bacteria in routine laboratories. Clin Microbiol Infect. 2018;24(11):1139–48.

111. Butler RN, Arora KK, Collins JG, Flanigan I, Lawson MJ, Roberts-Thomson IC, Williams JF. Pentose phosphate pathway in rat colonic epithelium. Biochem Int. 1990;22(2):249–60.
112. Batist G, Mekhail-Ishak K, Hudson N, DeMuys JM. Interindividual variation in phase II detoxification enzymes in normal human colon mucosa. Biochem Pharmacol. 1988;37(21): 4242–3.
113. Blachier F, Davila AM, Benamouzig R, Tome D. Channelling of arginine in NO and polyamine pathways in colonocytes and consequences. Front Biosci. 2011;16(4):1331–43.
114. Rolfe VE, Milla PJ. Nitric oxide stimulates cyclic guanosine monophosphate production and electrogenic secretion in Caco-2 colonocytes. Clin Sci (Lond). 1999;96(2):165–70.
115. Roediger WE, Babidge W. Human colonocyte detoxification. Gut. 1997;41(6):731–4.
116. Paone P, Cani PD. Mucus barrier, mucins and gut microbiota: the expected slimy partners? Gut. 2020;69(12):2232–43.
117. Johansson MEV, Sjövall H, Hansson GC. The gastrointestinal mucus system in health and disease. Nat Rev Gastroenterol Hepatol. 2013;10(6):352–61.
118. Nadalian B, Yadegar A, Houri H, Olfatifar M, Shahrokh S, Aghdaei HA, Suzuki H, Zali MR. Prevalence of the pathobiont adherent-invasive Escherichia coli and inflammatory bowel disease: a systematic review and meta-analysis. J Gastroenterol Hepatol. 2021;36(4):852–63.
119. Stevenson E, Minton NP, Kuehne SA. The role of flagella in Clostridium difficile pathogenicity. Trends Microbiol. 2015;23(5):275–82.
120. Chen L, Wang J, Yi J, Liu Y, Yu Z, Chen S, Liu X. Increased mucin-degrading bacteria leads to thinner mucus layer and aggravates experimental colitis. J Gastroenterol Hepatol. 2021;36 (10):2864–74.
121. Bergsrtom K, Shan X, Casero D, Batushanski A, Lagishetty V, Jacobs JP, Hoover C, Kondo Y, Shao B, Gao L, Zandberg W, Noyovitz B, McDaniel JM, Gibson DL, Pakpour S, Kazemian N, McGee S, Houchen SW, Rao CV, Griffin TM, Sonnenburg JL, McEver RP, Braun J, Xia L. Proximal colon-derived O-glycosylated mucus encapsulates and modulates the microbiota. Science. 2020;370(6515):467–72.
122. Christansen CB, Gabe MBN, Svendsen B, Dragsted LO, Rosenhilde MM, Holst JJ. The impact of short-chain fatty acids on GLP-1 and PYY secretion from the isolated perfused rat colon. Am J Phys. 2018;315(1):G53–65.
123. Lebrun LJ, Lenaerts K, Kiers D, Pais de Barros JP, Le Guern N, Plesnik J, Thomas C, Bourgeois T, Dejong CHC, Kox M, Hundscheid IHR, Khan NA, Mandard S, Deckert V, Pickkers P, Drucker DJ, Lagrost L, Grober J. Enteroendocrine L cells sense LPS after gut barrier injury to enhance GLP-1 secretion. Cell Rep. 2017;21(5):1160–8.
124. Lewis JE, Miedzybrodska EL, Foreman RE, Woodward ORM, Gay RG, Goldspink DA, Gribble FM, Reimann F. Selective stimulation of colonic L cells improves metabolic outcomes in mice. Diabetologia. 2020;63(7):1396–407.
125. Song Y, Koehler JA, Baggio LL, Powers AC, Sandoval DA, Drucker DJ. Gut-proglucagon-derived peptides are essential for releasing glucose homeostasis in mice. Cell Metab. 2019;30 (5):976–86.
126. Arora T, Akrami R, Pais R, Bergqvist L, Johansson BR, Schwartz TW, Reimann F, Gribble FM, Bäcked F. Microbial regulation of the L-cell tanscriptome. Sci Rep. 2018;8(1):1207.
127. Larraufie P, Doré J, Lapaque N, Blottière HM. TLR ligands and butyrate increase Pyy expression through two distinct but inter-regulated pathways. Cell Microbiol. 2017;19(2)
128. Tolhurst G, Heffron H, Lam YS, Parker HE, Habid AM, Diakogiannaki E, Cameron J, Grosse J, Reimann F, Gribble FM. Short-chain fatty acids stimulate glucagon-like peptide-1 secretion via the G-protein-coupled receptor FFAR2. Diabetes. 2012;61(2):364–71.
129. McKinley ET, Sui Y, Al-Kofahi Y, Millis BA, Tyska MT, Roland JT, Santamaria-Pang A, Ohland CL, Jobin C, Franklin JL, Lau KS, Gerdes MJ, Coffey RJ. Optimized multiplex immunofluorescence single-cell analysis reveals tuft cell heterogeneity. JCI Insight. 2017;2 (11):e93487.
130. Gallo RL, Hopper LV. Epithelial antimicrobial defence of the skin and intestine. Nat Rev Immunol. 2012;12(7):503–16.

Intestine Offers Board and Lodging for Intestinal Microbes on a Short- or Long-Term Stay

Abstract

The gut microbiota is a complex mixture of bacteria, archaea, viruses, and fungi. Approximately 150 different bacterial species can be identified at the individual level, but the heterogeneity of composition is important when comparing individuals between them. Ecological basis for such many species corresponds likely to the large number of substrates that are available for the global bacterial metabolic activity. Most studies have been performed using fecal samples, thus representing the microbiota composition representative of the distal part of the large intestine, and only a few analyses have been performed in the different parts of the intestinal tract. In the small intestine, the transit of the luminal fluid is rapid, and the concentration of bacteria is low and progressively increases from the duodenum to the ileum. In contrast, in the large intestine, the transit is considerably slowed down and the bacteria concentration is much higher. The intestinal tract is rapidly colonized after birth, and the intestinal microbiota composition found in neonates moves progressively within the first 3 years to the adult composition. Thereafter, the bacterial composition stays relatively stable. Interestingly, the composition of milk has an impact on the metabolic activity of intestinal bacteria in infants. The microbiota composition and metabolic activity depend mainly on environmental parameters, including alimentary parameters. The intestinal microbiota evolves in very old individuals, and the reasons for such changes are likely multifactorial.

The small and large intestines are inhabited by a complex mixture of microbes among which bacteria have been the subject of vast majority of studies. As introduced above, although most bacteria are in transit with a periodic excretion of large amounts of bacteria in feces, a small proportion of bacteria are adhering to the intestinal mucosa and thus are in close contact with this structure, and we will see that this contact has important implications for the host intestinal physiology [1–3].

Bacteria are dependent on the host for obtaining the substrates they need for their constant renewal. These substrates are either originating from the diet, or from the endogenous compounds. More precisely, endogenous compounds can be defined as those that are synthesized by the host and utilized by the host and the microbiota in a process of recycling. Then, the idea rapidly comes to mind that, depending on the dietary compounds that individuals are used to consume during their meals, the microbiota will be provided with different sources and quantities of substrates. As a matter of fact, as will be presented in the following chapters, the availability of dietary substrates provided by the hosts for their intestinal microbiota metabolic activity has important consequences for the communication between intestinal microbes, and for communication between bacteria and their host.

2.1   The Human Intestinal Microbiota: A Complex Mixture

The gut microbiota that is present in the intestinal content in mammals, including humans, is composed of a vast population of microbes. This population of microbes includes not only bacteria, but also archaea, viruses, and fungi. Archaea is a group of primitive prokaryotes that, based on their different characteristics, forms a separate group from bacteria and eukaryotes [4]. Archaea use various organic and inorganic substrates as energy sources [5]. They are detected in the mammalian digestive tract as part of the usual microbiota [6, 7] and are considered as essential constituents of the human microbiota [8]. Archaea colonize infants during the first year after birth [9]. Majority of the archaea in the human intestinal tract are methane-producing microbes, called methanogens, which are characterized by their ability to produce methane from hydrogen under aerobic conditions [4]. By removing hydrogen from the intestinal content, methanogens increase the fermentation process made by the intestinal bacteria, allowing accordingly more complete anaerobic degradation of the available organic substrates [10]. Then archaea are species that play a role in the optimal energy yield of the entire human intestinal microbiota [11].

The human intestinal virome, which establishes during the first years following birth, is defined as the population of viruses present in our intestine. This virome is personalized and rather stable and largely dominated by the so-called phages [12]. Phages, by infecting specific populations of intestinal bacteria, are one element that is implicated in the bacterial composition of the intestinal microbiota [13].

Fungi, although representing a minor component of the gut microbiota, may be involved in a competitive relationship with bacteria in the human gut [14]. Although 50 genera of fungi are commonly detected in the human intestine, 10 genera account for vast majority of detected organisms [15, 16]. The fungi most commonly detected in the human intestine include *Candida* and *Saccharomyces* [17]. Some clinical studies point to the role of *Candida* as one element in antibiotic-associated diarrhea [18].

Lastly, protozoans in the intestine, although classically not included as part of the microbiota itself, represent a heterogeneous group of eukaryotic organisms, with some of them being considered as parasites [19]. The term gut ecosystem usually refers to the biological community of microorganisms living in the environment of the gut.

Regarding bacteria, more than 1000 bacterial species have been reported to be present in our gastrointestinal tract, each individual harboring however only around 150 different species [20]. The corresponding collective bacterial genome represents more genes than the ones found in the human genome [21]. However, 90% of the intestinal bacteria belong to two phyla, namely Bacteroidetes and Firmicutes [22]. Ecological basis for this great number of species in the human intestinal microbiota resides likely in the fact that there is a very large number of different substrates that are available for bacterial metabolic activity. In the human body, it has been approximated that the total number of bacteria is in the same order of magnitude than the total number of human cells [23].

Of note, in most studies performed with volunteers, the bacteria composition has been determined in stool. Bacteria in stool represent the bacterial population present in the very distal part of the large intestine (that is rectum), and thus we are unfortunately left with few data regarding the bacterial population that are present in the more proximal parts of the large intestine, and in the small intestine. This is largely due to the technical difficulties in collecting luminal fluids in the intestine. This can be done however by intubation with a naso-intestinal sampling device in healthy volunteers [24]. This rather invasive method allows to recover the luminal fluid in different parts of the small intestine. Small intestinal luminal fluid can also be recovered in ileostomy effluent obtained from patients with colectomy and terminal ileostomy [25]. In other less technical words, the luminal fluid can be obtained from the distal small intestine in patients for which, for medical reasons, the colon has been resected. In animal models, it is possible to recover the luminal fluids in the different parts of the small and large intestines, but it requires euthanasia of the animals.

2.2   Bacteria Concentration, Composition, and Diversity During the Whole Life

Installation of the Intestinal Microbiota after Birth and Evolution of Its Composition During Development
Although it is commonly accepted that the intrauterine environment and fetus are germ-free before birth, this dogma has been somewhat challenged [26]. Indeed, experimental evidence show the presence of bacteria in the intrauterine environment [27, 28] that could possibly be due to the translocation of bacteria from the mother's gut to the intrauterine environment via the bloodstream. In support of this hypothesis, bacterial species found within the intestine, have been isolated from the umbilical cord blood between the mother and fetus. In addition, the meconium, that is the intestinal biological fluid that is accumulated in the fetus intestine during gestation, appears to be not completely sterile [29, 30]. However, it is only fair to recognize that the amplitude and the physiological implication of such a putative mother-to-fetus bacterial transfer remain unclear [31].

During and after birth, infants are exposed to large amounts of bacteria that originate from the mother and the surrounding environment. The large inter-individual variability observed between infants, as observed when analyzing the fecal microbiota during the first few months after birth [32], is likely due, in part at least, to the inter-individual variability of the intestinal microbiota composition between mothers [33].

The Intestinal Microbiota in Infants Moves Progressively to an Adult-like Microbiota

The intestine of the vaginally delivered newborn is rapidly colonized by an increasing number of bacteria which are largely those present in the vagina of the mother, and to a lesser extent by anal/fecal bacteria [34]. The intestinal microbiota of infants is overall very different when compared with the one present in adults [35]. The usual pattern of gut microbiota development in infants involves early colonization by facultative anaerobes. Indeed, within a few days, strictly anaerobic bacteria develop in the infant's intestine. Around 1 year of age, the infant microbiota composition is closer to the ones commonly found in adults [32], and at around 3 years of age, the fecal microbiota can be considered as relatively mature, being not vastly different that the ones characteristic of the adult stage [36, 37].

Analysis of the intestinal bacteria in meconium produced by newborns after birth revealed a strong correlation between the first microbial communities of either the mother's vagina in the case of vaginal delivery, or of the mother's skin in the case of cesarean section [38]. The differences in terms of intestinal microbiota between infant delivered by cesarean-section, and the ones that were vaginally delivered appear to rapidly decrease with time [39]. In adults, the mode of delivery apparently left no marked traces of the initial differences at the level of the intestinal microbiota [40]. Overall, the available data are compatible with the view that infants born by cesarean, and thus not directly exposed to the vaginal and enteric microbiota of the mother, are colonized by bacteria of the skin and of the hospital environment.

When comparing the influence of breastfeeding compared with formula-feeding on the fecal microbiota composition, relatively contradictory results have been obtained depending on the studies considered. Breastfed infants have been shown to be colonized by a less diverse intestinal microbiota than infants fed with bottle [41]. Breastfeeding appears to favor colonization by bifidobacteria [42]. This effect is presumably related to the different composition of the mother's milk when compared with formula, and notably to the presence in the mother's milk of peptides with antimicrobial properties such as immunoglobulin A, lactoferrin, lysozyme, and oligosaccharides [43, 44]. Regarding milk oligosaccharides, they favor the growth of certain bacterial species, including bifidobacteria which are known to be favored by breastfeeding [45, 46]. Interestingly, milk oligosaccharides have been shown to decrease the risk of neonatal diarrhea [47]. Higher levels of protein in infant formula and the lack in formula of oligosaccharides that are normally present in human milk promote a shift toward amino acid fermentation by the intestinal microbiota [42], thus demonstrating that the composition of milk can influence the metabolic activity of the intestinal bacteria.

During and after weaning, important modifications of the microbiota composition are observed [48–50]. Antibiotic exposure in the period following birth greatly influences the development of neonatal intestinal microbiota [51].

The human bacteriome, which is characteristic of one given individual, and is established in a rather stable way after the first years of life [36, 52], represents finally an enormous number of bacteria in the entire gut of the adult, around 10^{14} bacteria [53].

When comparing individuals between each other, high variability in microbiota composition is observed [54, 55].

Evolution of the Microbiota During Aging

In individuals over the age of 65, in a context of important inter-individual variability, the intestinal microbiota does not show vast modification in terms of relative composition, when compared with younger subjects, with however measurable modifications of specific bacterial species [56–59]. The reasons for observed changes in the microbiota composition appear unclear as several parameters are modified in the elderly, including dietary patterns in specific given situations [60], digestive secretion [61], nutrient absorption [62], and intestinal transit [63], all these changes being theoretically able to have an impact on the composition and on the metabolic activity of the intestinal microbiota. Of note, when comparing short-chain fatty acid (that are quantitavely major bacterial metabolites) concentrations in stools obtained from senior volunteers (56–65 years old) and older volunteers (65–95 years of age) with similar body mass index and dietary intakes, it was found that butyrate, acetate, and propionate are less concentrated in fecal samples obtained from older than from senior subjects [64].

In centenarians, the difference of microbiota composition appears relatively marked when compared with younger individuals [56]. Interestingly, the global metabolic activity of the intestinal microbiota toward the available substrates is modified in the elderly [65].

Supply of Substrates by the Host to the Intestinal Microbiota

The concentration of bacteria in the intestinal content is increasing when moving from the proximal to the distal small intestine, and further on in the large intestine [66]. The main parts of the intestine that are permanently colonized by a massive number of bacteria are the distal ileum and large intestine.

The Substrate Availability for Bacteria Is Different According to the Segments of the Intestinal Tract

In healthy individuals, the relatively rapid transit of the luminal content through the stomach and small intestine within 4–6 hours is not compatible with the development of a large concentration of bacteria in the proximal parts of the small intestine [67]. In contrast, the concentration of bacteria in the colon is as much as 10^9–10^{12} colony-forming unit (CFU)/g of content. Indeed, bacteria in healthy human feces represent more than 50% of the total solid part [68]. The spectacular increase in the concentration of bacteria in the large intestine is notably due to a much slower transit

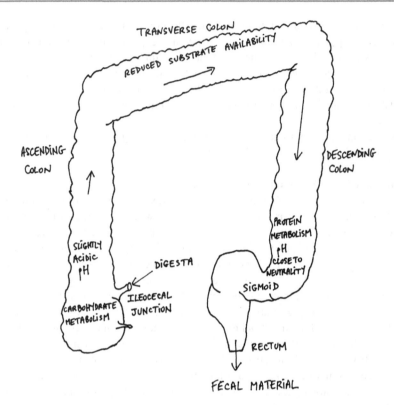

Fig. 2.1 Substrate availability in the human large intestine. In this scheme, major characteristics of substrate utilization in the different parts of the intestine are indicated

of the intestinal content in the large than in the small intestine, allowing intense metabolism of the available substrates supplied by the host. Indeed, in humans, the transit time of the content in the colon averages approximately 60 hours [69]. Transit time in the large intestine is much variable from an individual to another [70–72]. A longer transit time in the large intestine is associated with higher microbiota richness [73].

From a microbiological point of view, the large intestine can be viewed as an open system with substrates supplied by the host flowing in the small intestine, and for the undigestible part of the food, being transferred from the distal small intestine to the caecum through the ileo-caecal junction (Fig. 2.1). In healthy adults, the luminal contents of the distal parts of the intestinal tract are characterized by a low level of oxygen tension, thus explaining why they usually harbor large communities of obligate anaerobes [74].

The composition of the bacteria in the intestine, as well as the bacterial metabolic activity, are known to be influenced by dietary substrate availability, as will be developed in the following paragraphs. The small intestine, except in its distal part, contains a high concentration of nutrients in the post-prandial period, but low concentration of bacteria. Within the large intestine, the cecum and the proximal colon bacteria are supplied with a fluid that is relatively rich in undigested substrates, while, in contrast, the bacteria in the transverse and distal colon received a more limited amount of substrates, mainly because the readily available substrates have been already partly used in the more proximal parts of the large intestine [75] (Fig. 2.1).

The Microbiota Composition, Diversity, and Metabolic Capacity Depend Mainly on Environmental Parameters

Apart from the composition, diversity is another parameter that is often used to characterize the bacterial population [76]. Global differences in intestinal microbiota composition between groups of individuals have been described depending on the different geographical areas, apparently mainly because of different dietary habits [55, 77–84]. Importantly, microbiota that differ in terms of composition may share some degree of metabolic redundancy, thus yielding to not completely different bacterial metabolite profiles [85, 86].

Short-Term Dietary Modifications Can Modify Intestinal Microbiota Composition

In dietary intervention studies, the impact of dietary changes on the microbiota composition is measured in volunteers. In such an intervention, groups of individuals receive for given periods of time meals with different compositions. By doing so, it has been found that changes in numerous dietary parameters have indeed an impact on the microbiota composition, as will be detailed in the following paragraphs. The limitation of such studies is that the changes in microbiota composition is generally measured after several weeks of dietary intervention, and thus it is not known if the observed changes in microbiota composition observed remain for the long term. As the intestinal microbiota appears resilient, meaning that it can come back to an initial equilibrium after perturbation of its initial composition [87, 88], it is doubtful that the recorded changes will last in the long term.

To correctly interpret the data obtained from intervention in volunteers, it is useful to keep in mind that the initial differences in the bacterial composition between individuals (before the dietary intervention), may partly mask the differences attributable to changes in the food composition when comparing the control and experimental groups. From that point of view, the experimental procedure for dietary intervention in volunteers, in which a parallel design is used, displays some advantages. Indeed, in such a design, the volunteers are their own control, and thus the microbiota composition is analyzed in each participant before and after the experimental intervention after different periods of time, thus limiting the impact of inter-individual differences.

Importantly, as explained previously, almost all studies dedicated to the influence of food composition on the intestinal microbiota are originating from analysis of bacteria in the fecal material, thus ignoring the minor part of adhering bacteria present in the gut [3]. The vast part of bacteria is in faster transit, being finally evacuated in the feces after several days.

Host Genetic Background Is Playing a Minor Role in Fixing the Intestinal Microbiota Composition

As discussed above, the differences in intestinal microbiota composition between individuals, appear largely associated with eating habits but depend also on individual genetic factors. In humans, studies have been performed in monozygotic and dizygotic twins to determine if genetic factors in host can have an impact on the intestinal microbiota composition [89]. Indeed, it has been shown that a single host gene expression may be related to an effect on the diversity and population structure of the host gut microbiota [90]. Second evidence regarding the implication of host genetic factors for the intestinal microbiota composition is coming from a study showing that the Human Leukocyte Antigen (HLA) genotype plays a role in establishing infant's gut microbiota [91]. Human Leukocyte Antigen system is a highly polymorphic family of genes involved in immunity and responsible for identifying self-versus non-self [92]. In addition, indirect evidence has been supplied by studies showing that there is a greater similarity among the microbiota of the monozygotic twins than among microbiota of dizygotic twins [93, 94]. However, it is worth to underline that one study with monozygotic and dizygotic twins suggests that the overall heritability of the microbiome is low [55]. Thus, although genetic factors of the hosts presumably affect the intestinal microbiota composition, these genetic factors are likely to have less impact on the general population than the environmental factors [95], and notably the factors related to nutrition. Although it appears that the intestinal microbiota in one given individual is rather stable in the long-term perspective, episodes of imbalance of gut microbiota composition, called dysbiosis, are observed on some occasions, and such imbalance has been associated with alterations of intestinal epithelium physiology and metabolism, as will be developed in Chap. 4.

Microbiota Composition and Metabolic Activity toward Alimentary Compounds Can Be Modified by Pharmaceuticals, and Microbiota Can Metabolize Some Pharmaceuticals

Diverse pharmaceuticals can modify the microbiota composition in humans [96], and thus its overall metabolic activity toward compounds present in the alimentation. Thus, treatment with pharmaceuticals can modify the luminal concentrations of bacterial metabolites derived from dietary compounds, and then can modify the metabolic crosstalk between the intestinal microbes and their lodging host [97]. An illustrating example of such interaction is given by a study showing that individuals carrying an intestinal microbiota with high capacity for p-cresol production from tyrosine display reduced capacity to sulfonate acetaminophen (known as paracetamol in Europe), because of competition between absorbed p-cresol and acetaminophen for o-sulfonation in the liver [98]. This study is of prime importance as it demonstrates metabolic competition between a metabolite produced by the intestinal microbiota and a drug largely used for the treatment of fever and pain [99] (Fig. 2.2).

In turn, microbiota is known to metabolize some of the pharmaceuticals used for the treatment of diverse pathologies [100]. Microbes are equipped with a battery of enzymes which metabolize xenobiotics (notably pharmaceuticals in that case), and

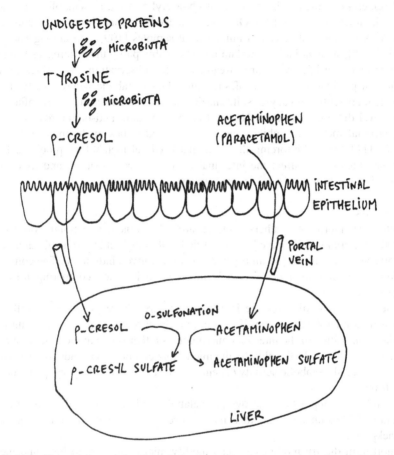

Fig. 2.2 An illustrative example of metabolic competition between a bacterial metabolite and a commonly used drug by the intestinal microbiota. P-cresol is a bacterial metabolite produced by the intestinal microbiota from tyrosine while acetaminophen is largely used for fever and pain treatment. These compounds compete for o-sulfonation in the liver cells

which are often different from the enzymatic activities of the host [101]. These bacterial enzymes include hydrolases, lyases, oxidoreductases, and transferases [102–104], all these activities being potentially able to modify pharmaceutical chemical structures, and thus their activities.

In fact, the human gut microbiota is known to transform more than 170 pharmaceuticals into metabolites with modified pharmacological properties when compared with the parent compounds [96, 101, 105]. For instance, sulfasalazine, a pharmaceutical compound used for inflammatory bowel disease medication [106], is absorbed by the small intestine up to 15–30%, while the rest of the compound is reduced by gut microbes into sulfapyridine and the active anti-inflammatory agent 5-aminosalicylic acid (5-ASA) [107, 108]. Then, various intestinal bacteria can metabolize 5-ASA into N-acetyl 5-ASA, a metabolite that lacks anti-inflammatory activity [97]. Of note, considerable variability in acetylation rate of 5-ASA has been observed in human fecal samples [109], and among bacterial species [110], such variability explaining presumably partly the differences between patients of the efficiency of drug treatment. This observation encourages further studies on possible implication of such microbial metabolic heterogeneity in the variable therapeutic efficacy of sulfasalazine among patients afflicted by inflammatory bowel diseases. Although N-acetyl ASA is without anti-inflammatory effect, this bacterial metabolite inhibits the growth of anaerobes, including *Clostridium difficile* [111], thus reinforcing the view that bacterial metabolites produced from pharmaceuticals may affect the intestinal microbiota composition, and its overall metabolic activity toward available substrates.

Key Points
- Intestinal microbiota gathers bacteria, archaea, viruses, and fungi. More than 1000 bacterial species have been identified within the gut microbiota, but around 150 species are found within the intestine of one given individual. The composition of the intestinal microbiota is much variable when comparing different individuals.
- The microbiota colonizes the intestine after birth. Three years after birth, the intestinal microbiota composition is not vastly different to the one in adult age. The composition of the intestinal microbiota is rather stable in the long term after first years of age. In the elderly, the microbiota composition gradually evolves.
- The bacterial metabolic activity depends on the ability of dietary and endogenous substrates.
- Microbiota composition and overall metabolic activity mainly depend on environmental parameters, and to a much lower extent on the individual genetic background.
- Short-term dietary modification can modify microbiota composition and metabolic activity.

Microbiota and the Host Intestinal Epithelium: Binding of Microbe Components to Epithelial Cells and Consequences
Apart from the metabolic crosstalk between the intestinal microbiota and the host, the microbiota and the host can exchange signals through the recognition of microbial components by the intestinal epithelial cells. These components can bind to a family of receptors called Toll-like receptors (TLRs) that are present in the intestinal epithelium. These microbial components, mainly of bacterial and viral origin, include the lipopolysaccharide (also called LPS) present in gram-negative bacteria, the bacterial flagellin, viral double- and single-stranded RNA, and various microbial lipids and lipoproteins [112–116]. This Toll-like receptor signaling is involved in multiple physiological processes including the control of renewal of the intestinal

epithelium, the epithelial barrier function, and the immune tolerance toward the gut microbiota, thus participating in gut homeostasis [117, 118]. Regarding metabolites produced by the intestinal bacteria from alimentary compounds, some of them, just like microbial components, can also bind to receptors present on host cells and modulate host physiological functions as will be detailed in the following chapters.

Microbiota, Production of Toxins, and the Host Intestinal Epithelium

Bacterial toxins, mostly made of different proteins or peptides produced by the metabolic activity of bacteria, are generally considered as a family of compounds that are distinct from the family of bacterial metabolites [119]. Emerging data point toward an effect of bacterial metabolites on some bacterial species that produce toxins.

Several compounds produced by intestinal bacteria are known to be deleterious for the host tissues. Just to take a first example of such toxins, toxin A and toxin B that are produced by *C. difficile* act on the colonic epithelium and immune cells, resulting in increased fluid secretion and inflammation [120]. Interestingly, the luminal environment can influence germination and growth from clostridial spores [121]. The secondary bile acid deoxycholate (that is synthesized in the large intestine by a subset of anaerobic bacteria from a minor proportion of unreabsorbed bile acids, see Chap. 3), induces germination of *C. difficile* spores but prevented the growth of this bacteria in its vegetative form [122]. Then, it appears in that example that the bacterial metabolite deoxycholate is acting on the biology of *C. difficile*.

Colibactin is another example of bacterial toxins that are produced by different bacterial species including several members of the *Enterobacteriaceae* family, and which induce chromosomal instability and DNA damage in colonic epithelial cells [123–126]. However, in that case, there is still no information on the effects, if any, of the impact of the luminal environment characteristics on the production of colibactin by bacterial species.

Key Points

- Not only bacterial metabolites, but also microbial components, mainly from bacterial and viral origin, can bind to receptors present in the intestinal epithelial cells, modulating its physiology and metabolism.
- Some bacterial metabolites can affect the biology of toxin-producing bacteria.

References

1. Lighthart K, Belzer C, de Vos WM, Tytgat HLP. Bridging bacteria and the gut: functional aspects of type IV pili. Trends Microbiol. 2020;28(5):340–8.
2. Palmela C, Chevarin C, Xu Z, Torres J, Sevrin G, Hirten R, Barnich N, Ng SC, Colombel JF. Adherent-invasive Escherichia coli in inflammarory bowel disease. Gut. 2018;67(3): 574–87.
3. Rolhion N, Darfeuille-Michaud A. Adherent-invasive Escherichia coli in inflammatory bowel disease. Inflamm Bowel Dis. 2007;13(10):1277–83.

4. Matijasic M, Mestrovic T, Paljetak HC, Peric M, Baresic A, Verbanac D. Gut microbiota beyond bacteria-mycobiome, virome, archeome, and eukaryotic parasites in IBD. Int J Mol Sci. 2020;21(8):2668.
5. Valentine DL. Adaptations to energy stress dictate the ecology and evolution of Archea. Nat Rev Microbiol. 2007;5(4):316–23.
6. Janssen PH, Kirs M. Structure of the archeal community of the rumen. Appl Env Microbiol. 2008;74(12):3619–25.
7. Raymann K, Moeller AH, Goodman AL, Ochman H. Unexplored archeal diversity in the great ape gut microbiome. mSphere. 2017;2(1):00026–17.
8. Gaci N, Borrel G, Tottey W, O'Toole PW, Brugère JF. Archaea and the human gut: new beginning of an old story. World J Gastroenterol. 2014;20(43):16062–78.
9. Wanpach L, Heintz-Buschart A, Hogan A, Muller EEL, Narayanamasy S, Laczny CC, Hugerth LW, Bindl L, Bottu J, Andersson AF, de Beaufort C, Wilmes P. Colonization and succession within the human gut microbiome by archea, bacteria, and microeukaryotes during the first year of life. Front Microbiol. 2017;8:738.
10. Samuel BS, Gordon JI. A humanized gnotobiotic mouse model of host-archeal-bacterial mutualism. Proc Natl Acad Sci U S A. 2006;103(26):10011–6.
11. Dridi B, Henry M, El Khéchine A, Raoult D, Drancourt M. High prevalence of Methanobrevibacter smithii and methanosphaera stadtmanae detected in the human gut using an improved DNA detection protocol. PLoS One. 2009;4(9):e7063.
12. Carding SR, Davis N, Hoyles L. Review article: the human intestine virome in health and disease. Aliment Pharmacol Ther. 2017;46(9):800–15.
13. Sausset R, Petit MA, Gaboriau-Routhiau V, De Paepe M. New insights into intestinal phages. Mucosal Immunol. 2020;13(2):205–15.
14. Balfour Sartor R, Wu GD. Roles for intestinal bacteria, viruses, and fungi in pathogenesis of inflammatory bowel diseases and therapeutic approaches. Gastroenterology. 2017;152(2):327–39.
15. Hallen Adams HE, Suhr MJ. Fungi in the healthy human gastrointestinal tract. Virulence. 2017;8(3):352–8.
16. Mukherjee PK, Sendid B, Hoarau G, Colombel JF, Poulain D, Ghannoum MA. Mycobiota in gastrointestinal diseases. Nat Rev Gastroenterol Hepatol. 2015;12(2):77–87.
17. Paterson MJ, Oh S, Underhill DM. Host-microbe interactions: commensal fungi in the gut. Curr Opin Microbiol. 2017;40:131–7.
18. Lacour M, Zunder T, Huber R, Sander A, Daschner F, Frank U. The pathogenic significance of intestinal Candida colonization. A systematic review from an interdisciplinary and environmental medical point of view. Int J Hyg Environ Health. 2002;205(4):257–68.
19. Burgess SL, Gilchrist CA, Lynn TC, Petri WA Jr. Parasitic protozoan and interactions with the host intestinal microbiota. Infect Immunol. 2017;85(8):e00101–17.
20. Salazar N, Valdés-Varela L, Gonzalez S, Gueimonde M, de Los Reyes-Gavilan CG. Nutrition and the gut microbiome in the elderly. Gut Microbes. 2017;8(2):82–97.
21. Ley RE, Peterson DA, Gordon JI. Ecological and evolutionary forcers shaping microbial diversity in the human intestine. Cell. 2006;124(4):837–48.
22. Eckburg PB, Bik EM, Bernstein CN, Purdom E, Dethlefsen L, Sargent M, Gill SR, Nelson KE, Relman DA. Diversity of the human intestinal microbial flora. Science. 2005;308(5728):1635–8.
23. Sender R, Fuchs S, Milo R. Are we really vastly outnumbered? Revisiting the ratio of bacterial to host cells in humans. Cell. 2016;164(3):337–40.
24. Calvez J, Benoit S, Piedcoq J, Khodorova N, Azzout-Marniche D, Tomé D, Benamouzig R, Airinei G, Gaudichon C. Very low ileal nitrogen and amino acid digestibility of zein compared to whey protein isolate in healthy volunteers. Am J Clin Nutr. 2021;113(1):70–82.
25. Chacko A, Cummings JH. Nitrogen losses from the human small bowel: obligatory losses and the effect of physical form of food. Gut. 1988;29(6):809–15.

26. Jimenez E, Marin ML, Martin R, Odriozola JM, Olivares M, Xaus J, Fernandez L, Rodriguez JM. Is meconium from healthy newborns really sterile? Res Microbiol. 2008;159(3):187–93.
27. Aagaard K, Ma J, Antony KM, Ganu R, Petrosino J, Versalovic J. The placenta harbors a unique microbiome. Sci Transl Med. 2014;6(237):237ra65.
28. Di Giulio DB, Romero R, Amogan HP, Kusanovic JP, Bik EM, Gotsch F, Kim CJ, Erez O, Edwin S, Relman DA. Microbial prevalence, diversity and abundance in amniotic fluid during preterm labor: a molecular and culture-based investigation. PLoS One. 2008;3(8):e3056.
29. Ardissone AN, de la Cruz DM, Davis-Richardson AG, Rechcigl KT, Li N, Drew JC, Murgas-Torrazza R, Sharma R, Hudak ML, Triplett EW, Neu J. Meconium microbiome analysis identifies bacteria correlated with premature birth. PLoS One. 2014;9(3):e90784.
30. Moles L, Gomez M, Heilig H, Bustos G, Fuentes S, de Vos W, Fernandez L, Rodriguez JM, Jimenez E. Bacterial diversity in meconium of preterm neonates and evolution of their fecal microbiota during the first month of life. PLoS One. 2013;8(6):e66986.
31. Milani C, Duranti S, Bottacini F, Casey E, Turroni F, Mahony J, Belzer C, Delgado Palacio S, Arboleya Montes S, Mancabelli L, Lugli GA, Rodriguez JM, Bode L, de Vos W, Geuimonde M, Margolles A, van Sinderen D, Ventura M. The first microbial colonizers of the human gut: composition, activities, and health implications of the infant gut microbiota. Microbiol Mol Biol Rev. 2017;81(4):e00036–17.
32. Palmer C, Bik EM, Di Giulio DB, Relman DA, Brown PO. Development of the human infant intestinal microbiota. PLoS Biol. 2007;5(7):e177.
33. Healey GR, Murphy R, Brough L, Butts CA, Coad J. Interindividual variability in gut microbiota and host response to dietary interventions. Nutr Rev. 2017;75(12):1059–80.
34. Houghteling PD, Walker WA. Why is initial bacterial colonization of the intestine important to infants' and children' health? J Pediatr Gastroenterol Nutr. 2015;60(3):294–307.
35. Matamoros S, Gras-Leguen C, Le Vacon F, Potel G, de La Cochetiere MF. Development of intestinal microbiota in infants and its impact on health. Trends Microbiol. 2013;21(4):167–73.
36. Rodriguez JM, Murphy K, Stanton C, Ross RP, Kober OI, Juge N, Avershina E, Rudi K, Narbad A, Jenmalm MC, Marchesi JR, Collado MC. The composition of the gut microbiota throughout life, with an emphasis on early life. Microbiol Ecol Health Dis. 2015;26:26050.
37. Yassour M, Vatanen T, Siljander H, Hämäläinen AM, Härkönen T, Ryhänen SJ, Franzoca EA, Vlamakis H, Huttenhower C, Gevers D, Lander ES, Knip M, DIABIMMUNE study Group, Xavier RJ. Natural history of the infant gut microbiome and impact of antibiotic treatment on bacterial strain diversity and stability. Sci Transl Med. 2016;8(343):343ra81.
38. Dominguez-Bello MG, Costello EK, Contreras M, Magris M, Hidalgo G, Fierer N, Knight R. Delivery mode shapes the acquisition and structure of the initial microbiota across multiple body habitats in newborns. Proc Natl Acad Sci U S A. 2010;107(26):11971–5.
39. Chu DM, Ma J, Prince AL, Antony KM, Seferovic MD, Aagaard KM. Maturation of the infant microbiome community structure and function across multiple body sites and in relation to mode of delivery. Nat Med. 2017;23(3):314–26.
40. Falony G, Joossens M, Vieira-Silva S, Wang J, Darzi Y, Faust K, Kurilshikov A, Bonder MJ, Valles-Colomer M, Vandeputte D, Tito RY, Chaffron S, Rymenans L, Verspecht C, De Sutter L, Lima-Mendez G, D'Hoe K, Jonckeere K, Homola D, Garcia R, Tigchelaar EF, Eeckhaudt L, Fu J, Henckaerts L, Zhernakova A, Wijmenga C, Raes J. Population-level analysis of gut microbiome variation. Science. 2016;352(6285):560–4.
41. Bezirtzoglou E, Tsiotsias A, Welling GW. Microbiota profile in feces of breast- and formula-fed newborns by using in situ hybridization (FISH). Anaerobe. 2011;17(6):478–82.
42. He X, Parenti M, Grip T, Lönnerdal B, Timby N, Domelöff M, Hernell O, Slupsky CM. Fecal microbiome and metabolome of infants fed bovine MFGM supplemented formula or standard formula with breast-fed infants as reference: a randomized controlled study. Sci Rep. 2019;9 (1):11589.
43. Bode L. Human milk oligosachharides: prebiotics and beyond. Nutr Rev. 2009;67(S2):S183–91.

44. Smilowitz JT, Lebrilla CB, Mills DA, Bruce German J, Freeman SL. Breast milk oligosaccharides: structure-function relationships in the neonate. Annu Rev Nutr. 2014;34: 143–69.
45. Andreas NJ, Kampmann B, Mehring Le-Doare K. Human breast milk: a review on its composition and bioactivity. Early Hum Dev. 2015;91(11):629–35.
46. Moubareck CA. Human milk microbiota and oligosaccharides: a glimpse into benefits, diversity, and correlations. Nutrients. 2021;13(4):1123.
47. Morrow AL, Ruiz-Palacios GM, Altaye M, Jiang X, Guerrero ML, Meinzen-Derr JK, Farkas T, Chaturvedi P, Pickering LK, Newburg DS. Human milk oligosachharide blood group epitopes and innate immune protection against campylobacter and calicivirus diarrhea in breastfed infants. Adv Exp Med Biol. 2004;554:443–6.
48. Davis MY, Zhang H, Brannan LE, Carman LJ, Boone JH. Rapid change of fecal microbiome and disappearance of Clostridium difficile in a colonized infant after transition from breast milk to cow milk. Microbiome. 2016;4(1):53.
49. Fallami M, Amarri S, Uusijarvi A, Adam R, Khanna S, Aguilera M, Gil A, Vieites JM, Norin E, Young D, Scott JA, Doré J, Edwards CA, The Infabio Team. Determinants of the human infant intestinal microbiota after the introduction of first complementary foods in infant samples from five European centers. Microbiology (Reading). 2011;157(5):1385–92.
50. Koenig JE, Spor A, Scalfone N, Fricker AD, Stombaugh J, Knight R, Angenent LT, Ley RE. Succession of microbial consortia in the developing infant gut microbiome. Proc Natl Acad Sci. 2011;108(S1):4578–85.
51. Tanaka S, Kobayashi T, Songjinda P, Tateyama A, Tsubouchi M, Kiyohara C, Shirakawa T, Sonomoto K, Kakayama J. Influence of antibiotic exposure in the early postnatal period on the development of intestinal microbiota. FEMS Immunol Med Microbiol. 2009;56(1):80–7.
52. Lozupone CA, Stombaugh JI, Gordon JI, Jansson JK, Knight R. Diversity, stability and resilience of the human gut microbiota. Nature. 2012;489(7415):220–30.
53. Valdes AM, Walter J, Segal E, Spector TD. Role of the gut microbiota in nutrition and health. BMJ. 2018;361:k2179.
54. Costello EK, Lauber CL, Hamady M, Fierer N, Gordon JI, Knight R. Bacterial community variation in human body habitats across space and time. Science. 2009;326(5960):1694–7.
55. Yatsunenko T, Rey FE, Manary MJ, Trehan I, Dominguez-Bello MG, Contreras M, Magris M, Hidalgo G, Baldassano RN, Anokhin AP, Health AC, Warner B, Reeder J, Kuczynski J, Caporaso JG, Lozupone CA, Lauber C, Clemente JC, Knights D, Knight R, Gordon JI. Human gut microbiome viewed across age and geography. Nature. 2012;486(7402):222–7.
56. Biagi E, Nylund L, Candela M, Ostan R, Bucci L, Pini E, Nikkïla J, Monti D, Satokari R, Franceschi C, Brigidi P, De Vos W. Through ageing, and beyond: gut microbiota and inflammatory status in seniors and centenarians. PLoS One. 2010;5(5):e10667.
57. Claesson MJ, Cusack S, O'Sullivan O, Greene-Diniz R, de Weerd H, Flannery E, Marchesi JR, Falush D, Dinan T, Fitzgerald G, Stanton C, van Sinderen D, O'Connor M, Harnedy N, O'Connor K, Henry C, O'Mahony D, Fitzgerald AP, Shanahan F, Twomey C, Hill C, Ross RP, O'Toole PW. Composition, variability, and temporal stability of the intestinal microbiota of the elderly. Proc Natl Acad Sci U S A. 2011;108(S1):4586–91.
58. O'Toole PW, Jeffery IB. Gut microbiota and aging. Science. 2015;350(6265):1214–5.
59. Odamaki T, Kato K, Sugahara H, Hashikura N, Takahashi S, Xiao JZ, Abe F, Osawa R. Age-related changes in gut microbiota composition from newborn to centenarian: a cross-sectional study. BMC Microbiol. 2016;16:90.
60. Govindaraju T, Sahle BW, McCaffrey TA, McNeil JJ, Owen AJ. Dietary patterns and quality of life in older adults. Nutrients. 2018;10(8):971.
61. Grassi M, Petraccia L, Mennumi G, Fontana M, Scarno A, Sabetta S, Fraioli A. Changes, functional disorders, and diseases in the gastrointestinal tract of the elderly. Nutr Hosp. 2011;26(4):659–68.
62. Drozdowski L, Thomson AB. Aging and the intestine. World J Gastroenterol. 2006;12(47): 7578–84.

63. Soenen S, Rayner CK, Jones KL, Horowitz M. The ageing gastrointestinal tract. Curr Opin Clin Nutr Metab Care. 2016;19(1):12–8.
64. Salazar N, Gonzales S, Nogacka AM, Rios-Covian D, Arboleya S, Guemonde M, de Los Reyes-Gavilan CG. Microbiome: effects of ageing and diet. Curr Issues Mol Biol. 2020;36: 33–62.
65. Woodmansey EJ, McMurdo MET, Macfarlane GT, Macfarlane S. Comparison of compositions and metabolic activities of fecal microbiotas in young adults and in antibiotic-treated and non-antibiotic-treated elderly subjects. Appl Environ Microbiol. 2004;70(10): 6113–22.
66. Gibson PR, Barrett JS. The concept of small intestine bacterial overgrowth in relation to functional gastrointestinal disorders. Nutrition. 2010;26(11–12):1038–43.
67. Gorbach SL. Population control in the small intestine. Gut. 1967;8(6):530–2.
68. Stephen AM, Cummings JH. The microbial contribution to human fecal mass. J Med Microbiol. 1980;13(1):45–56.
69. Stephen AM, Wiggins HS, Cummings JM. Effects of changing transit time on colonic microbial metabolism in man. Gut. 1987;28(5):601–9.
70. Bharucha AE, Anderson A, Bouchoucha M. More movement with evaluating colonic transit in humans. Neurogastroenterol Motil. 2019;31(2):e13541.
71. Miller LE, Ibarra A, Ouwehand AC. Normative values for colonic transit time and patient assessment of constipation in adults with functional constipation: systematic review with meta-analysis. Clin Med Insights Gastroenterol. 2017;11:1–8.
72. Wilson CG. The transit of dosage form through the colon. Int J Pharm. 2010;395(1–2):17–25.
73. Roager HM, Hansen LBS, Bahl MI, Frandsen HL, Carvalho V, Gobel RJ, Dalgaard MD, Plichta DR, Sparholt MH, Vestergaard H, Hansen T, Sicheritz-Ponten T, Bjorn Nielsen H, Pedersen O, Lauritzen L, Kristensen M, Gupta R, Licht TR. Colonic transit time is related to bacterial metabolism and mucosal turnover in the gut. Nature Microbiol. 2016;1(9):16093.
74. Rigottier-Gois L. Dysbiosis in inflammatory bowel diseases: the oxygen hypothesis. ISME J. 2013;7(7):1256–61.
75. Macfarlane GT, Cummings JH. The colonic flora, fermentation, and large bowel digestive function. In: Phillips SF, Pemberton JH, Shorter RG, editors. The large intestine: physiology, pathophysiology, and disease. New York: Raven Press; 1991.
76. Rutayisire E, Huang K, Liu Y, Tao F. The mode of delivery affects the diversity and colonization pattern of the gut microbiota during the first year of infants' life: a systematic review. BMC Gastroenterol. 2016;16(1):86.
77. Borenstein E, Kupiec M, Feldman MW, Ruppin E. Large-scale reconstruction and phyloge-netic analysis of metabolic environments. Proc Natl Acad Sci U S A. 2008;105(38):14482–7.
78. David LA, Maurice CF, Carmody RN, Gootenberg DB, Button JE, Wolfe BE, Ling AV, Devlin AS, Varma Y, Fischbach MA, Biddinger SB, Dutton RJ, Turnbaugh PT. Diet rapidly and reproducibly alters the human gut microbiome. Nature. 2014;505(7484):559–63.
79. De Filippo C, Di Paola M, Ramazzotti M, Albanese D, Pieraccini G, Banci E, Miglietta F, Cavalieri D, Lionetti P. Diet, environment, and gut microbiota. A preliminary investigation in children living in rural and urban Burkina Faso and Italy. Front Microbiol. 2017;8:1979.
80. Freilich S, Goldovsky L, Gottlieb A, Blanc E, Tsoka S, Ouzounis CA. Stratification of co-evolving genomic groups using ranked phylogenetic profiles. BMC Bioinformatics. 2009;10:355.
81. Gomez A, Petrzelkova KJ, Burns MB, Yeoman CJ, Amato KR, Vlckova K, Modry D, Todd A, Jost Robinson CA, Remis MJ, Torralba MG, Morton E, Umana JD, Carbonero F, Rex Gaskins H, Nelson KE, Wilson BA, Stumpf RM, White BA, Leigh SR, Blekhman R. Gut microbiome of coexisting BaAka pygmies and Bantu reflects gradients of traditional subsis-tence patterns. Cell Rep. 2016;14(9):2142–53.
82. Martinez I, Stegen JC, Maldonado-Gomez MX, Murat Eren A, Siba PM, Greenhill AR, Walter J. The gut microbiota of rural Papua new Guineans: composition, diversity patterns, and ecological processes. Cell Rep. 2015;11(4):527–38.

83. Salonen A, de Vos WM. Impact of diet on human intestinal microbiota and health. Annu Rev Food Sci Technol. 2014;5:239–62.
84. Schnorr SL, Candela M, Rampelli S, Centanni M, Consolandi C, Basaglia G, Turroni S, Biagi E, Peano C, Severgnini M, Fiori J, Gotti R, De Bellis G, Luiselli D, Brigidi P, Mabulla A, Marlowe F, Henry AG, Crittenden AN. Gut microbiome of the Hadza hunter-gatherers. Nat Commun. 2014;5:3654.
85. Moya A, Ferrer M. Functional redundancy-induced stability of gut microbiota subjected to disturbance. Trends Microbiol. 2016;24(5):402–13.
86. Thursby E, Juge N. Introduction to the human gut microbiota. Biochem J. 2017;474(11): 1823–36.
87. Fassarella M, Blaak EE, Penders J, Nauta A, Smidt H, Zoedental EG. Gut microbiome stability and resilience: elucidating the response to perturbations in order to modulate gut health. Gut. 2021;70(3):595–605.
88. Sommer F, Moltzau Anderson J, Bharti R, Raes J, Rosenthal P. The resilience of the intestinal microbiota influences health and disease. Nat Rev Microbiol. 2017;15(10):630–8.
89. Marietta E, Rishi A, Taneja V. Immunogenetic control of the intestinal microbiota. Immunology. 2015;145(3):313–22.
90. Spor A, Koren O, Ley R. Unravelling the effects of the environment and host genotype on the gut microbiome. Nat Rev Microbiol. 2011;9(4):279–90.
91. De Palma G, Capilla A, Nova E, Castillejo G, Varea V, Pozo T, Garrote JA, Polanco I, Lopez A, Ribes-Koninckx C, Marcos A, Garcia-Novo MD, Calvo C, Ortigosa L, Pena-Quintana L, Palau F, Sang Y. Influence of milk-feeding type and genetic risk of developing coeliac disease on intestinal microbiota of infants: the PROFICIEL study. PLoS One. 2012;7 (2):e30791.
92. Madden K, Chabot-Richards D. HLA testing in the molecular diagnostic laboratory. Virchows Arch. 2019;474(2):139–47.
93. Benson AK, Kelly SA, Legge R, Ma F, Low SJ, Kim J, Zhang M, Oh PL, Nehrenberg D, Hua K, Kachman SD, Moriyama EN, Walter J, Peterson DA, Pomp D. Individuality in gut microbiota composition is a complex polygenic trait shaped by multiple environmental and host genetic factors. Proc Natl Acad Sci U S A. 2010;107(44):18933–8.
94. Goodrich JK, Waters JL, Poole AC, Sutter JL, Koren O, Blekhman R, Beaumont M, van Treuren W, Knight R, Bell JT, Spector TD, Clark AG, Ley RE. Human genetics shape the gut microbiome. Cell. 2014;159(4):789–99.
95. Rothschild D, Kurilshikov A, Korem T, Zeevi D, Costea PI, Godneva A, Kalka IN, Bar N, Shilo S, Lador D, Vila AV, Zmora N, Pevsner-Fischer M, Israeli D, Kosower N, Malka G, Wolf BC, Avnit-Sagi T, Lotan-Pompan M, Weinberger C, Zhernakova A, Elinav E, Segal E. Environment dominates over host genetics in shaping the human gut microbiota. Nature. 2018;555(7695):210–5.
96. Weersma RH, Zhernakova A, Fu J. Interactions between drugs and the gut microbiome. Gut. 2020;69(8):1510–9.
97. Koppel N, Maini Redkal V, Balskus EP. Chemical transformation of xenobiotics by the human microbiota. Science. 2017;356(6344):eaag2770.
98. Clayton TA, Baker D, Lindon JC, Everett JR, Nicholson JK. Pharmacometabonomic identification of a significant host-microbiome interaction affecting human drug metabolism. Proc Natl Acad Sci U S A. 2009;106(34):14728–33.
99. Tan E, Braithwaite I, McKinley CJD, Dalziel SR. Comparison of acetaminophen (paracetamol) with ibuprofen for tnt of fever or pain in children younger than 2 years: a systematic review and meta-analysis. JAMA Netw Open. 2020;3(10):e2022398.
100. Zimmermann M, Zimmermann-Kogadeeva M, Wegmann R, Goodman AL. Mapping human microbiome drug metabolism by gut bacteria and their genes. Nature. 2019;570(7762):462–7.
101. Sousa T, Patterson R, Moore V, Carlsson A, Abrahamsson B, Basit AW. The gastrointestinal microbiota as a site for the biotransformation of drugs. Int J Pharm. 2008;363(1–2):1–25.

102. El Kaoutari A, Armougom F, Gordon JI, Raoult D, Henrissat B. The abundance and variety of carbohydrate-active enzymes in the human gut microbiota. Nat Rev Microbiol. 2013;11(7): 497–504.
103. Levin BJ, Huang YY, Peck SC, Wei Y, Martinez-Del Campo A, Marks JA, Franzosa EA, Huttenhower C, Balskus EP. A prominent glycyl radical enzyme in human gut microbiomes metabolizes trans4-hydroxl-l-proline. Science. 2017;355(6325):eaai8386.
104. Ryan A, Kaplan E, Nebel JC, Polycarpon E, Crescente V, Lowe E, Preston GM, Sim E. Identification of NAD(P)H quinone oxidoreductase activity in azoreductases from P. aeruginosa: azoreductases and NAD(P)H quinone oxidoreductases belong to the same FMN-dependent superfamily of enzymes. PLoS One. 2014;9(6):e98551.
105. Spanogiannopoulos P, Bess EN, Carmody RN, Turnbaugh PJ. The microbial pharmacists within us: a metagenomics view of xenobiotic metabolism. Nat Rev Microbiol. 2016;14(5): 273–87.
106. Damiao AOMC, de Azevedo MFC, Carlos AS, Wada MY, Silva TVM, Feitosa FC. Conventional therapy for moderate to severe inflammatory bowel disease: a systematic literature review. World J Gastroenterol. 2019;25(9):1142–57.
107. Azadkhan AK, Truelove SC, Aronson JK. The disposition and metabolism of sulphasalazine in man. Br J Clin Pharmacol. 1982;13:523–8.
108. Barberio B, Segal JP, Quraishi MN, Black CJ, Savarino EV, Ford AC. Efficacy of oral, topical, or combined oral and topical 5-aminosalicylates, in ulcerative colitis: systematic review and network meta-analysis. J Crohns Colitis. 2021;15(7):1184–96.
109. van Hogezand RA, Kennis HM, van Schaik A, Koopman JP, van Hees PA, van Tongeren JH. Bacterial acetylation of 5-aminosalicylic acid in fecal suspensions cultured under aerobic and anaerobic conditions. Eur J Clin Pharmacol. 1992;43(2):189–92.
110. Deloménie C, Fouix S, Longuemaux S, Brahimi N, Bizet C, Picard B, Denamur E, Dupret JM. Identification and functional characterization of arylamine N-acetyltransferases in Eubacteria: evidence for highly selective acetylation of 5-aminosalicylic acid. J Bacteriol. 2001;183(11):3417–27.
111. Sandberg-Gertzen H, Kjellander J, Sundberg-Gilla B, Järnerot G. In vitro effects of sulphasalazine, azodisal sodium, and their metabolites on Clostridium difficile and some other fecal bacteria. Scand J Gastroenterol. 1985;20(5):607–12.
112. Harris G, KuoLee R, Chen W. Role of toll-like receptors in health and diseases of gastrointestinal tract. World J Gastroenterol. 2006;12(14):2149–60.
113. Hayashi F, Smith KD, Ozinski A, Hawn TR, Yi EC, Goodlett DR, Eng JK, Akira S, Underhill DM, Aderem A. The innate immune response to bacterial flagellin is mediated by toll-like receptor 5. Nature. 2001;410(6832):1099–103.
114. Hemmi H, Takeuchi O, Kawai T, Kaisho T, Sato S, Sanjo H, Matsumoto M, Hoshino K, Wagner H, Takeda K, Akira S. A toll-like receptor recognizes bacterial DNA. Nature. 2000;408(6813):740–5.
115. Takeuchi O, Hoshino K, Kawai T, Sanjo H, Takada H, Ogawa T, Takeda K, Akira S. Differential roles of TLR2 and TLR4 in recognition of gram-negative and gram-positive bacterial cell wall components. Immunity. 1999;11(4):443–51.
116. Underhill DM, Ozinsky A, Smith KD, Aderem A. Toll-like receptor-2 mediates mycobacteria-induced proinflammatory signaling in macrophages. Proc Natl Acad Sci U S A. 1999;96(25): 14459–63.
117. Burgueno JF, Abreu MT. Epithelial toll-like receptors and their role in gut homeostasis and disease. Nat Rev Gastroenterol Hepatol. 2020;17(5):263–78.
118. Frosali S, Pagliari D, Gambassi G, Landolfi R, Pandolfi F, Cianci R. How the intricate interaction among toll-like receptors, microbiota, and intestinal immunity can influence gastrointestinal physiology. J Immunol Res. 2015;2015:489821.
119. Doxey AC, Mansfield MJ, Montecucco C. Discovery of novel bacterial toxins by genomics and computational biology. Toxicon. 2018;147:2–12.

120. Chandrasekaran R, Lacy DB. The role of toxins in Clostridium difficile infection. FEMS Microbiol Rev. 2017;41(6):723–50.
121. Abt MC, McKenney PT, Palmer EG. Clostridium difficile colitis: pathogenesis and host defense. Nat Rev Microbiol. 2016;4(10):609–20.
122. Sorg JA, Sonenhein AL. Bile salts and glycine as cogerminants for Clostridium difficile spores. J Bacteriol. 2008;190(7):255–2512.
123. Dubinski V, Dotan I, Gophna U. Carriage of colibactin-producing bacteria and colorectal cancer risk. Trends Microbiol. 2020;28(11):874–6.
124. Faïs T, Delmas J, Barnich N, Bonnet R, Delmasso G. Colibactin: more than a new bacterial toxin. Toxins (Basel). 2018;10(4):151.
125. Wami H, Wallenstein A, Sauer D, Stoll M, von Bünau R, Oswald E, Müller R, Dobrindt U. Insights into evolution and coexistence of the colibactin- and yersiniabactin secondary metabolite determinants in enterobacterial populations. Microb Genom. 2021;7(6):000577.
126. Wilson MR, Jiang Y, Villalta PW, Stornetta A, Boudreau PD, Carra A, Brennan CA, Chun E, Ngo L, Samson LD, Engelward BP, Garrett WS, Balbo S, Balskus EP. The human gut bacterial genotoxin colibactin alkylates DNA. Science. 2019;363(6428):eaar7785.

Metabolism of Dietary Substrates by Intestinal Bacteria and Consequences for the Host Intestine

3

Abstract

Numerous substrates from alimentary origin are metabolized by the gut microbiota. Their luminal concentrations are the net result of production, utilization, and absorption through the intestinal epithelium. Undigested proteins in the colon are degraded into amino acids by the bacteria which utilize them for their own protein synthesis and for other metabolic pathways, giving rise to numerous bacterial metabolites. Several among them, including formate, oxaloacetate, *p*-cresol, and indole are used for microbial communication. Others like lactate and hydrogen sulfide are used as fuels by colonocytes. Some like branched-chain fatty acids are involved in electrolyte transport. Indole and indole-related compounds represent beneficial bacterial metabolites for the colonic mucosa physiology, while bacterial metabolites like ammonia, *p*-cresol, and hydrogen sulfide affect energy metabolism in colonocytes when present in excess. Polyamines produced by intestinal bacteria are involved in bacterial growth and epithelium renewal, but excessive concentrations of putrescine exert deleterious effects on the intestinal barrier function. The intestinal microbiota produces compounds with neurotransmitter functions in the host. Norepinephrine for instance is involved in bacterial communication, while dopamine and tryptamine are involved in colon physiology. Consumption of high-protein diet by volunteers modifies the bacterial metabolite profile as well as the expression of genes involved in the epithelium renewal processes in rectal biopsies. Regarding indigestible polysaccharides, they are metabolized by the intestinal microbiota giving rise to short-chain fatty acid production. Among these latter, butyrate is highly metabolized in the mitochondria of colonocytes, allowing energy production and regulation of the cytosolic concentration of butyrate. Such regulation appears crucial for determining the effects of butyrate on gene expression. In addition, short-chain fatty acids are involved in different aspects of colon physiology, including notably the regulation of the intestinal immune system. Lower intake of indigestible polysaccharides shifts the metabolic activity of the intestinal bacteria to other

F. Blachier, *Metabolism of Alimentary Compounds by the Intestinal Microbiota and Health*, https://doi.org/10.1007/978-3-031-26322-4_3

sources of substrates like amino acids and lipids. Regarding lipids, few proportions of them are usually transferred from the small to the large intestine. The effects of lipid-derived bacterial metabolites on the colonic epithelium have been little investigated. The effects of phytochemicals, notably polyphenols, on the colon epithelium have been investigated, and beneficial effects of several polyphenol-derived bacterial metabolites have been reported in different experimental situations. Regarding vitamins of the B group, although the intestinal bacteria use primarily these compounds for their own metabolism and growth, a part of bacteria-derived vitamins B may remain available for the host. Lastly, it has been shown that several food additives and compounds produced during cooking processes are metabolized by the intestinal microbiota giving rise to bacterial metabolites with biological activities on the colonic epithelium.

As can be easily imagined, the rapid growth of bacteria within the large intestine content requires extensive amounts of substrates that are provided by the host. Approximately half of the bacterial biomass in the colon is lost via the feces every day [1], thus indicating an intense bacterial growth and associated anabolism (synthesis) that is possible only if the required substrates are available in sufficient quantities. The substrate requirement for bacterial growth in terms of quantities is much different depending on the concentration of bacteria in one given part of the intestine. In the colon, where the concentration of bacteria is very high, the substrate requirement is obviously much higher than in the proximal part of the small intestine. The situation in the distal part of the small intestine, thus in the distal ileum, is intermediate with an already high concentration of bacteria. In addition, in the large intestine, in contrast with the situation in the small intestine, there is a relatively long stasis of the luminal content, thus increasing the time during which bacteria are in contact with their substrates.

Production of Metabolites by Intestinal Bacterial Species Depends on Substrate Availability and on Bacterial Metabolic Capacities
The bacteria are thus dependent on the dietary substrates consumed by the host, but they also use endogenous substrates produced in the body. During metabolism of dietary substrates, intestinal bacteria can produce numerous metabolic end products. They can also produce intermediary products that can accumulate if the synthesis rate is greater than its further metabolism. These metabolites can accumulate in the luminal fluid. They can be in different forms, either as a gas that can be partly dissolved in the aqueous phase of the luminal content, dissolved in case of hydrophilic metabolites, or for hydrophobic compounds in poorly or not dissolved form. The metabolites generated by the microbiota can be either in free form or can be bound to various luminal compounds [2].

The concentration of a given bacterial metabolite in the intestinal content at a given time will be primarily the net result of its production by the synthesizing bacterial species, of its utilization by other bacteria in the consortium with the appropriate metabolic capacities, and of its absorption across the intestinal epithelium. In other word, for a given stable population of bacteria within the intestinal content, the substrate availability will be one of the main parameters in fixing amounts of bacterial metabolites produced. The absorption of the bacterial metabolites will depend on numerous parameters including the form in which bacterial metabolites are, either in free or bound form. The physical form of some bacterial metabolites, either cationic, anionic, or neutral will be also important, notably for transport and diffusion across the intestinal epithelium. The concentrations of bacterial metabolites available for microbial communication and for microbial-host communication will be the primary parameters that will determinate the effects of these compounds.

3.1   Metabolism of Proteins and Amino Acids by the Intestinal Microbiota and Impact on the Intestinal Epithelium Metabolism and Functions

In the small intestine, the digestion of proteins by the exocrine pancreas secretion releases peptides and amino acids. Although the intestinal bacteria have the capacity to synthesize amino acids for their own metabolism, not all bacterial species are able to synthesize all amino acids, and thus they depend on the amino acids provided by the host. The bacteria present in the small intestine can compete on some occasion with the host for amino acid utilization, but the part of amino acids provided by the host to the small intestine microbiota is very low compared to the amino absorbed by the host. Of note, in the pig model, it has been demonstrated that dietary and endogenous proteins from host are significant contributors for microbial protein synthesis in the distal small intestine [3]. Incidentally, the pig represents an irreplaceable experimental model for intestinal physiologists and nutritionists, as well as for gastroenterologists [4, 5]. The pig model is generally a more relevant model for extrapolation to humans than rodents [6]. Pigs are omnivorous, make spontaneous individual meals, and display similarities with humans regarding nutritional requirements [7, 8]. The pig model is advantageous since it allows the recovery of large numbers of enterocytes and colonocytes, even in neonates, thus allowing the test of dietary intervention on intestinal cell metabolism and physiology at different stages of development [9], as well as measurement of parameters related to digestion and absorption. Despite the numerous advantages in the utilization of the pig model, when comparing pigs and humans, differences are observed regarding gut anatomy [10] and microbiota composition. Xiao and collaborators have determined the gut microbiota characteristics in pig fecal samples [11] and found in pigs a total of 7.7 million non-redundant genes representing 719 metagenomic species. When comparing the functional pathways identified in pigs and humans, 96% of the functional pathways in humans are present in the catalogue of the pig microbiota community, suggesting that the pig represents a rather good model for extrapolation from pig fecal microbiota metabolic activity data to humans.

In the mammalian large intestine, most amino acids are not directly available for the microbiota to any large extent. This is because the process of amino acid absorption by the small intestine is efficient, so that small amounts of free amino acids are transferred from the small to the large intestine [12]. In the large intestine, the amino acids originate mainly from the proteins that have not been digested, or not fully digested in the small intestine [13] (Fig. 3.1). Indeed, although the process of protein digestion is an efficient process, with digestibility being in most cases equal or even higher than 90% [14–18], a significant amount of undigested, or not fully digested proteins is transferred through the ileocaecal junction from the small to the large intestine [19]. Some dietary proteins are more resistant to digestion by the proteases secreted in the intestine from the exocrine pancreas, and thus these undigested proteins are transferred in larger part from the small to the large intestine. These proteins include for instance gluten which is relatively resistant to digestion due to the presence of proline sequences [20]. The amounts of proteins transferred through the ileocecal junction represent, depending on the dietary habit and the individual metabolic profiles, between 6 and 12 g dietary and endogenous proteins [12, 21–23]. The amounts of undigested proteins can be usefully compared to the usual protein consumption in geographical areas like Europe and the USA, which averages approximately 85 g per day in adults [24, 25].

In the large intestine, the bacterial proteases and peptidases release amino acids from luminal proteins [26]. These amino acids, at difference with the situation in the small intestine, cannot be absorbed to any significant extent by the colonic epithelium of mammals [27], except during the neonatal period [28–30]. These amino acids are thus available for utilization by the intestinal bacteria for their own protein synthesis and utilization in other metabolic routes, with production of numerous metabolites [31] as will be detailed in the following chapters. This is an important aspect of the host-microbiota metabolic crosstalk where the host provides amino acids to bacteria.

In the Small Intestine, Bacteria can Provide Essential Amino Acids to the Host, But the Metabolic Crosstalk Between Host and Bacteria Appears Much more Orientated Towards the Supply of Amino Acids to Bacteria by the Host

On the opposite, there are some reasons to consider that bacteria may provide some amino acids to the host, and notably some among the so-called indispensable amino acids [32]. These amino acids are called indispensable because not synthesized, or not sufficiently synthesized, to provide enough amino acids in comparison with the requirement for protein synthesis and optimal physiological functions [33]. These indispensable amino acids must then be provided by the dietary supply. Indispensable amino acids gather 9 amino acids (isoleucine, leucine, valine, lysine, methionine, phenylalanine, threonine, tryptophan, and histidine) among the 20 amino acids. From a metabolic point of view, among the nine indispensable amino acids, two amino acids, namely lysine and threonine, are considered as strictly indispensable as they cannot be synthesized in the body, notably by reamination of their corresponding keto acids [34]. Among the 11 other amino acids, six of them (arginine, cysteine, proline, tyrosine, glutamine, and glycine) are classified as conditionally indispensable, because their biosynthesis capacity in the body tissues may

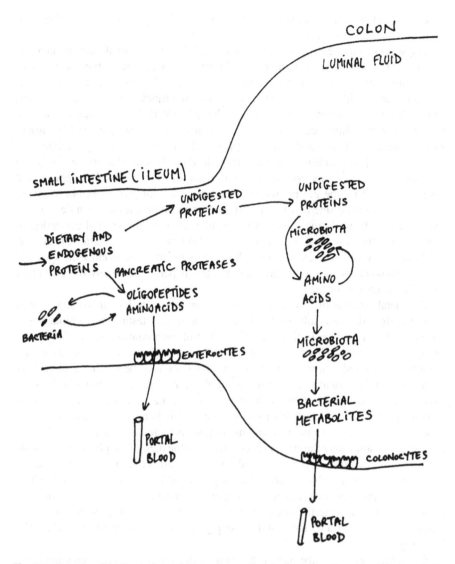

Fig. 3.1 Transfer of undigested protein from the small to the large intestine and utilization of this nitrogenous material by the intestinal microbiota. This scheme indicates transfer of undigested proteins from the ileum to the colon through the ileocecal junction. While pancreatic proteases in the luminal fluid of the small intestine release oligopeptides and amino acids that are absorbed by the enterocytes, amino acids released in the large intestine by the microbial metabolic activity cannot be absorbed through the colonocytes. In the large intestine, amino acids are mainly used for microbial protein synthesis and to produce bacterial metabolites among which several of them are absorbed in the portal blood

be insufficient compared to the requirement in specific physiological or pathophysiological situations [35, 36].

The supply of some indispensable amino acids by the intestinal microbiota to the host is likely to happen mainly in the small intestine since, as said above, colon does not absorb amino acids to any significant extent (Fig. 3.1). For instance, a study has examined the ability of germ-free rats (without microbes), and conventional rats (with microbes) to incorporate ^{15}N from $^{15}NH_4Cl$ into body lysine (an amino acid which is not transaminated in mammalian tissues), to determine if the ^{15}N enrichment found in the lysine is due to absorption of lysine synthesized by the intestinal microbiota [37]. The authors concluded from the data obtained that all the ^{15}N-lysine measured in the host was from microbiota origin. In the pig model, it appears that lysine produced by the gut microbiota is mainly used for protein synthesis in the splanchnic area (intestine and liver) [38]. In human infants, it has been determined that amino acids in plasma can derive from urea after hydrolysis and utilization of nitrogen by the intestinal microbiota [39]. However, in this latter study, the mechanisms by which amino acids synthesized by the microbiota enter the systemic pool of amino acids was not determined, but this is performed likely by absorption through enterocytes of the small intestine.

The metabolic fate of ^{15}N, given as urea or ammonia, to human volunteers was also investigated, and results show that threonine from intestinal microbiota origin appears in blood plasma [40], even if the extent of such contribution to the whole-body threonine metabolism could not be estimated in this study. These results are coherent with a study showing that genes involved in threonine synthesis are present in the core of functional genes expressed by the human gut microbiota [41]. However, the respective parts of threonine synthesized by the gut microbiota for its own protein synthesis or left available for protein synthesis in host is not known.

Overall, the supply to the host of amino acids by intestinal bacteria is likely to represent tiny amounts of amino acids compared to the dietary supply of amino acids through protein digestion [42]. Thus, although the respective parts of amino acids supplied by the host to the bacteria, and the part supplied by the bacteria to the host remain to be precisely determined, the flux of amino acid in the metabolic crosstalk between host and bacteria appears much more orientated towards the supply of amino acids to bacteria by the host than supply of amino acids by the bacteria to the host (Fig. 3.1).

Several indirect evidence support this latter proposition. Firstly, when comparing apparent absorption of some amino acids measured in vivo with the metabolic capacities of isolated enterocytes towards these amino acids, it appears that the microbiota is likely to participate in the metabolism of amino acids present in the luminal fluid. For instance, in the pig model, lysine is very little oxidized in the enterocytes [43], suggesting that this amino acid is mainly used in enterocytes for protein synthesis. However, when expressed as a percentage of the enteral tracer input, there is a substantial first-pass intestinal metabolism of lysine (35%) in young pigs [44], of which only 18% is recovered in the pig intestinal mucosal proteins. As lysine catabolism in the intestinal mucosa is quantitatively greater than the amino acid incorporation into mucosal proteins, it is tempting to propose that the intestinal microbiota, and maybe other cell types than the enterocytes in the intestinal mucosa, use lysine [45]. Regarding this latter point, it is noteworthy that the intestinal mucosa contains large numbers of immune cells in which amino acid catabolism appears to play a critical role in both innate and adaptive immunity in close relationship with the modulation of the gut barrier function [46].

The same reasoning can be made for other indispensable amino acids like methionine and phenylalanine which, although virtually not catabolized in isolated enterocytes [43], appear to be utilized in the intestine, and probably partly by the intestinal microbiota. Indeed, in the young pig model again, the net portal balance of methionine represents 48% of intake, suggesting that a relatively large part of the available methionine is consumed within the intestine [44, 47]. In accordance with these results, the parenteral methionine requirement (methionine given in blood) is approximately 70% of the enteral requirement (methionine given in the stomach) in pigs [48]. Regarding phenylalanine, it has been determined that, despite no measurable catabolism in enterocytes, there is marked first-pass metabolism of this amino acid in young pigs (35% expressed as a percentage of the enteral tracer), of which 18% is recovered in mucosal proteins [44].

In Both the Small and Large Intestine, Numerous Bacterial Species do not Possess the Capacities for the Synthesis of all the Amino Acids

Many of the metabolic pathways involved in amino acid synthesis in bacteria are conserved across the bacterial lineages, including those inhabiting the intestine [49]. However, major differences are observed when comparing the metabolic capacity for amino acid synthesis at the species and strain levels. For instance, whole genome analysis has shown that the gut bacterium *Clostridium perfringens* displays no metabolic capacity for the synthesis of glutamate, arginine, histidine, lysine, methionine, serine, threonine, aromatic and branched-chain amino acids [50]; and thus, depend on the host for the supply of these amino acids for their metabolic needs. Another gut bacterium, *Lactobacillus johnsonii*, appears incapable of carrying out synthesis of almost all amino acids due to a lack of complete anabolic pathways [51]. Other intestinal bacteria including *Campylobacter jejuni, Enterococcus faecalis*, and *Streptococcus agalactiae*, are able in mammals, including humans, to carry out the synthesis of only specific amino acids [52]. In contrast, other intestinal bacteria like *Clostridium acetobutylicum* are equipped with a complete set of genes for amino acid biosynthesis [53]. It is worth keeping in mind that the sole presence of genes implicated in amino acid synthesis within one given bacterium does not allow to conclude on their functionality in terms of capacity for amino acid synthesis. A typical example of such limitation is found for the common gut bacterium *Lactococcus lactis*, for which the genes allowing the synthesis of the 20 amino acids have been identified, but these bacteria require supplementation with isoleucine, valine, leucine, histidine, methionine, and glutamate for growth, since genes involved in the biosynthetic pathways of these amino acids have been demonstrated to lead to non-functional pathways due to point mutations [54, 55].

To describe the metabolic pathways that are involved in the synthesis of amino acids by the gut bacteria, it is useful to define families of amino acids that are linked

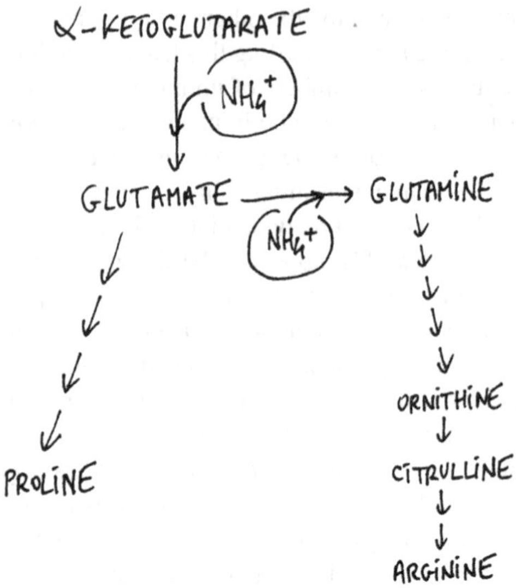

Fig. 3.2 Bacterial biosynthetic pathways of glutamate amino acid family. In this scheme, the utilization of ammonium for the biosynthesis of glutamate and glutamine is underlined since such utilization likely contributes to fix the ammonia concentration in the luminal fluid

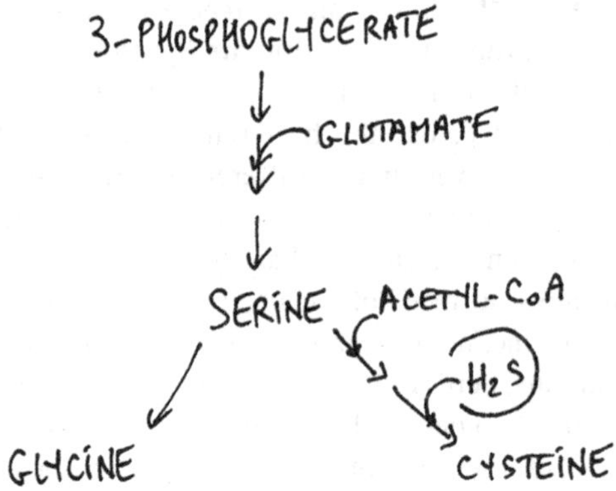

Fig. 3.3 Bacterial biosynthetic pathways of serine amino acid family. In this scheme, the utilization of hydrogen sulfide (H$_2$S) for cysteine synthesis is underlined since it likely contributes to the H$_2$S concentration in the luminal fluid

by the metabolic steps involved. These descriptions are also of interest as it allows to visualize several metabolites that are used during bacterial amino acid biosynthesis, and that are active on the intestinal ecosystem, like ammonia and hydrogen sulfide (see next paragraph). In the glutamate family of amino acids (glutamate, glutamine, proline, arginine), proline and arginine are derived in bacteria from glutamate; while glutamine is synthesized from glutamate and ammonia [56–58]. In the Fig. 3.2, the most common pathways are presented, but alternative pathways have been described in some bacterial species [49].

Regarding the serine family that is made by serine, glycine, and cysteine, serine is the precursor for the synthesis of glycine and cysteine in bacteria (Fig. 3.3) [59, 60]. As indicated in the figure, hydrogen sulfide (H$_2$S) is involved as co-precursor for the synthesis of cysteine in bacteria.

Fig. 3.4 Bacterial biosynthetic pathways of aspartate amino acid family. In this scheme, the utilization of ammonium for the biosynthesis of asparagine, and the utilization of hydrogen sulfide (H$_2$S) for the synthesis of methionine are underlined since it likely contributes to the concentrations of these bacterial metabolites in the luminal fluid

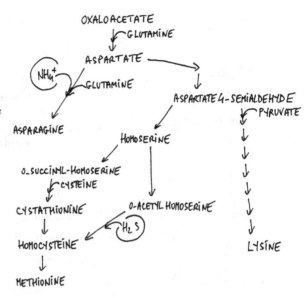

The aspartate family is made up by aspartate itself and by the four amino acids derived from this amino acid (asparagine, lysine, threonine, methionine), with ammonia and hydrogen sulfide being implied in some of the metabolic pathways (Fig. 3.4) [61–63].

In the pyruvate family (isoleucine, valine, leucine, alanine), pyruvate is the precursor of isoleucine, valine, and leucine in metabolic pathways involving several steps, while being converted to alanine in a single step (Fig. 3.5) [64]. Of note, ammonia appears to be involved in the metabolic pathways linking pyruvate and the four amino acids. The most common pathway for pyruvate conversion to alanine is the pathway that uses aspartate as co-precursor, but other metabolic routes have been described [49].

The next family is the aromatic family that gathers phenylalanine, tyrosine, and tryptophan (Fig. 3.6). The most common pathways for the synthesis of these amino acids in bacteria start with the condensation of the glycolytic intermediate phosphoenolpyruvate and the pentose phosphate pathway intermediate erythrose 4-phosphate, followed by metabolic steps involving ammonia [65].

Lastly, the histidine family is made solely by histidine. Histidine biosynthesis in bacteria is a complex ten-step pathway. The works performed on the histidine biosynthetic pathways have been mainly made using *E. Coli*, *Salmonella typhimurium*, and *Corynebacterium glutamicum*, demonstrating globally large conservation, with however some differences depending on the different species examined [54, 66].

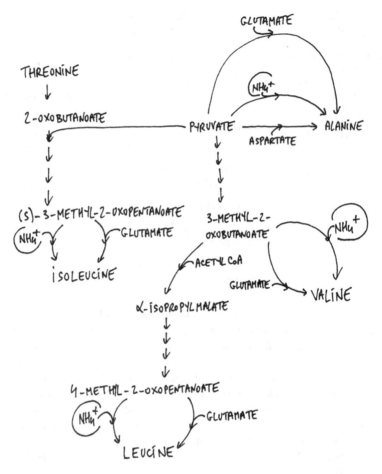

Fig. 3.5 Bacterial biosynthetic pathways of pyruvate amino acid family. In this scheme, the utilization of ammonium for the biosynthesis of alanine, isoleucine, valine, and leucine is underlined since it likely contributes to the ammonium concentration in the luminal fluid

Key Points
- In the small intestine, bacteria can use a part of amino acids provided by the host in addition to the ones they are able to synthesize. Conversely, amino acids produced by the bacteria can be provided to the host.
- The metabolic crosstalk between host and bacteria in small intestine is much more orientated towards the supply of amino acids from host to bacteria rather than supply of amino acids from bacteria to host.
- In the large intestine, bacteria release amino acids from undigested proteins and use them for protein synthesis. Bacteria utilize amino acids in numerous metabolic pathways, resulting in the production of a complex mixture of metabolites.
- Concentration of bacterial metabolites in the intestinal fluid depends on rate of production, rate of utilization, and absorption through the intestinal epithelium.

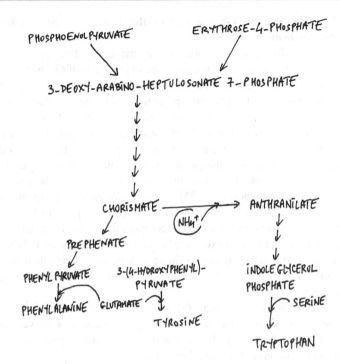

Fig. 3.6 Bacterial biosynthetic pathways of aromatic amino acid family. In this scheme, the utilization of ammonium for the synthesis of tryptophan is underlined since it likely contributes to the ammonium concentration in the luminal fluid

Amino Acid-Derived Bacterial Metabolites and Effects on the Intestinal Epithelium

The amino acid metabolism in intestinal bacteria is obviously not restricted to protein synthesis, and bacteria produce numerous metabolic end products from amino acids. Of considerable interest, among these bacterial metabolites, several of them are known from several decades to be also produced by the host cells, and to exert important physiological functions that are known to be disrupted in specific pathologies. Other metabolites are known to be exclusively produced by the intestinal bacteria, since the host tissues have no metabolic capacities in terms of enzymatic equipment to produce them. This distinction is important to consider when evaluating the impact of bacterial metabolite production on the physiology and metabolism of the host.

In the following paragraphs, the metabolites produced from amino acids by the intestinal bacteria will be individually presented together with what is known on their action on the intestinal epithelium, and on the consequences of such action on intestinal physiology and metabolism. Then, the overall effects of an increase in the amount of dietary protein on the colonic ecosystem will be presented together with the consequences of such increase for the colonic physiology.

Bacteria Degrade Protein Obtained from the Host and Use Amino Acids for Their Metabolism

Although the identification and biochemical characterization of the bacterial proteases and peptidases that are involved in the degradation of dietary and endogenous sources of proteins have been little studied, it appears that the intestinal microbiota is equipped to perform proteolysis of the available substrates [67, 68]. Some bacteria, such as lactic acid bacteria, have developed proteolytic systems to compensate for their reduced capabilities to synthesize amino acids [69]. These proteolytic systems include extracellular proteases that degrade proteins into smaller oligopeptide, and this is followed by their entry into bacteria via transporters. Finally, numerous intracellular peptidases degrade the peptides into shorter peptides and amino acids [70, 71]. Amino acids and their metabolites can be imported and exported from the bacteria via transmembrane proteins [72], allowing regulation of the concentrations of these compounds in the bacterial cells, and allowing presumably amino acid exchange between bacterial species. Efflux systems for some amino acids such as lysine, arginine, threonine, cysteine, leucine, isoleucine, and valine have been well-described in the bacteria E. coli and *Corynebacterium glutamicum* [73].

Bacteria may directly incorporate the available amino acids as substrates for protein biosynthesis or may use them as energy substrates or in various other metabolic pathways. Protein synthesis appears to be a major pathway for amino acid utilization by bacteria of the small intestine [74].

Low oxygen concentration is a characteristic of the intestinal luminal content, but due to some oxygen diffusion from the capillary network in the lamina propria situated below the intestinal epithelium, there is a zone adjacent to the intestinal mucosa where oxygen concentration is likely somewhat higher than in zones situated more distally [75]. Under aerobic environment, bacteria typically convert alpha-amino acids to their corresponding alpha-keto acids or saturated fatty acids via transamination or deamination, the alpha-keto acids being then used as energy substrates in the tricarboxylic cycle.

In healthy adults, the distal parts of the intestinal tract usually harbor large communities of obligate anaerobes [76] that live in the absence of oxygen [77]. In the absence of oxygen or other suitable electron acceptors, only strict or facultative intestinal anaerobic bacteria, such as Clostridia and Fusobacteria, can utilize amino acids as energy source, thus fermenting amino acids to short-chain fatty acids, hydrogen (H_2), carbon dioxide (CO_2), and ammonia, together with the production of other bacterial metabolites including H_2S, phenols, alcohols, and various organic acids [13, 78]. Several mechanisms for alpha-amino acid degradation have been described in anaerobic bacteria, including the Strickland reaction that is found in many proteolytic Clostridia. This latter reaction involves the coupled oxidation and reduction of amino acids to organic acids. Other fermentation pathways found in various Clostridia as well as *Fusobacterium* spp. and *Acidaminococcus* spp. involve single amino acids that act as electron donors as well as acceptors [57, 78]. The genus *Clostridium* contains unique amino acid degradation pathways, such as B12--dependent   aminomutases,   selenium   containing   oxidoreductases,   and oxygen-sensitive 2-hydroxyacyl-CoA dehydratases [79]. Amino acids can also be metabolized through decarboxylation reactions ultimately yielding amines and polyamines as products. Luminal parameters, such as pH, can influence the catalytic activity of different bacterial enzymes, such as deaminases and decarboxylases, thus affecting the production of specific end products [13].

Bacterial Metabolism of Amino Acids Depends on the Relative Availability of Substrates

Amino acids, depending on their structure and on the bacterial species considered, may be utilized for specific metabolism and functions. For instance, *Clostridium stricklandii* preferentially uses threonine, arginine, and serine for carbon and energy sources, but little utilizes glutamate, aspartate, and aromatic amino acids, and uses lysine as fuel only in stationary growth phase [79].

The ratio of available carbohydrates/proteins determines substrate utilization by the gut microbiota [80], and in humans, it has been shown that higher availability of complex carbohydrates (like plant fibers and resistant starch) lowers the process of protein fermentation [81–83]. Interestingly, bacteria appear to be able to modulate their metabolism according to what can be called the "energy context." Indeed, when fermentable carbohydrates like plan fibers, are abundant for intestinal bacteria, amino acids are mostly used for biosynthetic processes in anabolic reactions, but little for energy production. On the contrary, in a context of low availability of fermentable carbohydrates, several amino acids are largely used for energy production [49]. Then due to a high carbohydrate fermentation in the proximal colon, there is a progressive decrease of carbohydrate availability in more median and distal parts of the colon, resulting in higher protein degradation and amino acids utilization [84]. Longer transit time is also associated with higher level of protein fermentation [85, 86]. Therefore, the relative amounts of non-digestible carbohydrates and undigested/not fully digested proteins strongly influence the bacterial metabolism in the large intestine, and thus the flow of bacterial metabolites produced. Approximately half of the metabolic pathways used by the intestinal microbiota do not occur in the cells of the host, and the largest group of such metabolic pathways involves pathways related to amino acid metabolism [87].

Numerous Amino Acid-Derived Bacterial Metabolites are Active on the large Intestine Epithelium

Fermentation of amino acids derived from protein present in the luminal content by the colonic microbiota produces numerous metabolites (Fig. 3.7). Several of these metabolites have been shown to be involved in communication between microbes. Regarding the effects on the host tissues, bacterial metabolites have been shown to be either beneficial or deleterious for the host intestinal epithelium. As will be detailed, some bacterial metabolites appear beneficial at a given concentration, while being deleterious at higher concentrations. Relevant examples of the bacterial metabolites produced from amino acids, and their effects on the intestinal epithelial cells are presented below.

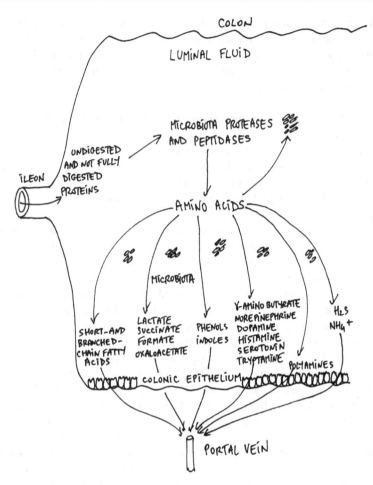

Fig. 3.7 Production of bacterial metabolites from amino acids in the large intestine. Amino acids released from the undigested dietary and endogenous proteins are used in the large intestine luminal fluid for microbial protein synthesis and to produce numerous bacterial metabolites that can be absorbed through colonic absorptive epithelial cells in the portal blood

Short-Chain Fatty Acids are Energy Substrates and Modulators of Gene Expression in Colonocytes

Some specific amino acids released from undigested proteins by the colonic microbiota contribute to the synthesis of the short chain-fatty acids (also known as SCFAs) butyrate, acetate, and propionate (Fig. 3.8) [88]. These compounds are recovered at millimolar concentrations in human colonic fluid, acetate being the most abundant, followed by propionate and butyrate as detailed in paragraph 3.2 [89]. Although it is well known that dietary substrates for short-chain fatty acid production are mainly fibers and resistant starches [90], isolated colonic bacteria growing in vitro on proteins as the only available carbon source have been

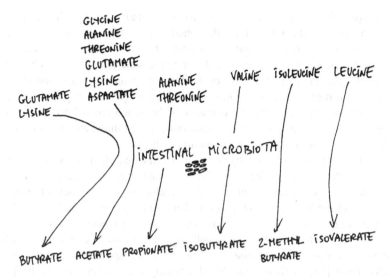

Fig. 3.8 Production of short-chain and branched-chain fatty acids from amino acids by the intestinal microbiota. Although the main substrates for short-chain fatty acid production are undigested polysaccharides, several amino acids are used by the intestinal microbiota for butyrate, acetate, and propionate production. In contrast, branched-chain fatty acids (that are isobutyrate, 2-methyl butyrate, and isovalerate) are produced exclusively from amino acids

demonstrated to produce short-chain fatty acids, as well as branched-chain fatty acids (also called BCFAs) [91]. Butyrate is produced by the gut microbiota from glutamate and lysine (Fig. 3.8). Acetate results from the metabolism of glycine, alanine, threonine, glutamate, lysine, and aspartate. Regarding propionate, it can be synthesized from alanine and threonine as precursors [19]. Of note, in rodent model, the consumption of a diet containing high amounts of proteins increases the total colonic content of butyrate, propionate, and acetate [92]. The short-chain fatty acids produced by the microbiota are used as energy substrates by the colonic epithelial cells and can modify gene expression in these cells as will be developed in the paragraph dedicated to dietary polysaccharides.

Branched-Chain Fatty Acids are Regulators of Electrolyte Movements Through the Colonic Epithelium

The branched-chain fatty acids isobutyrate, 2-methylbutyrate, and isovalerate are produced from indispensable branched-amino acids that are valine, isoleucine, and leucine respectively (Fig. 3.8). They are present in much lower quantities than the short-chain fatty acids in the mammalian large intestine content. In humans, the concentration of isobutyrate in the colon averages approximately 2.6 mM, while the concentration of isovalerate averages 3.4 mM [93]. Branched-chain fatty acids originate exclusively from the breakdown of proteins, and are not produced from carbohydrates, representing therefore good indicators of protein breakdown by the intestinal microbiota in the intestinal lumen [94]. Thus, five amino acids among the nine indispensable amino acids are used for short-chain and branched-chain fatty acid production by the microbiota. Branched-chain fatty acids are produced by numerous bacterial species including *Bacteroides, Clostridium,* and *Prevotella* [95]. The utilization of amino acids by the large intestine microbiota can be viewed as metabolic recycling of nutrients by bacteria in situation where the digestive capacity of the host is exceeded. Diets with a high proportion of proteins are thus expected to modify markedly the contents of branched-chain fatty acids in the colonic fluid when compared with diet containing more limited amounts of proteins, and this is exactly what is observed in mammals [92].

The effects of short-chain fatty acids and branched-chain fatty acids on the intestinal epithelium have been studied. Among branched-chain fatty acids, isobutyrate appears to exert regulatory roles on electrolyte movements through the colonic epithelium in experimental works. This capacity appears to be shared with short-chain fatty acids. Isobutyrate has been shown to change the movement of electrolytes through the absorptive cells of the colonic epithelium. Indeed, isobutyrate increases the expression of an apical transporter of Na^+ in colonic epithelial cells [96], activates the Na/H exchanger in colonic crypts [97], and stimulates sodium absorption in colonic biopsies [98]. Regarding chloride, isobutyrate reduces the secretion of this electrolyte in colonic biopsies [99, 100]. Incubation of isobutyrate at the apical side of the colon produced an alkalinization of the crypt lumen, thus indicating that this branched-chain fatty acid plays a role in fixing the characteristics of the microdomain surrounding the colonic crypt [101]. Isobutyrate appears in addition to be used as a fuel by colonocytes, notably when butyrate availability is deficient [102]. A mixture of the branched-chain fatty acids isovalerate and isobutyrate appears protective for the intestinal epithelium since it prevents the in vitro experimental disruption of the intestinal epithelial barrier [103].

Lactate, Succinate, Formate, and Oxaloacetate: Emerging Effects of These Organic Acids on the Intestinal Ecosystem

Lactate (isomers D and L), succinate, formate, and oxaloacetate are organic acids produced as intermediary metabolites during amino acid metabolism by the bacteria present in the intestine [19, 67]. These organic acids are found in the range of micromolar-to-low millimolar concentrations in the rat colonic content [92]. Lactate and formate can be detected in feces recovered from healthy donors, but high variability is recorded among individuals ranging from undetectable value up to millimolar concentrations [104]. In human colon, lactate concentration averages approximately 3.5 mM [93]. Numerous bacterial species produce lactate, including notably lactic acid bacteria such as *Lactobacillus* and *Enterococcus* [105]. Numerous bacteria including *Prevotellaceae* and *Veillonellaceae* are known succinate producers [106–108].

Several organic acids have been shown to be implicated in microbial communication. Formate, when secreted by the pathogenic bacteria *Shigella flexneri*, promotes virulence gene expression [109] (Fig. 3.9). Oxaloacetate was found to be the metabolite by which the bacteria *Escherichia coli* helps the parasite *Entamoeba*

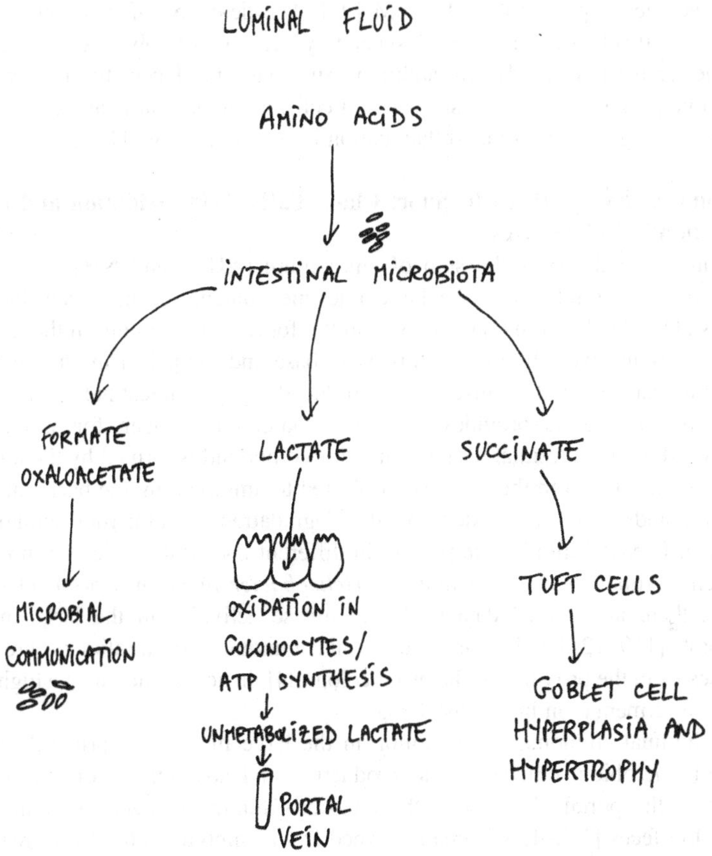

Fig. 3.9 Production of various organic acids by the intestinal microbiota and effects of these bacterial metabolites on microbial communication and colonic epithelial cell metabolism and physiology

histolytica to survive in the large intestine [110]. This result is of notable importance given the fact that this parasite can trigger a strong inflammatory response upon invasion of the colonic mucosa [111].

In rodent models, total amounts of L-lactate, D-lactate, succinate, and formate in the rat colonic fluid are several folds higher when the amount of protein is increased in the diet [92]. Among these organic acids, lactate and pyruvate are known to be transported in intestinal epithelial cells by members of the monocarboxylate transporter family, notably by the MCT1 isoform [112]. Of note, absorptive colonocytes are characterized by a high capacity for L-lactate oxidation, a capacity that is even higher than the capacity for butyrate oxidation (Fig. 3.9) [113]. Although the physiological relevance of such high oxidative capacity remains unknown, such metabolic capacity likely regulates the amount of luminal lactate released in the portal vein. Regarding succinate, the concentration of this metabolite in the human colon averages approximately 1.7 mM [93]. The detection of this metabolite by intestinal Tuft cells triggers hyperplasia and hypertrophy of goblet cells in the distal intestine epithelium [114]. In addition, succinate produced by the intestinal microbiota promotes the expansion of Tuft cells, an effect that is associated with a reduction of signs of intestinal inflammation in the mice model [115].

Ammonia in Excess Disturbs Short-Chain Fatty Acid Oxidation and Energy Production in Colonocytes

Ammonia, considered as the sum of ammonium (NH_4^+) and NH_3 is present at millimolar concentrations in the large intestine contents of mammals including humans [116, 117]. Ammonia is mostly in the form of ammonium in the intestinal luminal content given the pKa of this weak base and the pH of the luminal fluid. Ammonia can be both utilized and produced by the intestinal bacteria [20] (Fig. 3.10). Ammonia provides one main source of nitrogen for amino acids synthesized by the intestinal bacteria, and this compound is formed by the intestinal microbiota mostly from the conversion of urea to ammonia by the bacterial urease activities, and to a lower extent by the degradation of numerous amino acids [45, 118]. Urease activities are present in different bacterial species but not in the host cells. Although urease activity in *Helicobacter pylori* has been intensively studied, there are limited data regarding urease activities in the large intestine microbiota [119–121]. In humans, the luminal ammonia concentration progressively increases from the proximal to the distal colon [84], in accordance with a higher rate of protein fermentation in the distal colon.

The luminal ammonia concentration in the large intestine is primarily the net result of microbiota utilization and production and absorption from the luminal content to the portal blood, with the unabsorbed/unmetabolized ammonia being excreted in feces [122–128]. Urea is formed from ammonia in the liver urea cycle, and this cycle avoids accumulation of ammonia in the blood, and thus deleterious effects of excessive concentrations on the brain. Urea is excreted in the urine. However, a minor part of the circulating urea can diffuse from the blood to the intestinal lumen, serving there as a substrate for bacterial ureases, representing thus in an "host-bacteria cycle," a supply of ammonia for bacterial anabolism (Fig. 3.10). Urea transporters are expressed in the colonic mucosa where they likely participate in the transfer of urea from the circulation to the intestinal lumen [129]. Large amount of luminal ammonia can be absorbed through the large intestine absorptive epithelial cells in mammals, including humans [123, 128]. Ammonia in excess, either originating from the colonic content, or produced in colonocytes from glutamine utilization, inhibits in a dose-dependent manner the mitochondrial oxygen consumption in colonocytes [130], thus representing a metabolic troublemaker towards cell respiration, and thus energy production in this rapidly renewed structure. In addition, high millimolar concentrations of ammonia inhibit short-chain fatty acid oxidation in colonic epithelial cells [131, 132]. However, colonocytes are equipped with enzymatic activities that allow to partly detoxify ammonia in the mitochondria of colonocytes during its transfer from the luminal content to the bloodstream. This can be done by converting ammonia into citrulline in two

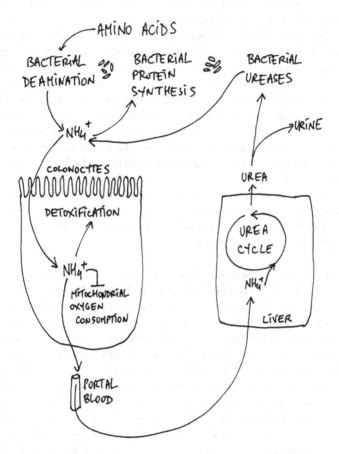

LUMINAL FLUID

AMINO ACIDS

BACTERIAL DEAMINATION

BACTERIAL PROTEIN SYNTHESIS

BACTERIAL UREASES

NH_4^+

URINE

COLONOCYTES

DETOXIFICATION

UREA

UREA CYCLE

NH_4^+

MITOCHONDRIAL OXYGEN CONSUMPTION

NH_4^+

LIVER

PORTAL BLOOD

Fig. 3.10 Production and utilization of ammonia by the intestinal microbiota and by the intestinal epithelial cells. In this scheme, ammonium production by the intestinal bacteria and utilization of this metabolite by bacteria are indicated. Detoxification of ammonium in the mitochondria of colonocytes allows to avoid accumulation of excessive concentrations that reduce mitochondrial oxygen consumption, and thus ATP synthesis

metabolic steps [116, 133], thus limiting the inhibitory effect of ammonia in excessive concentrations on colonocyte respiration (Fig. 3.10). Marked energy depletion in colonic epithelial cells may affect colonic epithelium renewal, and thus the intestinal barrier function [134]. Lastly, excessive concentrations of ammonia increase the paracellular permeability between colonocytes [135].

Phenol and Phenol Derivatives: Putative Deleterious Effect of Phenol in Excess for the Intestinal Epithelium

Phenylalanine catabolism by the intestinal bacteria produces phenyl-containing compounds like phenyl pyruvate, phenyl lactate, phenyl acetate, and phenyl propionate [136]. Tyrosine is used by the intestinal microbiota and leads to the production of hydroxyphenyl-containing molecules, including hydroxyphenyl pyruvate, hydroxyphenyl lactate, hydroxyphenyl propionate, and hydroxyphenyl acetate. Other bacterial metabolites are also formed from tyrosine like phenol, 4-ethylphenol, and p-cresol [137] (p-cresol will be presented below in a dedicated paragraph). In humans, the excretion and concentration of phenol in feces, that is approximately equal to 40 micromolar, can be diminished by a high resistant starch diet [81]. Phenols are the major products of aromatic amino acid metabolism in the distal bowel, and phenol concentration appears to be higher in the distal colon than in the proximal colon [138].

Little is known on the effects of phenol and phenol-related compounds on the intestinal epithelial cells. Phenol in excess has been shown to be cytotoxic towards colonocytes [139], and to increase epithelial permeability in in vitro model of human intestinal epithelial cells [20, 135, 140].

P-Cresol in Excess: A Bacterial Metabolite that Favors Clostridium Difficile Colonization and Affect Energy Production in Colonic Epithelial Cells

The phenolic compound p-cresol (4-methylphenol) is produced from the amino acid L-tyrosine, by the anaerobic flora of the large intestine [137] (Fig. 3.11). P-cresol cannot be synthesized by the host tissues and is thus exclusively originating from the metabolic activity of the microbiota. The p-cresol concentrations measured in the human colonic content is in the low millimolar range [81, 141–144]. In mammals, as expected, an increase of the dietary protein intake raises the fecal p-cresol concentration, due to an increased availability of the substrate L- tyrosine originating from undigested protein for bacterial metabolic activity [145, 146]. Conversely, the fecal excretion of p-cresol is diminished by a diet containing resistant starch [81, 147] or fiber [144]. The p-cresol concentration in the large intestine content has been measured in the pig model and averages 0.9 mM [148].

In vitro analysis has revealed that among bacteria present in the human gut, specific families of bacteria, namely *Fusobacteriaceae, Enterobacteriaceae, Clostridium*, and *Coriobacteriaceae,* are strong p-cresol producers [149]. The production of p-cresol by the intestinal bacteria can be diminished in vitro by other bacterial metabolites like lactic acid [150], thus representing an interesting illustration on the way a given bacterial metabolite can have an influence on the production of other bacterial metabolites (Fig. 3.11). Interestingly, the capacity of *Clostridium difficile* to produce p-cresol has been linked to its competitive advantage over other gut bacteria, including *Escherichia coli, Klebsiella oxytoca*, and *Bacteroides thetaiotaomicron* [151]. From a mouse model of *C. difficile* infection, it has been shown that excessive p-cresol production affects the gut microbiota biodiversity [151]. In addition, and importantly, by removing the availability of *C. difficile* to produce p-cresol, this bacterium is less able to recolonize the intestine after an initial

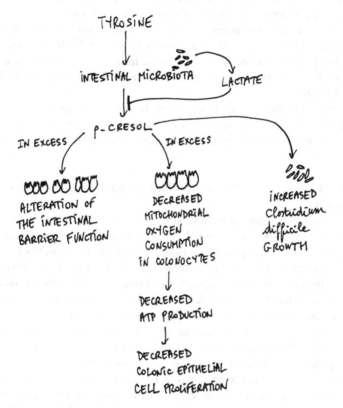

Fig. 3.11 Production of *p*-cresol by the intestinal microbiota and effects of this bacterial metabolite on microbial communication. Excessive concentrations of *p*-cresol in the luminal fluid affect energy metabolism in colonocytes, and more globally intestinal barrier function

infection (Fig. 3.11). This result is of major importance when considering that *Clostridium difficile* is a major cause of intestinal infection and diarrhea in individuals following antibiotic treatment [152]. As expected, *C. difficile* can tolerate concentration of *p*-cresol as high as 10 millimolar [153, 154], thus a much higher concentration than the one measured in the large intestine of mammals. P-cresol has been identified in volunteers as a fecal biomarker of *C. difficile* infection [155].

Although *p*-cresol is known to be absorbed by the intestinal epithelial cells, little is known on the way this bacterial metabolite is transported from the luminal side to the bloodstream. In the pig model with implanted catheters, the release of *p*-cresol by the intestine has been clearly shown [156], demonstrating trans-epithelial transfer.

P-Cresol can be Metabolized by the Colonic Epithelial Cells Giving Rise to the Production of Co-Metabolites

It appears that a part of p-cresol originating from bacteria can be metabolized by the colonic epithelial cells during its transcellular journey. Indeed, in volunteers, capacity of the colonic mucosa to conjugate phenolic compounds with sulfate (giving rise to p-cresyl sulfate) and glucuronide (giving rise to p-cresyl glucuronide) has been demonstrated [157, 158]. In an elegant study by Aronov and collaborators, comparison between p-cresyl sulfate levels in plasma from volunteers with and without colon strongly suggests that colon plays a role in the production of this latter compound [159].

The compounds that are originating from the microbial metabolic activity, and that are modified by the host are called co-metabolites since the production of such compounds is the result of a metabolic interplay between the intestinal microbiota and the host [160].

P-Cresol in Excess Affects Colonocyte Energy Metabolism and Epithelial Barrier Function

P-cresol may affect colonocyte metabolism. Indeed, at 0.8 mM, p-cresol diminishes mitochondrial oxygen consumption in human colonocytes when tested in vitro, and increases anion superoxide production thus indicating that this bacterial metabolite affects the mitochondrial function in these cells [161] (Fig. 3.11). This effect is associated with a reduction of the capacity of colonic epithelial cells to proliferate. In addition, in this latter study, it was demonstrated that pretreatment with p-cresol for one day results in an increased oxygen consumption by colonocytes, likely corresponding to an adaptive process aiming at counteracting the adverse effect of p-cresol on the cell respiration. Despite this tentative cellular metabolic compensation, after pretreatment of colonocytes for 3 days with increasing concentrations of p-cresol, the intracellular concentration of ATP is dose-dependently decreased, thus indicating energy deficiency in these cells under such experimental conditions. Using monolayers of human colonocytes, p-cresol tested between 1.6 and 6.0 millimolar concentrations dose-dependently increases the paracellular transport between colonocytes [162], suggesting that p-cresol at high concentrations, maybe because of energy depletion, can induce alteration of the intestinal barrier function (Fig. 3.11). This area of research is of major importance since alterations of normal barrier function have been detected in numerous clinical and experimental studies as an event that appears implicated in the etiology of inflammatory bowel diseases as will be detailed in Chap. 4.

Indole and Indole-Related Compounds: A Complex Family of Molecules Involved in the Intestinal Ecosystem

Intestinal bacteria convert tryptophan mainly into indole through the action of the bacterial enzyme tryptophanase, this latter enzyme being induced by tryptophan itself [163] (Fig. 3.12). Of note, in mammals, indole is exclusively originating from the bacterial metabolic activity since host cells do not have the metabolic capacity to produce this compound [164]. Indole is synthesized from various Gram-positive and

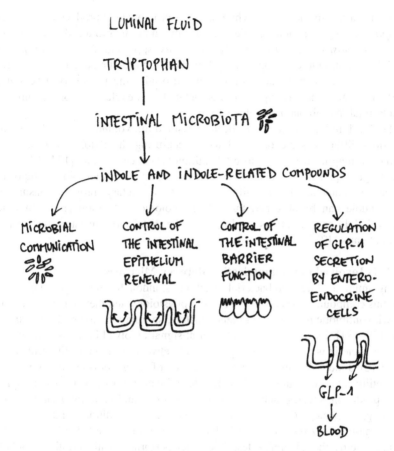

Fig. 3.12 Production of indole and indole-related compounds by the intestinal microbiota and effects of these bacterial metabolites on microbial communication and metabolism and physiology of colonic epithelial cells

Gram-negative bacterial species including *Escherichia coli*, *Proteus vulgaris*, *Clostridium* spp., and *Bacteroides* spp. [164–166]. Tryptophan can also be converted directly or indirectly by the intestinal microbiota into several indole-related compounds, including indole-3- pyruvate, indole-3- lactate, indole-3-propionate, indole-3-acetamide, indole acrylate, indole acetaldehyde, indole-3-acetate, indole-3-aldehyde, 3-methyl-indole (skatole), and indole-3-acetaldehyde [163, 167–171]. However, the precise metabolic pathways involved in the stepwise conversion of tryptophane into these minor indole-related compounds by the intestinal microbiota in the large intestine remain unclear. Among these numerous compounds, skatole will be the object of one dedicated paragraph which is presented below.

Information is available regarding the bacterial species involved in the production of indole-related compounds. Indole, as presented above, is synthesized from various bacterial species [165], while indole-3-propionate is produced by *Clostridium sporogenes* [172]. Other indole-related compounds are produced by *Bacteroides* species, *Peptostreptococcus* spp. *Lactobacillus* spp. and *Bifidobacterium* spp. [166]. In rodent, indole-3-propionate is found only in blood recovered from animals with an intestinal microbiota, but not in blood originating from germ-free animals [173], suggesting that, as expected, this metabolite is exclusively originating from the intestinal microbiota metabolic activity.

The fecal indole content has been measured in volunteers, but huge inter-individual differences are observed when measuring the indole fecal content in volunteers, ranging from 0.30 to 6.64 millimolar concentrations [174, 175]. Such large range is presumably due to differences in intestinal microbiota composition between volunteers, and to different levels of dietary protein consumption [176]. Indole can be absorbed through the colonic epithelium [177]. In rodent models, indole concentrations are higher in the distal large intestine that in more proximal parts [178].

Indole Is Involved in Communication Between Microbes

Among amino acid-derived bacterial metabolites, indole has raised growing interest since this compound is involved in bacterial physiology and metabolism in relationship with antibiotic resistance, virulence factors, sporulation, and biofilm formation [164, 179–181]. Indeed, indole is a bacterial signal involved in the communication between bacteria within the same species and between bacteria of different species (Fig. 3.12). Indole diminishes related virulence of *L. monocytogenes* by reducing cell motility and aggregation [182]. Indole influences host cell invasion by non-indole-producing species such as *Salmonella enterica* and *P. aeruginosa*, as well as the fungal species *Candida albicans* [183]. In addition, indole reduces enterohemorrhagic *Escherichia coli* adherence to epithelial cells [184], although in this latter study the adherence tests were not performed with intestinal epithelial cells. In another study, indole was shown to display bacteriostatic effects on lactic acid bacteria, while affecting their survival [185]. Of note, indole mitigates cytotoxicity by *Klebsiella* spp. by suppressing toxin production [186].

Indole and Indole-Related Compounds Exert an Overall Beneficial Effect on the Intestinal Mucosa

Several indole-related compounds, including indole-3-acetate, indole-3-propionate, indole-3-aldehyde, indole-3-acetaldehyde, and indole acrylate bind to the aryl hydrocarbon receptor [187], which is present in different cell types of the host including cells present in the intestinal mucosa, notably intestinal epithelial and immune cells [188]. The binding of these compounds to AhR participates in the maintenance of the intestinal mucosa homeostasis by acting on the control of the intestinal epithelium renewal, its barrier function, and activity of several intestinal immune cell types [189]. AhR can be also activated by dietary compounds [189], but bacterial metabolites appear to play a preponderant role on the AhR activation [190]. Only few commensal bacteria have been identified as able to produce AhR ligands, such as *Peptostreptococcus russellii* [191] and *Lactobacillus* spp. [192].

Protective effects of microbiota-derived aryl hydrocarbon receptor agonists on the intestinal mucosa have been suggested from both experimental and clinical data, with presumed consequences for the risk of metabolic syndrome [193]. In this latter study, metabolic syndrome was found to be associated with an impaired capacity of the gut microbiota to produce from tryptophan aryl hydrocarbon agonist receptors, such impaired capacity being paralleled by increased gut permeability and decreased secretion of GLP-1 from enteroendocrine cells [193], this hormone being notably known to control glycemia by stimulating insulin secretion [194]. In this study, the respective potential roles of the different indole-related compounds in these processes were not determined.

Exposure of human enterocytes to indole increases the expression of genes involved in the intestinal epithelial barrier function and mucin production [195] (Fig. 3.12). Also, oral administration of indole-containing capsule to rodent resulted in an increased expression in colonocytes of genes coding for tight junction proteins between epithelial cells [196]. In accordance with these results, indole increases transepithelial resistance in in vitro experiments using colonocyte monolayers [197], thus reinforcing the view that indole ameliorates the basal barrier function. Thus, indole and several related compounds exert beneficial effects on the intestinal mucosa in different experimental situations.

Indole at Excessive Concentration May Affect the Energy Metabolism in Colonocytes

However, indole when used at a 2.5 millimolar concentration affects the respiration of colonocytes by diminishing mitochondrial oxygen consumption [197], and thus mitochondrial ATP production. This latter effect was paralleled by a transient oxidative stress in colonocytes, which was followed by an increased expression of antioxidant enzymes, presumably as an adaptive process against the deleterious effect of indole exposure at excessive concentration [197].

Indole Modulates Hormone Secretion by the Enteroendocrine Cells of the Intestinal Epithelium

In in vitro experiments performed with immortalized and primary mouse colonic enteroendocrine L cells, indole modulates the secretion of glucagon-like peptide-1 (GLP-1) [198] (Fig. 3.12). Since GLP-1 slows down gastric emptying, stimulates insulin secretion by pancreatic beta cells, and diminishes appetite [194], it would be of major interest to study in vivo the effects of indole on these physiological parameters.

Skatole: A Bacterial Metabolite Involved in Communication Between Bacteria that Emerges as a Deleterious Molecule for Colonocytes When Present in Excess

Skatole is another tryptophan-derived bacterial metabolite [80, 169, 199, 200]. *Lactobacillus*, *Clostridium*, and *Bacteroides* are skatole producers [166, 201, 202]. Skatole displays a strong inhibitory effect on enterohemorrhagic *Escherichia coli* biofilm formation [203]. Biofilms are structures formed by the colon microbiota that line on the mucosal surface and these structures represent a player in the modulation, either positive or negative depending on the context, of the colon epithelial barrier function [204, 205].

Fecal skatole concentration in healthy individuals is usually low averaging 0.04 mM [200]. Skatole can be absorbed through the mammalian gut epithelium [206], being then released in the circulation [207]. Skatole concentration increases in the large intestine after a high meat diet consumption, or when luminal fermentation increases due to longer intestinal stasis [200].

Little is known on the effects of skatole on intestinal cells, but high concentrations of skatole induce cell death in human colonocytes [208].

Hydrogen Sulfide Is Produced by Intestinal Bacteria from Cysteine and Sulfate
Hydrogen sulfide (H_2S) is produced by different bacterial species from both dietary and endogenous compounds [209] (Fig. 3.13). H_2S is produced through cysteine catabolism, and by sulfate-reducing bacteria. Cysteine-degrading bacteria include *Fusobacterium, Clostridium, Escherichia coli, Salmonella, Klebsiella, Streptococcus, Desulfovibrio*, and *Enterobacter*, which convert cysteine to H_2S through the enzymatic activity of cysteine desulfhydrase [210–213]. Sulfate-reducing bacteria include *Desulfovibrio, Desulfobacter, Desoulfobulbus*, and *Desulfotomaculum*. *Desulfovibrio* is the dominant genera of sulfate-reducing bacteria, and it includes *Desulfovibrio piger* and *Desulfovibrio desulfuricans* [214, 215]. Although sulfate was first believed to be poorly absorbed by the human gastrointestinal tract, it appears that a net absorption of sulfate is measurable. The amount of dietary sulfate as well as the small intestine absorption capacity represent the main parameters affecting the amount of sulfate reaching the colon [216]. In volunteers, the ingestion of sulfate supplement increases the fecal sulfide production rate [217]. It appears that current concentrations of sulfate in the large intestine are generally adequate to support growth of sulfate-reducing bacteria.

H_2S appears to act as a protective agent in bacteria against the action of antibiotics and oxidative stress [218, 219] (Fig. 3.13). Conversely, the capacity of some bacteria to cope with surrounding H_2S may be related to some specific metabolic characteristics. For instance, in *Escherichia coli*, cytochrome bd-type O_2 oxidases, that are relatively insensitive to sulfide, allow bacterial oxygen consumption and growth in the presence of this compound [220].

Overall, it appears that H_2S that is not bound to luminal compounds, thus either as gas or dissolved in the aqueous phase of the colonic content, represents the chemical form that will diffuse through the mucous layers, and then into the colonocytes (Fig. 3.13). Several dietary compounds like zinc and proanthocyanidins (contained in plants) bind H_2S, thus reducing its concentration in free form [221–223]. Then, these dietary compounds that are not fully absorbed by the small intestine epithelium [224, 225] could represent a potential way to regulate the concentration of free H_2S in the large intestine luminal content. In germ-free mice, free H_2S concentrations are reduced in plasma and intestinal tissues when compared with animals with an intestinal microbiota [226], suggesting that the gut microbiota is a purveyor of this compound for the host.

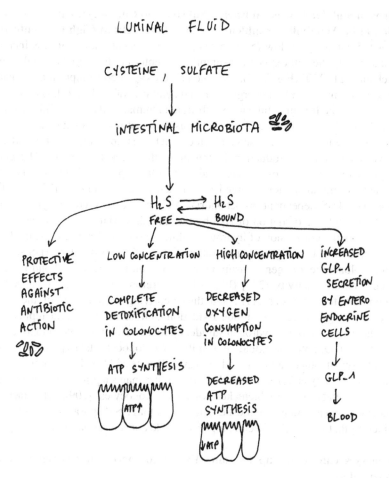

Fig. 3.13 Production of hydrogen sulfide (H_2S) by the intestinal microbiota and effects of this bacterial metabolite on antibiotic action and on metabolism and functions in intestinal epithelial cells. H_2S at low luminal concentrations is totally oxidized in colonocytes allowing mitochondrial ATP synthesis. However, this bacterial metabolite at high luminal concentrations seriously affects energy metabolism in colonocytes

The concentration of total H_2S, in its free or bound form, measured in the colonic luminal content is rather divergent according to the techniques used, ranging from high micromolar to low millimolar concentrations [227]. In addition to the exogenous source of H_2S, the colonocytes can produce endogenously a tiny amount of H_2S [228]. The part played by endogenous synthesis in H_2S intracellular concentration in colonocytes is presumably of quantitatively minor importance when compared with the part played by H_2S originating from exogenous luminal supply [227].

Hydrogen Sulfide: A Mineral Energy Substrate in Colonocytes at Low Concentration but a Metabolic Troublemaker for These Cells at High Concentrations
In isolated colonocytes, low (5 to 65 micromolar concentrations) of H_2S increases instantaneously the cell oxygen consumption by entering the mitochondrial respiratory chain [221, 229] (Fig. 3.13). This increased oxygen consumption is associated with an inner mitochondrial energization and synthesis of ATP [223]. H_2S has been established as the first mineral energy substrate in human cells [230]. This discovery has challenged the previous concept that considers that mammalian cells are exclusively dependent on carbon-based molecules, such as simple sugars, fatty acids, and amino acids, for energy production. Interestingly, this capacity to oxidize H_2S can be viewed as reminiscence of an ancestral metabolic capacity [231] that has been identified in animal models adapted to sulfide-rich environments [232], such as worms. This H_2S-dependent mitochondrial ATP synthesis in colonocytes is made possible through the oxidation of H_2S by the mitochondrial sulfide oxidation unit, which allows the conversion of hydrogen sulfide into thiosulfate [233] (Fig. 3.13).

In deep contrast, at higher concentrations (above 65 micromolar), H_2S severely inhibits colonocyte oxygen consumption by inhibiting the mitochondrial cytochrome c oxidase activity [229] (Fig. 3.13), therefore decreasing the capacity of mitochondria to synthesize ATP, and thus depriving colonocytes of their main source of energy. Thus, the sulfide oxidation unit in mitochondria represents a metabolic way to detoxify hydrogen sulfide, up to a certain threshold, and to recover energy from it. Of note, the oxidation of sulfide in colonocytes takes precedence over the oxidation of other carbon-based energy substrates [234]. In other words, the oxidation of H_2S by colonocytes for detoxification remains a priority given its action as energy metabolism troublemaker towards colonocytes [209]. The fact that colonocytes appear among the most efficient cells for H_2S disposal is not surprising, considering that these cells face the highest sulfide concentration in the body [234].

Colonocytes can Adapt up to Some Extent to Increased Hydrogen Sulfide Concentration
Several processes, that can be viewed as adaptive responses, appear to limit the deleterious effects of excessive luminal H_2S on colonocytes. Indeed, when human colonocytes are exposed to high sulfide concentrations, they spectacularly increase their capacity to utilize glucose in the glycolytic pathway [235] and then use another metabolic pathway for ATP synthesis, even if this way is much less efficient than the mitochondrial oxidation of fuels. Also, in rats fed with a high-protein diet in which the amount of H_2S increases in the colonic content, the expression of the gene corresponding to sulfide quinone reductase (also known as SQR), the first and rate-limiting enzyme for H_2S detoxification, increases in colonocytes [229]. In such situation, the water content in the colon luminal fluid increases as well [92]. Although the mechanisms that allow such water retention in the colon are not clear yet, such retention allows to limit the increase of H_2S concentration that faces the colonic epithelium. Thus, the estimation of the maximal luminal concentration of H_2S that can be safely handled by the colonic epithelial cells refers to the metabolic capacity of these cells to adapt to fluctuation in H_2S luminal concentration.

Hydrogen Sulfide can Increase Hormone Secretion in Enteroendocrine Epithelial Cells

Apart from the effect of H_2S on colonic absorptive cells, this metabolite has been shown to regulate endocrine function of the gut. Indeed, H_2S stimulates in vitro the secretion of GLP-1 by the enteroendocrine L-cells [236] (Fig. 3.13). GLP-1 facilitates control of glycemia by stimulating insulin secretion by pancreatic beta cells while suppressing glucagon secretion [237]. In the study by Pichette and collaborators, the authors, by using the prebiotic chondroitin sulfate, which increases the abundance of the sulfate-reducing bacteria *D. piger* and sulfate moiety in the distal intestine, measured an increased H_2S production. Such increased production was associated with metabolic improvements through increased GLP-1 secretion during oral glucose tolerance tests.

Polyamines: A Family of Molecules Produced by the Intestinal Microbiota with Important Effects on the Intestinal Mucosa Metabolism and Physiology

Polyamines are a family of small aliphatic amines that are present in bacterial and mammalian cells [238]. These compounds can be synthesized by intestinal bacteria and by host cells [239, 240] (Fig. 3.14), but several polyamines synthesized by bacteria are not produced by mammalian cells [240]. At physiological pH, polyamines are positively charged and bound to negatively charged biomolecules including DNA, and RNA as well as specific proteins [241, 242]. This latter capacity is related to their modulating effects on numerous functions associated with these compounds in bacteria and host cells [243–245].

Colonic Bacteria are a Main Source of Luminal Polyamines

Colonic bacteria represent a main source of luminal polyamines which include putrescine, spermidine, spermine, agmatine, and cadaverine. These polyamines are synthesized from their respective amino acid precursors. These precursors are ornithine/arginine and methionine for the "classical" polyamines that are putrescine, spermidine and spermine, and which are the most largely studied compounds among the polyamines (Fig. 3.14). Arginine is the precursor of agmatine in bacteria, while lysine is used for cadaverine synthesis in bacterial cells [239, 245].

Gram-negative bacteria such as *Escherichia coli* are well known to produce large amounts of putrescine, spermidine, agmatine, and cadaverine in minimal media [246]. Among the colonic bacteria, many members among *Bacteroides*, *Lactobacillus*, *Clostridium*, and *Bifidobacterium* produce various polyamines under colonic-type environmental conditions [247, 248].

The capacity of intestinal bacteria to produce polyamines depends on the characteristics of their luminal environment, including the availability of precursors furnished by the host [249], and the pH of the luminal fluid [247, 250]. The intestinal microbiota composition appears to represent another major parameter involved in determining the concentrations of the different polyamines in the intestinal luminal fluid [251]. Polyamines, including putrescine, spermidine and spermine, play a major role in bacterial growth [243, 252], and the capacity of bacteria to grow is partly linked to their capacity to synthesize polyamines (Fig. 3.14). In contrast, the

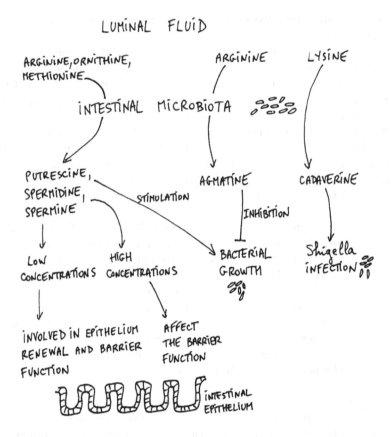

Fig. 3.14 Production of polyamines by the intestinal microbiota and effects of these bacterial metabolites on bacterial and intestinal epithelial cell growth and on intestinal barrier function

polyamine agmatine has been shown to represent a growth-inhibitory compound for some bacterial species [253]. In addition to endogenous synthesis, bacteria like *Escherichia coli* are equipped with several transport systems that allow them to import polyamines [243, 245]. Bacteria are also able to excrete polyamines in the extracellular medium [254], a process that allow to regulate the concentration of these compounds in bacterial cells.

Polyamines have been measured in intestinal contents at concentrations from micro- to millimolar concentrations [255]. The polyamines contained in the intestinal contents originate not only from the metabolic activity of the intestinal bacteria, but also from dietary supply [256], and from fully mature epithelial cells that are desquamated in the luminal content releasing their intracellular content. Indeed, enterocytes isolated from the mammalian small intestine contain mostly spermine, and to a lower extent spermidine and spermine [257–260]. The concentration of polyamines in absorptive epithelial cells of the small and large intestine is the net result of intracellular synthesis and degradation, uptake from the extracellular medium and release outside the cells [261–263]. Since the circulating concentration of putrescine, spermidine, and spermine in human serum is below 1.0 μM [264], thus very low, and since the capacity for the synthesis of these compounds from amino acid precursors is hardly detectable in intestinal epithelial cells from the small and large intestine [133, 258, 265], it can be deduced that the polyamines contained within absorptive epithelial cells are likely originating almost exclusively from the polyamines present in the intestinal content.

This deduction must be somewhat nuanced according to the stages of development considered. Indeed, and surprisingly, in the pig model, at birth, the absorptive epithelial cells of the small intestine already contain relatively high contents of spermine and spermidine, and putrescine [262]. Given that the intestinal fluid at birth contains almost no bacteria, this result is intriguing. It can be partly explained by the relatively high catalytic activity of the enzyme ornithine decarboxylase, the first enzyme in the polyamine-producing pathway [262]. This metabolic capacity is rapidly lost after birth since in the enterocytes isolated from 2 day old suckling piglets, the ornithine decarboxylase activity represents only 2% of the activity at birth. At that stage of development, the enterocytes from the host appear to be thus entirely dependent on the exogenous luminal source of polyamines [262].

In the proximal small intestine, due to low concentration of bacteria and rapid transit, the polyamines in the luminal content are likely mainly originating from the dietary supply. The small intestine displays large capacity for the absorption of these compounds [266], thus allowing to supply cells with polyamines in peripheral tissues. In contrast, in the colon, the polyamines in the luminal content result mostly from the metabolic activity of the bacteria towards the amino acid precursors [267].

Luminal Polyamines Exert Beneficial Effects on the Intestinal Mucosa, But Excessive Concentrations are Deleterious

Polyamines play important roles on the intestinal epithelium, and more largely on the intestinal mucosa physiology. Putrescine, spermidine and spermine are involved in fluid secretion by colonic crypts [268], and in post-prandial colonic motility [269]. Dietary supplementation with spermidine reinforces the intestinal barrier function in mice [270] (Fig. 3.14). Putrescine stimulates DNA synthesis in intestinal epithelial cells [271], and polyamines appear required for intestinal epithelium renewal [272–275]. A mixture of putrescine, spermidine and spermine has been found to be necessary for normal post-natal development of the small intestine and colon mucosa [276]. Microbial putrescine represents a stimulant for the proliferation of colonic epithelial cells [277], and thus putrescine is presumably involved in the rapid renewal of the intestinal epithelium. Interestingly, the polyamines putrescine and agmatine exert opposite effects, with putrescine being strictly necessary for proliferative colonic epithelial cells [278], and agmatine displaying strong anti-mitotic effect [279] (Fig. 3.14). Regarding cadaverine, experimental evidence strongly suggests that this polyamine plays a role in diminishing the pathogenesis of *Shigella* infections [245]. From these latter studies, it appears that the respective concentrations of the different polyamines are a crucial determinant of their overall effect on the colonic epithelium and ecosystem.

However, the reported effects of the polyamines are not exclusively beneficial, and the effects observed depend on the extracellular concentrations. Indeed, supplementation in mice with exogenous putrescine in excess disrupts tight junction permeability between epithelial cells [280] (Fig. 3.14). This latter effect is paralleled by a capacity of excessive putrescine to increase gut permeability and inflammatory cytokine concentrations in the colonic tissues. In addition, putrescine supplementation increases the attachment of bacteria to the colonic epithelium. Spermidine and spermine have also been shown to increase or decrease adhesion of bacteria to the mucus layer of the host, but the effects observed are complex, and still difficult to interpret in terms of consequences, since the effects observed depend on the bacterial strain studied, and on the age of host, either infants or adults [281].

Acetaldehyde: A Compound that can be Produced by the Intestinal Microbiota from Ethanol with Deleterious Effects on the Intestinal Epithelium

Ethanol can be produced in small amounts by the intestinal microbiota [282, 283]. Among bacteria, *Escherichia* are ethanol producers [284]. In the rat model, by increasing the amount of protein in the diet, the ethanol content in the colonic luminal fluid is increased several folds [92], suggesting overall capacity of the intestinal bacteria to produce ethanol as an accumulated intermediary metabolite during amino acid catabolism (Fig. 3.15). The intestinal microbiota metabolizes ethanol, either originating from alcohol consumption or, to a much minor extent by the microbiota metabolic activity, to acetaldehyde [285, 286] (Fig. 3.15). Indeed, several bacterial species catalyze such conversion, but this has been mainly studied regarding the bacteria in the oral cavity [287]. Acetaldehyde concentration is increased in the colon of rodents after alcohol administration, and these concentrations are lower in germ-free animals when compared with animals with a conventional intestinal microbiota [288], pointing out the role of the intestinal microbiota for acetaldehyde production from ethanol. In accordance with these results, the treatment of rodents with antibiotics decreases the intracolonic production of acetaldehyde from ethanol [289, 290]. Acetaldehyde injected in the lumen of the pig colon is metabolized to acetate [291]. Thus, acetaldehyde concentration in the colon appears to depend on the synthesis from ethanol and transformation to acetate by the colonic flora (Fig. 3.15). Ethanol chronic consumption modifies the concentrations of short-chain fatty acids in the large intestine. Indeed, in a rodent model, 8-weeks consumption of ethanol was found to modify such profile in the large intestine with decreased concentration of butyrate and increased concentration of acetate [292]. In human fecal samples, chronic alcohol consumption is associated with decreased propionate concentration, while butyrate and acetate concentrations are little affected [293].

Little is known on the effect of ethanol and acetaldehyde on the colonic epithelial cells. In in vitro experiments with human colonocytes, ethanol was found to alter the basal colonic barrier function [294] (Fig. 3.15). In volunteers, intraduodenal administration of ethanol increases both small and large intestine intestinal permeability [295]. Acetaldehyde has shown some deleterious effects on human intestinal goblet cells, decreasing the mitochondrial activity, decreasing the ATP cellular

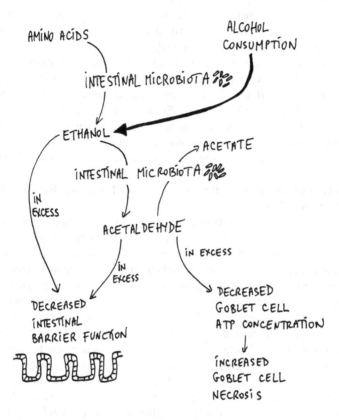

Fig. 3.15 Production of acetaldehyde by the intestinal microbiota from alcohol consumption and from ethanol synthesis by the intestinal microbiota, and effects of this bacterial metabolite on intestinal barrier function

concentration, increasing reactive oxygen species production, and finally increasing goblet cell necrosis [296], thus raising the view that acetaldehyde may adversely affect one component of the intestinal barrier function (Fig. 3.15). In the human colonic mucosa, acetaldehyde in excess reduces the expression of several proteins necessary for normal tight junction and adherens junction functions [297].

Bacterial Metabolites that are Known to be Neuroactive in the Host are Involved at First Place in Communication Between Bacteria, but also Emerge as Potential Modulator of Intestinal Physiology

Recent research works have shown that bacteria isolated from the mammalian gut, including humans, have the capacity to synthesize compounds that are well known to be produced by the host and which display neuroactive properties. Many of these bacterial metabolites result from the bacterial metabolism of amino acids. There are some indications that these metabolites play roles in the adaptation of bacteria to changes in their environment, indicating a long and complex evolutionary history for the functions of these compounds in the living world. However, there are still little information on the mechanisms that would link the intestinal bacterial metabolic activity and its effects on the intestinal wall.

Gut microbes synthesize several metabolites with neurotransmitter functions in the host [298]. These bacterial metabolites include gamma-aminobutyrate produced by *Lactobacillus* spp.; *Bifidobacterium* spp.; and *Lactococus lactis*, norepinephrine produced by various bacteria including *Escherichia* spp., and *Bacillus* spp., dopamine produced by *Bacillus* spp., histamine produced by numerous bacterial genera, and serotonin produced by various bacteria among which *Streptococcus* spp.; *Escherichia* spp.; *Enterococcus* spp., *Lactococcus*, and *Lactobacillus* [299, 300].

Gamma-Aminobutyric Acid Produced by the Intestinal Bacteria Is Likely Absorbed in the Blood

Gamma-aminobutyric acid (also known as GABA) is an amino acid which is not present in protein and known to act as the major inhibitory neurotransmitter in the central nervous system [301]. Gamma-aminobutyric acid is also found in the enteroendocrine cells of the intestinal epithelium [302]. Gamma-aminobutyric acid production is made by decarboxylation of glutamate via the enzyme glutamate decarboxylase. This enzyme is part of the glutamate decarboxylase system that is found in several bacteria genera, including *Lactobacillus* and *Bifidobacterium* [303–305]. This system is implicated in bacterial acid tolerance by maintaining intracellular pH homeostasis [306] (Fig. 3.16). Other factors, in addition to luminal acidic stress have been demonstrated to activate the glutamate decarboxylase system, including sodium, polyamines, and hypoxia. The newly formed gamma-aminobutyric acid can be metabolized to succinate, while the un-metabolized part is exported from the bacteria and released in the intestinal content [307]. This export is made by glutamate/gamma-aminobutyric acid antiporters that are found in numerous Gram-negative bacteria, including *Escherichia*, *Shigella*, *Brucella*, as well as Gram-positive bacteria like *Listeria*, *Lactobacillus*, *Lactococcus*, *Clostridium*, and *Bifidobacterium* [306]. The effects of bacterial gamma-aminobutyric acid on the intestinal wall, if any, remain to be explored. In germ-free animals, reduced levels of gamma-aminobutyric acid are measured in the intestinal content and in the serum when compared with animals inoculated with fecal microbiota [308], thus suggesting that gamma-aminobutyric acid produced by the microbiota can be absorbed in the blood.

Norepinephrine Is Involved in Communication Between Intestinal Bacteria

Norepinephrine (also called noradrenaline) is synthesized from the amino acids tyrosine and plays physiological roles as central and peripheral neurotransmitter [309]. Norepinephrine acts also as hormone after being released in the blood after secretion from the adrenal medulla cells, playing numerous physiological roles including regulation of intestinal motility [310, 311]. Norepinephrine from bacterial origin can affect the growth and virulence factors of some anaerobic bacteria such as

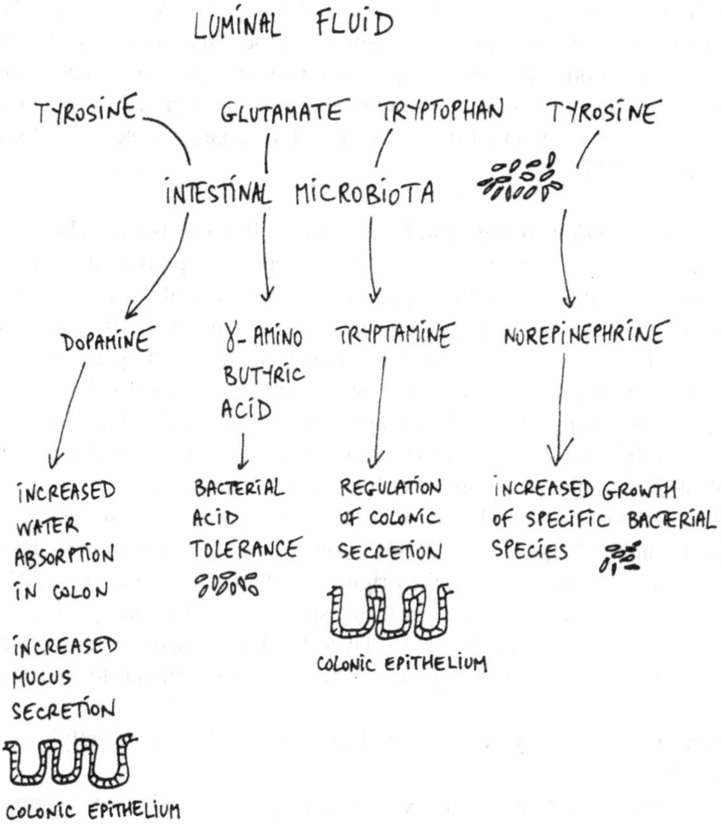

LUMINAL FLUID

TYROSINE   GLUTAMATE TRYPTOPHAN   TYROSINE

INTESTINAL MICROBIOTA

DOPAMINE   Y-AMINO   TRYPTAMINE   NOREPINEPHRINE
            BUTYRIC
            ACID

INCREASED   BACTERIAL   REGULATION   INCREASED GROWTH
WATER   ACID   OF COLONIC   OF SPECIFIC BACTERIAL
ABSORPTION   TOLERANCE   SECRETION   SPECIES
IN COLON

INCREASED   COLONIC EPITHELIUM
MUCUS
SECRETION

COLONIC EPITHELIUM

Fig. 3.16 Production of compounds with known neuroactive properties by the intestinal microbiota and effects of these compounds on microbial communication and on colonic physiology

Clostridium Perfringens [312, 313]. The pathogens *Klebsiella pneumoniae, Pseudomonas aeruginosa, Enterobacter cloacae, Shigella sonnei,* and *Staphylococcus aureus* all display increased growth in vitro in the presence of norepinephrine [314] (Fig. 3.16). Germ-free mice show marked reduction of norepinephrine in the cecal lumen, and colonization with microbiota restored the basal cecal level [315]. The effects of bacterial norepinephrine on the host intestinal wall, if any, are not known.

Dopamine Increases Water Absorption Through the Colonic Epithelium and Promotes Mucus Secretion

Dopamine is produced in two steps from tyrosine. Dopamine is a neurotransmitter synthesized in central and peripheral nervous system [316]. Bacterial tyrosinases which catalyze the conversion of tyrosine to L-dihydroxyphenylalanine (DOPA), the direct precursor of dopamine, are widely found in many bacterial genera [317]. Several bacterial species produce dopamine including *Bacillus subtilis, Escherichia coli,*

Klebsiella pneumoniae, *Proteus vulgaris*, and *Staphylococcus aureus* [299]. Of major interest, the administration of dopamine in the lumen of the colon increased colonic water absorption in mice [315], indicating that dopamine in the colonic fluid can increase water movement through the colonic epithelium in mammals (Fig. 3.16). In addition, in in vitro experiments, dopamine promotes mucus secretion in rat distal colon [318].

Histamine Production by Intestinal Bacteria: Is It Toxic for the Host?
Histamine and its receptors have been firstly described as parts of the immune and gastrointestinal systems, but their presence in the central nervous system and implication in behavior has gained increased attention [319]. Histamine is synthesized by mast cells, basophils, platelets, histaminergic neurons, and enteroendocrine cells [320], but intestinal absorptive epithelial cells of the small intestine display no capacity for histamine production from histidine [321]. Production of histamine from the decarboxylation of the amino acid histidine has been demonstrated in numerous Gram-positive and Gram-negative bacterial strains [322]. In *Lactobacillus* spp., histamine is readily exported from bacteria using the histidine/histamine antiporters [323]. Excessive histamine production by bacteria in food stored improperly can be at the origin of histamine intoxication [324]. Histamine intoxication can affect the gastrointestinal tract physiology by provoking diarrhea and vomiting [325]. The part played by the excessive microbiota-derived histamine in the toxic effects of this compound on the host remains to be determined.

Serotonin Produced by Intestinal Bacteria Is Likely Partly Absorbed in the Blood
Serotonin (also known as 5-hydroxytryptamine) is not only a neurotransmitter produced in specialized cells of the human brain, but also acts as hormone secreted by intestinal enteroendocrine cells [326]. Serotonin is known to regulate intestinal secretion and motility, as well as visceral sensitivity [327]. The effects of bacterial serotonin on the host intestinal wall, if any, is not known. In germ-free animals, there is a reduction of serotonin concentration in blood compared to conventional animals with microbiota, and serotonin concentration can be restored via recolonization of mice with microbiota [173]. These latter results suggest that a part of the serotonin produced by the intestinal microbiota can be absorbed through the intestinal epithelium in blood. Gut microbiota promotes serotonin synthesis in enteroendocrine cells of the colonic epithelium [328]. Thus, bacteria can interact with the host to induce the endogenous production of serotonin, such induction being able to affect intestinal motility [328].

Tryptamine Produced by Intestinal Bacteria Participates in the Regulation of Secretion in Colon
Tryptamine is a monoamine like 5-hydroxytryptamine (serotonin) that is produced by the intestinal microbiota from tryptophan. Tryptamine can be produced by different bacterial species, notably *Clostridium sporogenes* [329]. In the intestinal

tract, tryptamine activates the 5-HT4 receptor expressed in the colonic epithelium, and such activation controls colonic secretion [330] (Fig. 3.16).

Key Points
- The fecal and/or colonic luminal contents in numerous amino acid-derived bacterial metabolites (including branched-chain fatty acids, lactate, ammonia, phenol, p-cresol, indole, skatole, hydrogen sulfide, polyamines, and acetaldehyde) are depending on the characteristics of the diet consumed.
- Several amino acid-derived bacterial metabolites, like formate, oxaloacetate, p-cresol, indole, hydrogen sulfide, certain polyamines, gamma-aminobutyric acid, and norepinephrine, are involved in communication between microbes, bacterial colonization, and bacterial growth.
- Several amino acid-derived bacterial metabolites including branched-chain fatty acids, lactate, succinate, indole, dopamine, and tryptamine exert beneficial effects on the intestinal epithelium physiology and metabolism in terms of regulation of electrolyte movement, epithelial renewal, and energy supply.
- Several amino acid-derived bacterial metabolites in excess like ammonia, phenol, p-cresol, skatole, hydrogen sulfide, putrescine, and acetaldehyde exert deleterious effects on intestinal epithelial cells in terms of energy metabolism, epithelial permeability, and cell viability.
- Detoxification pathways against some deleterious amino acid-derived bacterial metabolites, like ammonia and hydrogen sulfide, have been identified in colonocytes.

Effects of the Amounts of Dietary Proteins on the Intestinal Ecosystem and Consequences

The in vitro and in vivo tests of individual effects of amino acid-derived bacterial metabolites on the colonic epithelial cells are a good starting point, as presented above, to identify the metabolites that are beneficial or deleterious for these cells. Such an identification can indeed help in defining a safe luminal environment that helps in maintaining the colorectal epithelium in "healthy state." However, in real life situation, the intestinal epithelium is facing simultaneously, behind its protective mucous layers, a multitude of compounds. The amounts and/or concentrations of these compounds can vary markedly according to dietary changes in the short- and long-term period of times [331, 332].

Regarding the digestibility of dietary protein, animal protein is overall higher than the digestibility of plant protein [333], thus implying that on a basis of an equal amount of dietary protein, consumption of plant protein versus animal protein will likely result in a higher amount of protein being transferred from the small to the large intestine. Among plant proteins, some of them, like for instance proteins in rapeseed, are characterized by markedly lower digestibility [334] when compared to other dietary proteins. Then, both the amounts of dietary proteins consumed and the sources of proteins will affect the luminal content in terms of bacterial metabolite composition. In addition, the content of amino acids in each dietary protein will change the availability of the different amino acids to be used as substrates by the large intestine bacteria, and then will presumably change the production rates of the various metabolites.

Of note, the amount and type of dietary protein consumed over the world can be markedly different between geographical areas. For instance, in the western world, the amounts of proteins in the diet are largely above the recommended intake, considering the metabolic and physiological requirement [335], while in some developing countries, the consumption can be below the requirement [336]. The protein requirement has been determined to be 0.83 g/kg body weight/day in adults [335]. Of note, in several developing countries, the ratio of proteins originating from plant over proteins from animal origin is well above the ratio calculated for Western countries [336].

After these preliminaries, and considering that amino acid-derived bacterial metabolites, as explained above, have been shown to be active on colonic epithelial cells in preclinical in vitro and in vivo studies, what are the known consequences of an increase of the protein content in the diet on the large intestine ecosystem?

High-Protein Diet are Used for Body Weight Reduction and Muscle Mass Augmentation

The amount of dietary protein ingested can be markedly increased by consuming the so-called high-protein diets. These high-protein diets are belonging to the numerous types of weight-loss diets that are currently proposed and consumed [337] by millions of overweight and obese individuals among populations in Europe and USA who wish to decrease their body weight [337–339], and also by athletes and exercisers [340] who wish to increase their muscle mass and physical performance [341]. These high-protein diets represent a heterogeneous group of diets with different compositions [342] but are all characterized by a higher proportion of protein (25–30% of total energy intake) among the two other dietary macronutrients that are carbohydrates and fat, when compared with the generally recommended macronutrient proportion. This recommended proportion is approximately 10–15% daily energy from protein, 55–75% from carbohydrates, and 15–30% energy from fat [343, 344]. One of the main rationales for the consumption of high-protein diets for body weight reduction is that it is recognized that, on a basis of equal energy content, dietary proteins are more satiating than carbohydrates and fats [345], thus diminishing calorie uptake. Regarding the consumption of high-protein diets by athletes, the main rationale for such utilization is to provide more amino acids (considering that usual amino acid supply is insufficient) following exercise to build more muscle proteins, thus participating in muscle hypertrophy [346]. The impact of high-protein diet on gut health, considering the metabolic activity of the intestinal microbiota, remains an important but emerging topic.

Short-Term High-Protein Diet Consumption does not Change the Large Intestine Microbiota Composition But do Modify Production of Bacterial Metabolites

As said above, it has been determined that based on regular western diet, approximately 12 g of undigested, or not fully digested dietary and endogenous proteins are transferred from the small to the large intestine [21]. The amount of nitrogenous material is increased nearly proportionally when the amount of dietary protein consumed increases [347]. Relatively few human intervention studies have examined the effects of changing the amount of dietary protein in the diet on the gut microbiota composition and metabolic activity. In addition, in several studies, the dietary protein intake was not the only parameter modified between the groups of volunteers, since energy and/or fiber intake were also different. These two latter parameters are known to affect the gut microbiota composition [348–350], rendering correct interpretation of the results obtained difficult. Two studies have used high-protein diet without modification of dietary fiber and energy intake [351, 352]. These two studies did not detect any measurable changes in the fecal and rectal biopsies-associated microbiota composition after 2 to 3 weeks consumption of the high-protein diet.

However, with regards to the metabolic activity of the intestinal microbiota, several intervention studies in humans have shown that high-protein diet induces a shift from carbohydrates to protein degradation by the gut microbiota, as judged from the analysis of bacterial metabolites in feces [145, 351, 353, 354]. The effects of high-protein diet appear rather homogeneous despite differences in the experimental design, emphasizing the importance of substrate availability -amino acid availability in this case- for determining the changes in the luminal environment within the large intestine. Most of the studies performed reported, as expected, that high-protein diet consumption induces increase in feces of amino acid-derived branched-chain fatty acids, such as the bacterial metabolites isobutyrate, isovalerate, and 2-methylbutyrate [351, 354, 355]. In contrast, a decrease of the short-chain fatty acid butyrate was consistently found after high-protein diet consumption [351, 354, 356, 357]. However, several of these studies were characterized by a decrease in fiber content in the diet consumed by the volunteers receiving the high-protein diet, knowing that fibers are major substrates for short-chain fatty acid production by the intestinal microbiota. However, such a reduction in the fecal concentration of butyrate was found even when volunteers from the high-protein diet group consumed similar amounts of dietary fibers and calories than volunteers in the normo-protein group, indicating that the increased amount of protein in the diet plays a significant role in such a decrease [351]. As indicated in the next paragraph, butyrate is well known to be a major luminal oxidative substrate, and a regulator of histone acetylation, and thus of gene transcription in human colonocytes [89, 358], raising the view that decreased butyrate concentration in the large intestine fluid would be presumably detrimental for the rectal mucosa homeostasis.

Two studies in volunteers receiving a high-protein diet found a marked increase in fecal ammonia concentration [145, 353], while two others did not [351, 354], likely due to different experimental protocols. As ammonia is efficiently absorbed through the colon epithelium [123], the fact that ammonia concentration was not increased in the fecal fluid does not mean that ammonia production by the intestinal microbiota was not increased in the colonic fluid after high-protein diet consumption. Indeed, when recovering the colonic content in rodent models after high-protein diet ingestion, the ammonia concentration in this content was markedly increased when compared to animals receiving a normo-protein diet [116, 130]. High-protein diets were found in addition to increase the fecal concentration of several S-containing bacterial metabolites, including hydrogen sulfide [145, 359].

Analysis of Bacterial Metabolites and Co-Metabolites in Urine: A Useful Way to Decipher Intestinal Microbiota Metabolic Activity
Urinary metabolomics analysis is much useful to identify the bacterial metabolites and co-metabolites which have been produced by the gut microbiota, absorbed from the lumen to the bloodstream through the intestinal epithelium (with or without metabolism in colonocytes), possibly further metabolized by the host in the liver and other organs outside the splanchnic area, and finally excreted in the urine where they accumulate (Fig. 3.17). High-protein diet consumption increases the concentration of several amino acid-derived bacterial metabolites and co-metabolites in the urine of volunteers. For instance, high-protein diet ingestion results in an increased urinary excretion of the bacterial metabolite phenol [353]. Regarding co-metabolites, *p*-cresyl sulfate have been repetitively found to be more excreted in the urine after high-protein diet consumption than after a normo-protein diet ingestion [145, 351, 360]. This latter compound results from the metabolism of the bacterial metabolite *p*-cresol in the liver [361]. An increased excretion of *p*-cresyl sulfate in urine is an indication of an increased absorption of the bacterial metabolite *p*-cresol across the large intestine epithelium, and further metabolism by host tissues, and thus presumably of the production of this compound by the intestinal microbiota. This increased urinary excretion of *p*-cresyl sulfate after high-protein diet consumption is an important parameter to be considered, given the deleterious effects of excessive *p*-cresol on colonocytes (see the paragraph "Amino acid-derived bacterial metabolites and effects on the intestinal epithelium"). Another co-metabolite, indoxyl sulfate, was found to be increased in the morning spot urine after high-protein diet consumption [351]. Since indole, the bacterial precursor for the synthesis of indoxyl sulfate in the liver, has been shown to contribute to the maintenance of the colonic barrier function [195, 196], an increased bacterial production of indole may be beneficial for the host. This potential beneficial effects on the intestinal mucosa must however be confronted to the deleterious effects of indoxyl sulfate in excess on kidney cells [362] (see Chap. 5).

The Production of Metabolites Produced from Amino Acids by the Gut Microbiota Is Different When Consuming Animal or Plant Proteins
The situation is further complicated by the fact that the changes observed in the urinary metabolome after high-protein diet consumption are different according to the nature of protein used as supplement. Indeed, when volunteers ingest for three weeks a high-protein diet obtained by supplementing the diet with casein, their urine

Fig. 3.17 Synthesis of co-metabolites by the host from bacterial metabolites. In this scheme, the case of bacterial metabolite synthesized from amino acids, and conversion of these metabolites into co-metabolites by the host tissues are presented, as well as excretion of these compounds in the urine

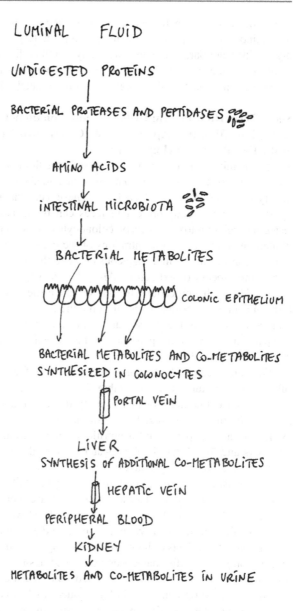

LUMINAL   FLUID

UNDIGESTED PROTEINS

BACTERIAL PROTEASES AND PEPTIDASES

AMINO ACIDS

INTESTINAL MICROBIOTA

BACTERIAL METABOLITES

COLONIC EPITHELIUM

BACTERIAL METABOLITES AND CO-METABOLITES
SYNTHESIZED IN COLONOCYTES

PORTAL VEIN

LIVER
SYNTHESIS OF ADDITIONAL CO-METABOLITES

HEPATIC VEIN

PERIPHERAL BLOOD

KIDNEY

METABOLITES AND CO-METABOLITES IN URINE

contains, in reference to the urine obtained from the normo-protein control, increased concentration of urea (synthesized in the liver urea cycle from ammonia). Increased concentrations of the bacterial metabolites and co-metabolites isobutyrate, 3-hydroxybutyrate, 3-hydroxy isovalerate, *p*-cresyl sulfate, indoxyl sulfate and phenylacetyl glutamine were also measured in the urine [351]. However, when dietary supplementation was done with soy protein, although the increase in the bacterial metabolites and co-metabolites was like that recorded using casein supplementation, no increase of p-cresyl sulfate in urine was recorded [351]. Thus, the amino acid composition in the different dietary proteins, together with their digestibility, are parameters that are likely to affect differently the luminal environment of the intestinal epithelium in terms of bacterial metabolite composition, and with different impact on the colonic epithelium metabolism and physiology.

Short-Term High-Protein Diets Consumption do not Induce Fecal Water Toxicity, But Modify Expression of Genes Involved in the Maintenance of the Rectal Mucosa Homeostasis

To test the mixture of compounds, present in the luminal content of the distal part of the intestine, one possibility is to recover the so-called fecal water before and after a dietary intervention in volunteers [351]. These fecal water extracts which contain the hydrophilic compounds present in feces can be separated from the bacteria and tested for their cytotoxic effects on colonocytes. However, this technique has several limitations. Firstly, the concentrations of the compounds in the mixture are not the original concentrations in the aqueous phase of the luminal fluid, because of dilution during the process of extraction. Also, the fecal hydrophobic compounds are not extracted by such a procedure. In addition, the respective parts of compounds in free form and in form bound to fecal components most probably do not mimic the original ratios. With these reservations in mind, this fecal water test allows useful comparison of cytotoxic potential of the intestinal luminal content in different dietary situations [363].

When an isocaloric high-protein diet was given for two weeks to healthy human subjects, the mixture of water-soluble components recovered from the feces shown no increased cytotoxicity or genotoxicity towards human colonocytes when compared to normo-protein diet [352]. Similarly, in a study by Benassi-Evans et al. [364], the authors performed a nutritional intervention with high-protein diet for 52 weeks and found that the fecal water recovered from the volunteers consuming such diet was not more genotoxic that the ones recovered from control volunteers consuming isocaloric normo-protein diet. In the study by Beaumont et al. [351], supplementation of the regular diet with either casein or soy protein for 3 weeks did not result in higher cytotoxic potential of the fecal water when compared with the results obtained from isocaloric normo-protein diet volunteers.

Regarding the effects of high-protein diet ingestion on gene expression in the gut mucosa, only few studies have been performed. Using a six week- dietary intervention protocol, a study in rodents showed that casein-containing high-protein diet modifies gene expression in the colonic mucosa when compared with an isocaloric normo-protein diet [365]. Another study in rodents showed that a two week-intervention protocol with whole milk protein- containing high-protein diet down-regulates in colonocytes gene expression notably in relationship with cell metabolism, cell signaling, DNA repair, and cellular adhesion when compared with the situation in colonocytes isolated from animals that received an isocaloric normo-protein diet [366]. On the contrary, such diet was found to up-regulate the expression of genes involved in intestinal barrier function [366]. These experimental studies allow to establish the new proof of concept according to which increasing the amount of dietary protein in the diet results in a modification of gene expression in the colonic mucosa, and more specifically, in the epithelial layer. A randomized double-blind controlled study with overweight volunteers reported that 3 week-dietary supplementations with either casein or soy protein resulted in small amplitude changes in the expression of numerous genes in the rectal mucosa, notably for genes involved in the homeostatic processes of epithelial renewal, such as cell cycle and cell death [351].

Low-Protein Diet May Affect Intestinal Microbiota Composition and Metabolic Activity

In deep contrast with the consumption of diet rich in proteins, thus above the requirement, in Western countries, a part of the world population is affected by an amount of dietary protein sometimes largely below the requirement, thus affecting children growth and health status [367–369]. The normal pattern of intestinal microbiota evolution is disrupted in children suffering from undernutrition [370, 371], even if the part due to insufficient protein consumption remains unknown. Little is known on the impact of low-protein diet on the intestinal microbiota composition and metabolic activity, as well as consequences in terms of intestinal physiology. In the young pig model, low-protein-high-carbohydrate diet given for one month alters the fecal microbiota composition [372], while in mice, low-protein diet leads to a greater abundance of urease-producing bacterial species in feces [373], thus modifying the microbiota composition, and presumably the overall microbiota metabolic capacity.

Key Points

- High-protein diets are used by individuals who want to decrease their body weight in relationship with the associated satiating effect of such diets. Short-term consumption of such diets induces a shift from carbohydrate to protein degradation by the intestinal microbiota with increased amino acid-derived bacterial metabolites and co-metabolites recovered in the urine.
- The production of bacterial metabolites is different when supplementing the regular diet with either animal or plant proteins.
- Short-term consumption of high-protein diet causes little change in fecal and rectal mucosa-associated microbiota composition and diversity.
- Short-term consumption of high-protein diet modifies the expression of genes involved in the maintenance of rectal mucosa homeostasis.
- Children suffering from undernutrition display an altered pattern of intestinal microbiota. Low protein diets alter the fecal microbiota composition.

3.2   Metabolism of Polysaccharides by the Intestinal Microbiota and Impact on the Intestinal Epithelium Metabolism and Functions

Polysaccharide-Derived Bacterial Metabolites and Effects on the Intestinal Epithelium

Indigestible Polysaccharides Are Substrates for the Large Intestine Microbiota Giving Rise to Short-Chain Fatty Acids

Among the carbohydrates present in the diet, only monosaccharides can be directly absorbed by the small intestine enterocytes which use dedicated transporters for such a purpose, while disaccharides (like sucrose, lactose, and maltose) need to be firstly digested before absorption of the resulting monosaccharides (glucose, fructose, galactose) [374]. Indigestible carbohydrates in the diet belong to a family of compounds mostly, but not exclusively, present in plant sources that include resistant starch, non-starch polysaccharides, and indigestible oligosaccharides [375]. Resistant starch cannot be fully digested in the small intestine and are usually divided into different groups according to their relative resistance to digestion [376]. The non-starch polysaccharides, usually called fibers, include soluble and insoluble fibers [377]. The indigestible carbohydrates are transferred to the large intestine where they are used as primary carbon and energy source for bacterial growth [378]. In the process of fermentation of the dietary non-digestible carbohydrates, the human gut microbiota produces large amount of short-chain fatty acids, namely acetate, propionate, and butyrate [379]. Both dietary fibers and resistant starch are substrates for the production of short-chain fatty acids in the large intestine [380–384].

Non-digestible carbohydrates are not the unique sources of short-chain fatty acids, since, as indicated in the preceding paragraphs, several amino acids are used by the intestinal microbiota as substrates for such production. In addition to these precursors, dietary fructose has been identified as being converted to acetate by the intestinal microbiota [385]. Numerous butyrate-producing species are belonging to *Ruminococcaceae*, *Lactospiraceae*, and *Clostridiaceae* [386]. Other bacterial species, including *Faecalibacterium prausnitzii*, are also able to produce butyrate [387, 388], as well as *Eubacterium rectale* and *Roseburia*, and many clostridium species [389].

These compounds are recovered at millimolar concentrations in human colonic fluid. Acetate is the most abundant, averaging 54 mM in human colon, while propionate and butyrate represent one-third of acetate concentration, thus approximately 18 mM [84, 89, 390–393]. Due to the pKa of the short-chain fatty acids, and to the neutral or slightly acidic pH in the large intestine fluid, most of the short-chain fatty acids are in anionic, dissociated form in the large intestine content [394]. These compounds are transported in colonocytes through different processes. Specific transporters, notably the monocarboxylate transporters MCT1 and SLC5A8, are responsible for the entry of short-chain fatty acids inside colonocytes [112, 358, 395, 396]. Butyrate can also enter the colonocytes by using the SCFA/HCO$_3$

Fig. 3.18 Effects of the bacterial metabolite butyrate on the colonic epithelium metabolism and physiology

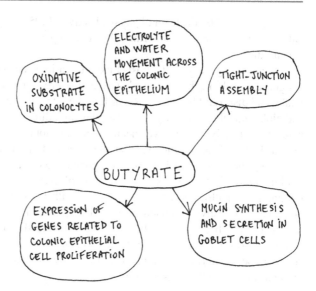

exchanger [397], while the minor amount of short-chain fatty acids in undissociated lipid-soluble form can enter colonocytes by diffusion [398]. Of note, from in vitro experiments with colonocytes and luminal membranes obtained from colonocytes, it has been demonstrated that butyrate transport by the MCT1 transporter can be reduced by L-lactate and propionate [399, 400], suggesting that competition between different organic acids present in the luminal fluid for entry into colonocytes may happen in vivo.

Short-Chain Fatty Acids Are Luminal Fuels for Absorptive Colonocytes
Butyrate, acetate, and propionate, after their entry inside absorptive colonocytes, are highly oxidized in these cells [401–403], thus serving as luminal fuels for colonocytes (Fig. 3.18) and explaining the micromolar circulating concentrations of these compounds in blood as compared to luminal millimolar concentration [390]. Thus, short-chain fatty acids utilization by the host cells represents a way to recycle energy from compounds produced by the intestinal microbiota from undigested (or not fully digested) dietary compounds like fibers and proteins.

Oxidation of Short-Chain Fatty Acids Brings a Minor Amount of Energy to the Host
However, such energy recycling appears to represent only a minor part of total energy consumption by the human body. Indeed, in humans, short-chain fatty acids contribute at the most to approximately 10% of the caloric requirement [404]. Then, the supply of these metabolites by microbiota to the host, and their subsequent oxidation represents a modest contribution in terms of energy supply, and this is particularly true when calorie intake is largely above the requirements. As a matter of fact, overweight and obesity are mainly driven by the amount of energy ingested

from the diet, and by the level of physical activity [405, 406], according to the energy balance status, and considering the interdependency of energy intake and expenditure.

Genetic background of the hosts linked to various metabolic pathways also plays a significant role in the risk of obesity and overweight at a given level of energy intake and disposal [407–409]. Regarding the intestinal microbiota, its composition and global metabolic activity appear to play an indirect role in the process of weight gain. The underlying putative mechanisms are notably related to the capacity of several bacterial metabolites to alter the intestinal epithelial barrier function, thus increasing the transfer of bacterial components and bacterial metabolites from the luminal fluid to the bloodstream, several of these latter being able to activate inflammatory pathways that are associated with both obesity and risk of adverse outcomes in obesity-associated diseases [410]. The situation is complicated by the fact that some bacterial metabolites, like butyrate that represents a fuel for the colonic absorptive epithelial cells, have been shown to modify the secretion of enteroendocrine hormones involved in the control of food intake [411], as detailed in the last paragraph of this section.

Short-Chain Fatty Acids Regulate Electrolyte and Water Movement Across the Colonic Epithelium
Butyrate, and to a minor extent acetate and propionate, have been shown to play a role in the movement of electrolytes across the colonic epithelium, thus participating in the regulation of absorption of electrolytes and water by the colonic absorptive cells [96–98, 398, 412]; as well as electrolyte secretion [100, 413] (Fig. 3.18).

Short-Chain Fatty Acids Stimulate Tight Junction Assembly
In in vitro experiments performed with monolayers of human colonocytes, butyrate stimulates tight junction assembly, thus enhancing the epithelial barrier function [414]. In another in vitro study performed with intestinal epithelial cells grown as monolayers, butyrate increased epithelial barrier function in a process involving increased expression of gene coding for claudin-1, a component of the tight junctions involved in intestinal barrier function [415] (Fig. 3.18).

Butyrate Oxidation in Colonocytes Regulates Its Intracellular Concentration: A Way to Regulate Gene Expression in Differentiated Cells
In colonocytes, a mixture of the three short-chain fatty acids regulates the expression of different genes involved in different aspects of their functions [416]. Among these compounds, butyrate appears to be the most effective in such regulation [417, 418]. Of notable interest, butyrate increases the expression of genes involved in the control of proliferation of cancerous colonic epithelial cell models, thus reducing their growth rate [419–421]. These results have led to the proposition that butyrate mitochondrial metabolism in these cells is not exclusively related to energy production, but also acts as a regulator of butyrate concentration, and thus presumably its effect on nuclear gene expression, and associated decreased growth [422, 423] (Fig. 3.19). In other words, when butyrate is highly metabolized in the

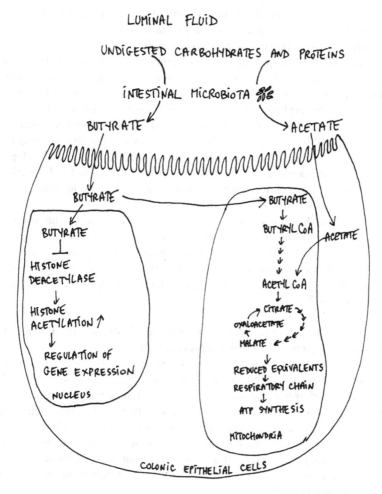

Fig. 3.19 Metabolism and mechanisms of action of butyrate and acetate in cancerous colonic epithelial cell models. In this scheme, the metabolism of butyrate in the mitochondria of cancerous colonocytes to regulate its cytosolic concentration (and thus its effects on gene expression in the nucleus) is presented

mitochondria of cancerous colonocytes, this bacterial metabolite little affect cell growth, while when this compound is little metabolized, it accumulates in the cytosol of cells and modifies gene expression resulting in reduced growth rate. Of note, reduced growth rate corresponds to a lower requirement of ATP synthesis, while sustained growth rate is associated with higher requirement of energy, notably for protein and ADN synthesis [424]. The mechanisms involved in the regulation of gene expression by butyrate in colonocytes include notably the capacity of butyrate to increase the acetylation of specific nuclear histone proteins [425]. Histone hyperacetylation in colonocytes represents an epigenetic mechanism driven by the environment that regulates gene expression in these cells [426].

Regarding the oxidation of butyrate to regulate its intracellular concentration, the central article of Roediger published in 1982 reinforces indirectly this latter concept [403]. As a matter of fact, although well oxidized in the mitochondria of the colonocytes, butyrate very modestly increases oxygen consumption in these cells [403]. This can be explained by the suppression of the oxidation of endogenous substrates in colonocytes in the presence of butyrate [403]. In other worlds, when butyrate is available from the microbial activity to colonocytes, these cells use butyrate as a preferential fuel for ATP synthesis, and thus use less endogenous substrates (to maintain energy homeostasis), such butyrate metabolism allowing regulation of its concentration within colonocytes, and thus its effect on gene expression.

Of note, and in the same line of thinking, it has been shown that the oxidation of butyrate by differentiated colonocytes in the surface epithelium and in the upper part of the colonic crypts regulates the concentration of butyrate in vicinity of proliferating epithelial stem/progenitor cells in the lower part of the crypts, thus protecting them from the deleterious effects of butyrate at excessive concentrations [427].

Short-Chain Fatty Acids Stimulate Mucin Synthesis in Goblet Cells and Hormone Secretion by Enteroendocrine Cells

Less data is available regarding the effects of short-chain fatty acids on non-absorptive intestinal epithelial cells. Regarding the mucous-secreting goblets cells, butyrate, and propionate increase the expression of the gene coding for the mucin MUC2 in human goblet cells [428], and acetate and butyrate induce mucin secretion in the colon after intraluminal administration [429]. In biopsies recovered from colonic resection samples, butyrate is able to increase the synthesis of mucin, and this stimulating effect is dependent on the oxidation of butyrate [430], thus suggesting that energy provision through butyrate metabolism in biopsies likely participate in the process (Fig. 3.18).

The role of short-chain fatty acids on hormone secretion by the enteroendocrine cells has been investigated. Although butyrate receptors have been identified in endocrine cells present in the colonic crypts [431, 432], the physiological significance of the presence of such receptors in the large intestine remains unclear [433]. Propionate and butyrate have been shown to stimulate in vitro the production and secretion of PYY in human enteroendocrine cells partly through binding to free fatty acid receptors (FFAR) [434]. When delivery of propionate to the colon in volunteers was performed by oral ingestion of inulin-propionate, an increased PYY and GLP-1 plasma concentrations were measured [435]. An infusion of acetate in the distal colon of human subjects resulted in an increased circulating concentration of PYY [436]. These results are of interest as they suggest that the three short-chain fatty acids increase in different preclinical and clinical studies the secretion of the satiety stimulators PYY and GLP-1 [437, 438]. Thus, it remains intriguing that in one hand, as stated in the previous paragraph, short-chain fatty acid oxidation by the host contributes to a modest extent to the calorie supply, while in the other hand, these bacterial metabolites increase the secretion of the satiety stimulators PYY and GLP-1 by enteroendocrine cells.

Key Points
- Indigestible polysaccharides are used by the large intestine bacteria to produce short-chain fatty acids which are used as energy substrates by colonocytes.
- Butyrate is used as energy substrate within colonocytes, allowing regulation of its intracellular concentration, and thus its effects on gene expression.
- Butyrate oxidation by differentiated colonocytes regulates the concentration of this compound near the proliferating stem cells, avoiding marked inhibition of their proliferation in case of excessive concentration.
- Short-chain fatty acids regulate electrolyte and water movement across the colonic epithelium and favor the maintenance of epithelial barrier.
- Short-chain fatty acids stimulate mucus synthesis and secretion by goblet cells and stimulate the secretion of the satiety hormones PYY and GLP-1 by colonic enteroendocrine cells.

Effects of the Amount of Dietary Polysaccharide on the Intestinal Ecosystem and Consequences

Higher Consumption of Fibers Is Generally Associated with Favorable Health Outcomes in Terms Of Intestinal Ecosystem and Intestinal Physiology

Systematic analysis of observational studies indicates that higher dietary fiber consumption is associated with favorable health outcomes [439]. Notably, higher consumption of fibers is associated with lower concentrations of serum inflammatory biomarkers [440, 441]. However, it is worth noting that higher consumption of fibers may be associated with higher intake of some vitamins [442], some minerals [443], and numerous phytomicronutrients [444], all these compounds being able to be part of the favorable outcomes observed in individuals with higher fiber intake. With these reservations in mind, what part of the beneficial effects of indigestible poly- saccharide on health outcomes could be attributable to changes in the intestinal ecosystem, and notably in reference to bacterial metabolite production from alimen- tary compounds?

Mechanistic studies and clinical trials with isolated or mixtures of different fibers have demonstrated regulatory effects on gastrointestinal physiology notably in terms of digestion and absorption of dietary compounds, transit time, stool consistency, and gut microbiota composition and metabolic activity [445–447].

When indigestible polysaccharides are in short supply, as often observed in Western countries [448], the large intestine microbiota may switch to other sources for growth such as amino acids originating from undigested proteins or fats [93, 378]. Such a metabolic switch has consequences for the composition of the luminal large intestine fluid, resulting in decreased butyrate concentration in feces [354, 357, 449].

In experimental and clinical studies, it has been consistently shown that supplementation with indigestible saccharides, in case of consumption of diet rich in protein, beneficially lowers the production of amino acid-derived bacterial metabolites with potential toxicity against the large intestine epithelium [81, 144, 147, 450–454]. These results are corroborated by in vitro fermentation with stools obtained from donors [68].

Very Low Dietary Fiber Consumption May Adversely Affect the Protective Mucus Layer in the Large Intestine

There is evidence that high fiber consumption contributes to an efficient protective mucus barrier upon the colonic epithelial layer [455]. It has been proposed that when the availability of fibers for the intestinal microbiota is limited, due to lower consumption, the proportion of mucus-degrading bacteria increases, therefore potentially affecting the characteristics of the mucus layer [456].

More generally, when the amount of microbiota-accessible carbohydrates is limited in comparison with what bacteria require for growth, bacteria become dependent on alternative sources represented notably by mucin glycans [457–459], therefore possibly altering mucus structure. Incidentally, some adhering bacterial species in the gastrointestinal tract are equipped with proteases that can directly target mucins [460], thus changing their structure, or reducing their viscoelasticity, allowing increased transfer of luminal bacteria through mucus [461]. The full understanding of how such a use of mucin glycans by the colonic bacteria modify the mucus properties is complicated by the fact that limited degradation of mucus is needed for the maintenance of a protective barrier function [462]. In other words, only marked utilization of mucin glycans by specific bacteria would be susceptible to affect the mucus layer properties.

Key Points
- Dietary fiber consumption is associated with beneficial regulatory effects on digestion, absorption, and intestinal transit time.
- The amounts of dietary fibers consumed have an impact on the microbiota composition and metabolic activity.
- Lower dietary indigestible polysaccharide consumption shifts the intestinal microbiota metabolic activity to other sources of substrates like amino acids and fats.
- Very low dietary fiber consumption affects the structure of the protective mucus gel.

3.3   Metabolism of Lipids by the Intestinal Microbiota

Lipid-Derived Bacterial Metabolites and Effects on the Intestinal Ecosystem

In a Context of Balanced Diet Consumption, Few Lipids are Transferred from the Small to the Large Intestine

A very small proportion of dietary fat, usually less than 5% in condition of regular diet consumption, reaches the large intestine [170, 463]. Bacteria in the gut degrade triglycerides and phospholipids into glycerol and fatty acids [464]. Triglycerides represent a large part of fat present in the diet, while phospholipids, notably phosphatidylcholine, represent only a minor part of dietary lipids. Of note, phosphatidylcholine can also be synthesized from endogenous bile acids [465]. Regarding the glycerol moiety, it serves as a precursor for the synthesis by *Lactobacillus reuteri* of   reuterin,   a   mixture   of   different   compounds   that   include 3-hydroxypropionaldehyde (3-HPA), 3-HPA hydrate, 3-HPA dimer, and acrolein [466–468]. The bacterial metabolite reuterin is known to exert broad-spectrum antimicrobial effect against several bacteria of the intestinal microbiota including *Clostridium difficile* and *Escherichia Coli* [469, 470] (Fig. 3.20). Among the metabolites present in reuterin, acrolein has been shown to exert deleterious effect towards intestinal epithelial cells, affecting tight junction proteins, and damaging intestinal barrier function, resulting in increased intestinal permeability [471] (Fig. 3.20).

Regarding phosphatidylcholine, the intestinal microbiota converts the choline moiety of the phospholipid into the bacterial metabolite trimethylamine [472] (Fig. 3.20). This trimethylamine production is performed by several species present in the intestinal microbiota including members of *Clostridium*, *Eubacterium* spp., and *Proteobacteria* [473, 474]. No information on the effects of trimethylamine on the intestinal epithelium is available at the time of the writing of this book. As will be developed   in   the   paragraph   devoted   to   the   gut-cardiovascular   system, trimethylamine is however well known to be absorbed through the intestinal epithelium and then metabolized to trimethylamine oxide in the liver [475].

Concerning the fatty acids released from triacylglycerol and phospholipids in the luminal fluid of the large intestine, there are several indications that free fatty acids display strong antimicrobial properties in vitro [476, 477], but the impact of such property for the control of the bacterial growth in the large intestine in real life situation remained to be determined.

In regards to specific bacterial metabolites that are derived from lipids, fecapentaenes, that are present in the human feces, are produced by the intestinal microbiota from polyunsaturated ether phospholipids (Fig. 3.20) and represent potent mutagens towards colon epithelial cell DNA [478], as will be detailed in Chap. 4. Fecapentaenes are produced by several species belonging to *Bacteroides* spp. [479].

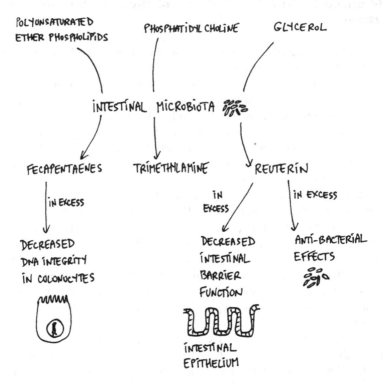

Fig. 3.20 Production of bacterial metabolites from lipids and glycerol moiety of lipids and effects of these compounds on the intestinal ecosystem. In this scheme, the effects of excessive concentrations of bacterial metabolites on DNA integrity in colonocytes, on the intestinal barrier function, and on bacteria are indicated

Key Points
- Generally, few lipids are transferred from the small to the large intestine.
- The glycerol moiety of triglycerides and phospholipids is metabolized by the gut microbiota into reuterin, a mixture of bacterial metabolites with antimicrobial properties.
- Among the compounds in reuterin, acrolein in excess exerts adverse effects on intestinal epithelial permeability.

Effects of the Amount of Dietary Lipids on the Intestinal Ecosystem and Consequences

Evidence from animal studies has shown that high-fat diets modify the intestinal ecosystem both in the distal part of the small intestine and in the large intestine.

High-Fat Diets Modify the Intestinal Microbiota Composition and Ileal Permeability

In the ileum, four weeks of high-fat diet given to rodents results in the colonization of the space between villous by a dense microbiota, a situation in contrast with what is observed in animals that receive a standard diet, since in that latter case the intervillous space is free of bacteria [480]. This may be related to the effects of fatty acids, which in addition to their reported antimicrobial effects, appear to act on the adhesion of bacteria to mucus. As a matter of fact, some lipids, depending on their chemical structure, have been shown to affect the adhesion of bacteria to the mucus layer. Indeed, in in vitro studies, polyunsaturated fatty acids inhibit adhesion of *Lactobacillus* to mucus, while gamma-linolenic acid and arachidonic acid promote adhesion of *Lactobacillus* to mucus [481].

Overall, the cecal and fecal microbiota composition are markedly modified by high-fat diets, notably regarding the relative abundance of *Lactobacillus*, *Bifidobacterium*, and *Bacteroides* [482, 483]. This intestinal dysbiosis is associated with numerous adverse events including decreased expression of genes involved in antimicrobial peptides in ileal epithelial cells, retention of mucin in goblet cells, decreased electrolyte secretion, upregulation of the pro-inflammatory cytokine tumor necrosis factor-alpha (TNF-α), decreased expression of genes coding for tight junction protein, and finally alteration of the ileal permeability [480, 483–486].

Secondary Bile Acids Can Decrease Clostridium Difficile Colonization But are Cytotoxic in Excess for the Intestinal Epithelial Cells

Bile acids are end products generated from the metabolism of cholesterol [487]. Cholic acid and chenodeoxycholic acid are the major primary bile acids synthesized in the liver, and then conjugated with taurine or glycine [488]. Bile salts are stored, together with phosphatidylcholine and cholesterol, in the gallbladder and secreted into the intestinal tract to facilitate nutrient digestion and absorption, notably regarding the lipid part of the diet [489]. In the enterohepatic circulation of bile acids, approximately 95% of bile acids are absorbed in the ileum and transported back to the liver via portal circulation where it controls the level of synthesis of new bile acids. The minor proportion of primary bile acids that have not been absorbed in the small intestine can be used by the intestinal microbiota for the synthesis of the secondary bile acids, namely mainly deoxycholic acid and lithocholic acid (Fig. 3.21). Many bacterial species, like *Clostridium*, *Lactobacillus*, *Bifidobacterium*, *Eubacterium*, *Escherichia*, and *Bacteroides* can deconjugate primary bile acids, thus preventing their reabsorption by the small intestine, and increasing their transfer to the large intestine where they are thus used by the intestinal microbiota for secondary bile acid production [490]. The average fecal excretion of deoxycholic acid and lithocholic acid in healthy volunteers is within the 100–200 mg/day range [491].

High-fat diets increase bile acid concentration in the small and large intestine to optimize fat absorption [492]. Accordingly, the total fecal bile acid concentration is increased by such diets. Bile acids are known for their general antimicrobial action [493]. Deoxycholic acid (also called DCA), the main secondary bile acid present in

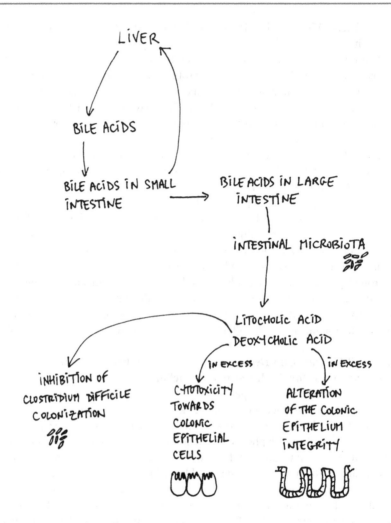

Fig. 3.21 Synthesis of the secondary bile acids by the intestinal microbiota and effects of these compounds on the colonic ecosystem. The effect of the secondary bile acid deoxycholic acid on *Clostridium difficile* colonization is indicated, as well as the effects of excessive concentrations of this bacterial metabolite on the colonic epithelium

human feces [494], has been shown to be strongly associated with the inhibition of *Clostridium difficile* colonization in animal models and in clinical practice [495, 496] (Fig. 3.21). This is of major interest since, as previously stated, *Clostridium difficile* is central for the induction of diarrhea and colitis associated with antibiotic consumption [497, 498].

Regarding the effects of deoxycholic acid on the host intestinal epithelium, this bacterial metabolite is known to be cytotoxic for the epithelium of the small intestine [499] and the large intestine [500]. Indeed, in in vitro and in vivo experiments performed in the mice model, it has been shown that deoxycholic acid in excess disrupts the epithelial integrity both in the small and large intestine [501, 502] (Fig. 3.21).

Key Points
- Unabsorbed bile acids are metabolized by the gut microbiota into the secondary bile acids deoxycholic and lithocholic acids.
- Deoxycholic acid inhibits *Clostridium difficile* colonization.
- Deoxycholic acid in excess is cytotoxic for colonic epithelial cells.

3.4   Metabolism of Purine Nucleotide by the Intestinal Microbiota

The bacteria present in human stool both produce hypoxanthine in the purine nucleotide degradation pathway, and degrade this compound [503], suggesting that the concentration of hypoxanthine in the distal large intestine depends on the net result of such production and degradation. Hypoxanthine has been identified as being able to increase ATP intracellular concentration in intestinal epithelial cells, thus acting presumably as an additional fuel in these cells, while promoting reinforcement of the intestinal barrier function [504].

3.5   Metabolism of Phytochemicals by the Intestinal Microbiota and Impact on the Intestinal Epithelium Metabolism and Functions

Bioactive phytochemicals are compounds contained in plants that display biological effects on various tissues and cells of the body. These bioactive phytochemicals gather an enormous number of different molecules, including notably phenolic structures, a group of compounds composed of more than 8000 currently known chemical structures [505]. Among this heterogeneous group of compounds from plant origin, polyphenols have been the object of intensive research in the last decades because of their reported beneficial effects observed in animal models, and in clinical studies [506]. The fact that relatively high intakes of polyphenols from dietary sources have been measured (for instance, approximately 820 mg/day in France [507]) is another reason for explaining the research effort made on the physiological effects of these compounds. Of note, less attention has been paid on the toxic effects that can be observed in case of consumption of excessive doses of polyphenols in the context of dietary supplementation [508].

The beneficial effects of polyphenols have been observed in context of type 2 diabetes [509], inflammatory situations [506, 510], and dysfunctions of the cardiovascular system [511]. Polyphenolic compounds are classically divided in high molecular weight tannins (that contain notably the proanthocyanidins [512]), and low molecular weight polyphenols. Most polyphenols are present in food as glycosides that are polyphenols conjugated to various sugars including glucose, galactose, rhamnose, and rutinose [513]. In the small intestine, low molecular weight polyphenols are partially absorbed, either directly, or after metabolic transformation during their transfer through enterocytes [514, 515].

The situation is much different for the high molecular weight tannins, since these dietary compounds are almost not absorbed in the small intestine and are thus transferred from the small to the large intestine [516]. Studies in germ-free animals, as well as in vitro incubation of fecal samples with different polyphenols, provide evidence that dietary polyphenols are overall highly metabolized by the colonic microbiota [513, 517–520]. Microbial species involved in metabolic transformation of dietary polyphenols include *Bacteroides distasomis*, *Bacteroides uniformis*, *Bacteroides ovatus*, *Enterococcus casseliflavus*, *Eubacterium cellulosolvens*, and *Eubacterium ramulus* [521].

The Intestinal Microbiota Produces Bacterial Metabolites from Polyphenols
The metabolic activity of the large intestine microbiota towards low molecular weight polyphenols and high molecular weight tannins generates numerous bacterial metabolites that are often organic acids, but also hydroxylated forms of polyphenols [522–530]. Human dietary intervention trials and in vitro metabolic studies with fecal samples incubated with different dietary plant polyphenols reveal large inter-individual variations in absorption, metabolism, and excretion of these compounds, which can be related in part to differences in the gut microbiota composition and metabolic activity [513, 531–533]. Such inter-individual variations in the microbial metabolism of polyphenols can be illustrated considering the microbial metabolism of the phytoestrogen daidzein, a polyphenolic compound that belongs to the flavo-noid group [534]. In most subjects, intestinal microbiota converts daidzein to o-desmethylangolensin, a metabolic pathway that involves *Clostridium* species. However, a minor proportion of the subjects, about 30% of volunteers, convert daidzein to (S)-equol through the metabolic activity of a wide range of bacterial species among the intestinal microbiota [513, 535].

Several Polyphenol-Derived Bacterial Metabolites Exert Beneficial Effects on the Intestinal Ecosystem
Some data are available regarding the effects of the polyphenol-derived bacterial metabolites on the host intestinal mucosa. The bacterial metabolite 3,4-dihydroxyphenylacetic acid (DOPAC) has been shown to possess capacity for free radical scavenging [536] (Fig. 3.22). In addition, DOPAC reduces the secretion of pro-inflammatory cytokines in mononuclear cells [537]. Thus, this bacterial metabolite displays interesting beneficial potential considering its capacity to reduce events associated with the inflammatory processes. Regarding the other bacterial metabolite 3,4-dihydroxybenzoic acid (also called protocatechuic acid (PCA)), it has been shown in several preclinical experiments with animal models to reduce the severity of chemically induced colitis. Indeed, in mice, PCA given intraperitoneally reduces the severity of colitis as attested by reduced damages to colonic mucosa, lower neutrophil infiltration, attenuated oxidative stress, and lower expression of

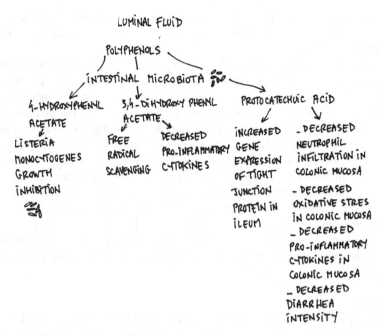

Fig. 3.22 Metabolism of polyphenols by the intestinal microbiota and effects of bacterial metabolites on the intestinal ecosystem. The effects of the 3 bacterial metabolites on bacterial growth and their protective action in case of inflammatory episodes are indicated

genes coding for the pro-inflammatory cytokines IL-6, TNF-alpha, and IL-1beta in colonic tissues [538] (Fig. 3.22). However, in this latter study, the way of administration of the bacterial metabolite PCA was by peritoneal injection and not by the intestinal luminal way. Nevertheless, other studies have confirmed the potentially beneficial effect of PCA towards intestine. In the pig model, dietary PCA supplementation increased the expression of several tight junction proteins in the ileal epithelium [539], while dietary supplementation with this bacterial metabolite in rodents with colitis ameliorates the incidence of diarrhea and bleeding, as well as the histological aspect of the colon mucosa. In addition, this compound lowers the infiltration of neutrophils in the colonic mucosa, while decreasing the expression of the inducible form of nitric oxide synthase (iNOS) [540] (Fig. 3.22). This latter decrease of iNOS is of interest, since iNOS overexpression has been found in colonic samples obtained from patients with inflammatory bowel diseases [541–543]. Overexpression of iNOS in the colonic mucosa results in excessive production of nitric oxide which reacts with reactive oxygen species, forming strong oxidant like peroxynitrite [544].

The effects of 4-hydroxyphenylacetate (HPA) on the intestinal ecosystem have been also studied. Of note, this bacterial metabolite is not only produced from polyphenols [545], but also from aromatic amino acids [546] (see paragraph 3.1). This latter bacterial metabolite inhibits the growth of the food-borne pathogen

Listeria monocytogenes (Fig. 3.22), an effect associated with alteration of the bacteria morphology, and decrease of the expression of several virulence genes [547].

Beneficial effects of polyphenols have been in addition related to their capacity to decrease the concentration and production of deleterious metabolites by the large intestine microbiota. For instance, proanthocyanidin-containing polyphenol plant extract reduces the production of human fecal samples of H_2S and ammonia [548]. Furthermore, in rats, proanthocyanidin-containing polyphenol extract attenuates the increase of H_2S concentration that is provoked in the large intestine by consumption of a high-protein diet [549]. These results are overall of interest as H_2S and ammonia inhibit the respiration of colonocytes when present in excess [2]. Lastly, proanthocyanidin-containing polyphenol extracts from plant origin, as well as the bacterial metabolites derived from proanthocyanidins, namely 3-phenylpropionic acid, 3,4-dihydrophenylpropionic acid, and 4-hydroxyphenylacetic acid are all able to prevent the alteration of barrier function provoked by excessive concentrations of the tyrosine-derived bacterial metabolite *p*-cresol [162].

Key Points
- Among phytochemicals, polyphenols can be transferred from the small to the large intestine where they are metabolized by the intestinal microbiota giving rise to bacterial metabolites.
- Among these bacterial metabolites, several of them display beneficial effects on the colonic ecosystem, inhibiting *Listeria monocytogenes*, exerting free radical scavenging effect, and reducing the concentration of ammonia and hydrogen sulfide which are deleterious for the colonocytes when present in excess.

3.6   Osmolality and pH of the Intestinal Luminal Fluid and Effects on the Intestinal Epithelium

In healthy humans, as stated above, diet can modify, at least transiently, the colonic microbiota composition [356] and/or its metabolic activity [2]. Consequently, bacterial metabolite concentrations in the colonic fluid can evolve depending on the overall net result of bacterial metabolism, absorption through the colonic epithelial absorptive cells, and on changes in water and electrolyte movement through the colonic epithelium, thus finally being able to affect the osmolality of the luminal content [550, 551].

Hyperosmotic Load Affects the Intestinal Epithelium Metabolism
Exposure of the mucosa of the small intestine to hyperosmotic load (treatment with hyperosmolar media) resulted in increased transepithelial resistance and alterations in absorptive epithelial cell tight-junction structure [552, 553]. In addition, hyperosmotic exposure of the intestinal epithelial cells results in the production of pro-inflammatory cytokines [554, 555]. In in vitro experiments with human

colonocytes, hyperosmolar media slow down their proliferation, in association with a transient reduction of cell mitochondrial oxygen consumption and decreased intracellular ATP content [556]. The barrier function of colonocyte monolayer was also transiently affected in this latter study since increased paracellular apical-to-basal permeability was measured. In addition, hyperosmotic stress induces secretion of the pro-inflammatory cytokine IL-8. By measuring expression of genes involved in energy metabolism, electrolyte permeability, and intracellular signaling, different response patterns to hyperosmotic stress occurred, depending on its intensity and duration, highlighting notably cellular adaptive capacities towards changes in luminal osmolarity [556].

Changes in the Luminal pH Affect the Transport and Diffusion of Several Bacterial Metabolites in Intestinal Epithelium Cells and Thus Their Effects on These Cells

Regarding the luminal pH, several bacterial metabolites, either acidic or alkaline, can affect this parameter. Conversely, the luminal pH can affect the acid/base ratio of several bacterial metabolites.

The pH of the human cecal content is slightly acidic ranging from 5.7 to 6.8, while ranging from 6.1 to 7.5 in the distal colon and rectum [394]. The pH at the colonic mucosal surface in healthy subjects is averaging 7.1 in cecum and proximal colon, whereas it ranges from 7.2 to 7.5 in the distal colon and rectum [557]. The luminal pH depends notably on the respective concentrations of a complex mixture of acids and bases in the large intestine content, among which hydrogen and bicarbonate secretion by the colonic mucosa represents important determinants [394]. Short-chain fatty acids participate as organic acids, together with other microbiota-derived acidic metabolites, like branched-chain fatty acids, lactate, succinate, and hydrogen sulfide. Ammonia intervenes in the luminal pH as a weak base. Overall change in the luminal pH can affect the uptake of luminal compounds by colonocytes, and then their action on these cells. For instance, hydrogen sulfide (H_2S) dissociates in solution, yielding hydrosulfide anion with a pKa equal to 7.04 [209]. Thus, when the luminal pH is more acidic, the hydrogen sulfide/hydrosulfide anion ratio in the large intestine increases, with H_2S, unlike hydrosulfide anion, easily penetrating biological membranes by diffusion [558]. A lower luminal pH will then increase H_2S entry into colonocytes. The same reasoning can be made regarding butyrate and ammonia. Butyrate, according to its pka value of 4.7, exists predominantly in the anionic dissociated form at the normal colonic pH, this anionic form being transported by the monocarboxylate transporter isoform 1 (MCT1), which is present in the colonocyte brush-border membranes [89, 559]. Then at a more acidic pH, the concentration of the anionic form will decrease and the diffusible undissociated form will increase, resulting presumably in lower uptake of this compound through the MCT 1 transporter by colonocytes, and increased diffusion of the undissociated lipid-soluble form. Lastly, ammonia (considered as the sum of NH_4^+ and NH_3) with a pKa equal to 9.02 is mainly present as the NH_4^+ ammonium form in the colonic fluid and transported in absorptive colonocytes through dedicated transporters [125]. A more alkaline luminal pH will displace the equilibrium in favor of NH_3, knowing that this latter compound is highly diffusible across colonocyte membranes [560]. Then, according to the pH of the colonic luminal fluid, the uptake of several bacterial metabolites by colonocytes will be modified, and consequently their effects on colonocytes will be different [2].

Beyond the influence of luminal pH on the transport of bacterial metabolites inside colonocytes, the modification of this parameter may per se affect the microbiota metabolic activity. For instance, low luminal pH reduces the synthesis of secondary bile acids by the intestinal microbiota [561], and the activity of luminal bacterial proteases in the colon [562].

Key Points
- Hyperosmolarity of the colonic luminal fluid affects colonocyte energy metabolism and epithelial barrier function.
- Changes in the pH of the colonic luminal fluid affect the acid/base ratio of several bacterial metabolites, thus modifying their entry into colonic epithelial cells, and thus their effects on these cells.
- Changes in the pH of the colonic fluid affect the microbiota metabolic activity.

3.7 Bacterial Metabolites and the Intestinal Immune System

The intestine represents the largest compartment of the immune system. The intestinal epithelium is continually exposed to a large variety of antigens and immuno-modulatory agents originating from the diet and from components of the intestinal microbiota, together with metabolites produced by its metabolic activity [563]. The intestinal epithelial cells and numerous immune cells express a series of receptors that recognize bacterial cell wall components, as well as bacterial metabolites, and such recognition has an impact on the intestinal immune system activity [564, 565]. Intestinal epithelial cells are known to represent crucial mediators of intestinal homeostasis that enables the establishment of an immunological environment permissive to colonization by bacteria [566]. The intestinal immune system includes lymphoid tissues together with populations of scattered innate and adaptive effector cells [567].

Different types of immune cells are present within the intestinal epithelium, within the lamina propria situated below the epithelium, and within the organized intestinal lymphoid tissue known as Peyer's patches [568, 569]. These structures are continuously exposed to an enormous diversity of microbial and food antigens [570]. Intestinal intraepithelial lymphocytes are, like their name indicate, located between intestinal epithelial cells [571]. They include a variety of cell types with different effector functions, including cells with cytotoxic, helper, and regulating properties [572]. The lamina propria contains numerous dendritic cells. These antigen-presenting cells can extend dendrites between intestinal epithelial cells without disrupting the intestinal barrier function, and sense antigens from the luminal content [224]. In healthy conditions, dendritic cells participate in the induction of tolerance towards the residing microbiota antigens [573, 574]. The intestinal

B cells, as part of the adaptive immune system, are one of the predominant cell populations in the lamina propria of the small and large intestine [575]. Lamina propria is also rich in T cells [576]. Differentiated T cells share some functional properties with the innate lymphoid cells of the lamina propria. Intestinal epithelial cells can also produce several cytokines that affect immune cells functions [577–579]. Conversely, intestinal epithelial cells respond to cytokines produced by immune cells in the lamina propria [566].

In the large intestine, immunoglobulins A (IgA), together with the presence of many regulatory T cells play a role in modulating the interactions between the host and its intestinal microbiota [580, 581]. Conversely, the intestinal microbiota contributes to shape the intestinal immune system [582].

The focus of this chapter is to review the actions played by the bacterial metabolites on the intestinal immune system. Diverse bacterial metabolites such as short-chain fatty acids and several amino acid-derived compounds can affect the immune response through interaction with cells present in the host intestinal tract [583].

Several Bacterial Metabolites Regulate the Intestinal Immune Response

Short-chain fatty acids appear to be able in in vitro experimental models to act on several immune cells in ways that are associated with anti-inflammatory actions. In vivo studies with experimental models have also been performed, but most of them have used oral provision of short-chain fatty acids, thus in a very different situation than the "real life situation" where these compounds are mostly provided from the luminal content through the metabolic activity of the intestinal microbiota. As will be detailed in the paragraph 4.1, although the anti-inflammatory effects of short-chain fatty acids are suggested by several in vitro studies, the roles of these effects in clinical situations remain quite controversial.

The short-chain fatty acid butyrate has the capacity to decrease the production of the pro-inflammatory mediators Il-6 and IL-12 by intestinal macrophages, the most abundant immune cell type in the intestinal subepithelial lamina propria [584] (Fig. 3.23). Macrophages differentiated in the presence of butyrate display enhanced antimicrobial activity, suggesting increased resistance to enteropathogens [585]. In addition, the three short-chain fatty acids butyrate, propionate, and acetate are all able to promote the production of the regulatory anti-inflammatory interleukin IL-22 by T cells and innate lymphoid cells [586] (Fig. 3.23). These short-chain fatty acids promote T cell differentiation into both effector and regulating T cells, promoting either immunity or immune tolerance depending on the immunological environment [587]. The three short-chain fatty acids regulate the metabolism and gene expression in B cells, promoting pathogen-specific antibody response [588]. Butyrate and propionate both diminish the production of the pro-inflammatory cytokine TNF-α by activated neutrophils [589] (Fig. 3.23).

Dopamine and norepinephrine are bacterial metabolites derived from amino acids that have been shown to interfere with the intestinal immune system since these compounds alter the bacterial internalization in intestinal Peyer's patches [590].

Fig. 3.23 Effects of the bacterial metabolite butyrate on immune cells. IL-6 and IL-12 are pro-inflammatory interleukins, while IL-22 is a regulatory anti-inflammatory interleukin

By using a combined metabolomics and bacterial genetic strategy, histamine was identified as a bacterial metabolite produced by *Lactobacillus reuteri* that suppresses the production of the pro-inflammatory TNF-alpha by isolated Toll-like receptor 2-activated human monocytoid cells [591].

Another amino acid-derived bacterial metabolite, namely spermine was shown to inhibit pro-inflammatory M1 macrophage activation through the suppression of the production of pro-inflammatory cytokines [583, 592]. As presented in the paragraph 3.1, the bacterial metabolite indole, which is produced from tryptophan, together with several indole derivatives, are compounds that act on several intestinal immune cell functions.

Finally, the secondary bile acid 3 β-hydroxydeoxycholic acid, which is produced by the intestinal microbiota, promotes the large intestine generation of peripheral regulatory T cells, and acts on dendritic cells to diminish their immune-stimulatory properties, thus contributing to the immunological balance in the large intestine [593].

Key Points
- Short-chain fatty acids exert immune-regulatory roles towards the intestinal immune system.

- Several bacterial metabolites including dopamine, norepinephrine, histamine, spermine, indole, and the secondary bile acids 3β-hydroxydeoxycholic acid exert regulatory roles on the intestinal immune system.

3.8   Minerals and the Intestinal Microbiota: Consequences for the Intestinal Ecosystem

The intestinal microbiota uses a significant amount of minerals provided by the host as essential elements for their metabolic needs and growth [594]. In addition, the intestinal microbiota, by itself, and through its metabolic activity, and thus through the production of metabolites, can affect directly and indirectly the absorption of minerals by the small and large intestine as will be presented below. Conversely, mineral deficiency and oral supplementation with several minerals can affect the gut microbiota composition [595].

Acetate and Propionate Increase Calcium Absorption in the Large Intestine
Calcium is the most abundant minerals in the human body, with approximately 99% being present in bones. The absorption of calcium by the intestine is central for maintaining bone health and calcium homeostasis [595]. Although in healthy humans, calcium is mostly absorbed in the small intestine, a minor proportion of calcium can be effectively taken up by the large intestine epithelial cells [596]. Rectal infusion of the microbiota-derived short-chain fatty acids acetate and propionate increased calcium absorption from the distal colon and rectum in volunteers [597].

Dietary Magnesium Deficiency Alters Intestinal Microbiota Composition and Metabolic Activity
Magnesium, as a co-factor of more than 300 enzymatic reactions, is involved in a wide variety of physiological and metabolic functions [598]. The intestinal absorption of magnesium occurs predominantly in the small intestine, but some uptake is ensured by the large intestine epithelium [599, 600]. Dietary magnesium deficiency was shown to alter the gut microbiota composition in a rodent model [601]. Such deficiency was associated with a modification of the luminal environment of the colonic epithelium, since marked increase of formate concentration was recorded in the colon of mice receiving low dietary intake of magnesium [602].

Propionate Increases Iron Absorption in the Large Intestine
Iron is involved in the synthesis of hemoglobin, myoglobin, and numerous enzymes involved in physiological and metabolic functions in the body [603, 604]. The absorption of iron from heme is much more efficient than the absorption of non-heme iron [605]. Non-absorbed iron that has not been used by the bacteria of the small intestine is transferred to the large intestine through the ileocaecal junction. Although iron absorption occurs mainly in the small intestine of mammals, a significant part of iron can be absorbed through the absorptive epithelial cells of the large intestine [606]. Iron is an essential element for almost all intestinal bacteria

[594]. Oral iron supplementation is used in case of iron-deficiency which represents a global health concern affecting children, women, and the elderly [607]. Due to limited iron absorption in the small intestine, increase transfer of iron from the small to the large intestine luminal content may occur notably in case of fortification or supplementation, affecting the gut microbiota composition with a reduction of the relative number of lactic bacteria including *bifidobacteria* and *lactobacilli*, while increasing enterobacteria including *Escherichia coli* [608, 609]. Little is known on the role of bacterial metabolites on the absorption of iron. Lower luminal pH in the colon, notably in case of increased organic acid production by the microbiota, increase the solubility of iron and iron uptake by colonocytes [610, 611]. In rodents, the short-chain fatty acid propionate increases iron absorption by the large intestine epithelium [612]. Two other bacterial metabolites, namely 1,3-diaminopropane (produced from the amino acids glutamate and aspartate [613]), and reuterin both downregulate expression of key elements of iron transport through absorptive enterocytes [614], thus demonstrating the involvement of the microbiota metabolic activity in systemic iron homeostasis.

The Intestinal Microbiota Utilizes a Significant Part of Dietary Zinc
Zinc plays major roles in different physiologic and metabolic processes notably regarding activity of different brain structures, and of the immune and endocrine systems [615–617]. Zinc is used to a significant level by the intestinal microbiota which use this element for its metabolic needs [594]. By comparing germ-free rodents with no intestinal microbiota, and conventional animals colonized by an intestinal flora, it has been determined that the dietary zinc requirement is lower in germ-free animals than in the conventional counterpart [618], thus pointing out the significant utilization of zinc by the intestinal microbiota. In rodents, zinc deficiency during pregnancy alters the fecal microbiota composition [619]. In a rodent model of *Clostridium difficile* infection, excessive dietary zinc consumption alters the intestinal microbiota composition and decreases resistance capacity against *Clostridium difficile* infection [620].

Key Points
- Short-chain fatty acids increase calcium and iron absorption through the large intestine epithelium.
- The bacterial metabolites 1,3-diaminopropane and 3-hydroxypropionaldehyde decrease iron transport through enterocytes.
- Gut microbiota diverts part of the available zinc for its own utilization, but excessive zinc intake lowers resistance towards *Clostridium difficile* infection.

3.9   Vitamins and the Intestinal Microbiota: Consequences for the Intestinal Ecosystem

The intestinal microbiota possesses the metabolic capacity for synthesizing some vitamins, notably vitamin K and vitamins of the B group [621–623]. Although the bacteria synthesize vitamin K and vitamins B for their own metabolism and growth, and for the growth of other microbial members of the intestinal community, a part of produced vitamins is available to the host [513, 624–627]. However, the contribution of intestinal bacteria for vitamin supply to the host appears more significant for the vitamins of the B group than for vitamin K [623]. Some bacterial species have the complete biosynthetic pathways to produce B vitamins, while others must acquire them from exogenous sources. However, both kind of bacterial species express transporters for B vitamins or B vitamin precursors [628, 629]. It has been shown that the colonic epithelial cells can absorb a wide variety of B vitamins [630]. This is of considerable interest since humans cannot synthesize vitamins of the B group (except for niacin), and thus depend on exogenous sources such as diet and intestinal microbiota. The vitamins of the B group that are synthesized by the microbiota play a central role in numerous metabolic pathways operating in the host tissues, notably those involved in energy production [631]. Of equal importance, vitamins of the B group are involved in the metabolism of methionine, this amino acid being the precursors of several metabolites with biological activity, and notably of S-adenosylmethionine which is a methyl donor involved in more than 60 methylation reactions in the body, including DNA and histone methylation. These 2 latter methylation processes represent an important component for the regulation of gene expression in cells [632].

Key Point
- Intestinal bacteria participate to the supply of vitamins of the B group and vitamin K to the host.

3.10   Food Additives and Compounds Produced during Cooking Processes and the Intestinal Microbiota: Consequences for the Intestinal Ecosystem

Xenobiotics is a general term that includes the chemicals to which an organism is exposed, with the notion that these chemicals are extrinsic to the normal metabolism of that organism [633, 634]. Thus, xenobiotics include dietary compounds, industrial chemicals, and pharmaceuticals [635]. In this chapter, restriction will be given to the presentation of typical examples of some food additives, and compounds formed in the cooking process, since the metabolism of dietary nutrients and micronutrients by the intestinal microbiota has been presented in previous chapters.

The metabolism of food additives and compounds generated during cooking processes by the gut microbiota remains an emerging field of research with considerable implications for the assessment of the toxicology risk and for the understanding of the mechanisms of action of these compounds.

These latter compounds can be active on the microbiota itself, and/or on the host's cells [636, 637]. They can be metabolized by the gut microbiota which produces diverse metabolites. Indeed, the gut microbiota is equipped with a broad repertoire of enzymatic activities, often not present in host tissues, which can lead to modification of xenobiotics and production of metabolites which can display biological effects, either beneficial or deleterious, depending on the overall context [638]. The amplitude of the metabolic activity of the intestinal microbiota towards these compounds will, like for any exogenous and endogenous substrates present in the luminal fluid, depends on the bioavailability of compounds in each range of concentrations and on the intestinal transit time.

The Intestinal Microbiota Metabolizes Food Additives
The intestinal microbiota interacts with some food additives like artificial sweeteners, emulsifiers, and preservatives [635]. For instance, gut microbes convert the artificial sweetener cyclamate into cyclohexylamine, and can metabolize the two other sweeteners stevioside and xylitol [639, 640]. However, the metabolic pathways involved in this conversion remain to be identified. The gut microbiota increases its metabolic capacity to metabolize cyclamate after prolonged exposure [641], suggesting that long-term ingestion of this sweetener may select and/or induce specific metabolic activities among the intestinal microbiota. However, the metabolism of sweeteners by the human intestinal microbiota in humans in real-life situation, and the consequences of such a metabolism for the intestinal epithelium remain to be studied.

Sulfite is a food preservative that acts as antioxidant [642, 643]. Food antioxidants are defined as chemical compounds added to some food to retard autoxidation [642]. Sulfite is absorbed by enterocytes in the duodenum [644]. Intestinal bacteria belonging to *Bilophila* and *Veillonellaceae* reduce sulfite into hydrogen sulfide. Hydrogen sulfide, as explained in paragraph 3.1, can be used as a mineral fuel in mitochondria of colonocytes at low concentration, but severely impairs mitochondrial oxygen consumption and ATP production in these cells at higher concentration [645].

Nitrite is also used as a food preservative, notably because nitrite inhibits the growth of microorganisms, notably *Clostridium botulinum* strain [646], which produces different toxins responsible for deleterious effects, notably in terms of neurotoxic effects [647]. Nitrite can react with amines in the large intestine of mammals, allowing the synthesis of N-nitroso compounds [648]. Among these N-nitroso compounds, nitrosamine synthesis can be promoted by commensal bacteria through catalysis of amine nitrosation [649]. Among nitrosamines, several of them, when present in excess, have been shown to provoke DNA damages in cells [650–652] as will be detailed in Chap. 4.

Compounds Formed During the Cooking Processes Can Be Modified by the Intestinal Metabolic Activity

The numerous cooking processes used around the world generate a myriad of new compounds that were not initially present in food and modify the structure of food components [653–655]. Typical examples of compounds that are formed during cooking are heterocyclic amines that are formed during charring of meat and fish, and that are poorly absorbed by the small intestine [651, 656]. These compounds can be altered by the gut microbiota activity. For instance, by comparing in a rodent model animals with and without intestinal microbiota, it was shown that the DNA damage provoked in colonic cells by the heterocyclic amine 2-amino-3-methylimidazol (4,5-f) quinolone (IQ) was more ample in conventional animals than in germ-free animals [657], indicating that the intestinal microbiota, likely by metabolizing IQ into secondary metabolites, play a role in the degree of genotoxic alterations in colonocytes. Also of major interest, the gut microbiota can hydrolyze IQ-glucuronide conjugates, such metabolites being formed within the liver from IQ present in the colonic fluid and excreted via the bile into the intestinal lumen [658]. This metabolic activity of the intestinal microbiota increased the time of exposition of host 'cells to IQ, thus presumably increasing the genotoxic insults in colonic epithelial cells.

Key Points
- Food additives, like sulfite and nitrite, can be metabolized by the intestinal microbiota giving rise to bacterial metabolites like hydrogen sulfide and nitrosamines which show adverse effects on the intestinal epithelium when present in excess.
- Compounds produced during some cooking processes, like heterocyclic amines, can be metabolized by the intestinal microbiota, giving rise to metabolites that can be involved in the overall toxicity of these compounds.

References

1. Stephen AM, Cummings JH. The microbial contribution to human fecal mass. J Med Microbiol. 1980;13(1):45–56.
2. Blachier F, Beaumont M, Andriamihaja M, Davila AM, Lan A, Grauso M, Armand L, Benamouzig R, Tomé D. Changes in the luminal environment of the colonic epithelial cells and physiopathological consequences. Am J Pathol. 2017;187(3):476–86.
3. Libao-Mercado AJO, Zhu CL, Cant JP, Lapierre H, Thibault JN, Sève B, Fuller MF, de Lange CFM. Dietary and endogenous amino acids are the main contributors to microbial protein synthesis in the upper gut of normally nourished pigs. J Nutr. 2009;139(6):1088–94.
4. Blachier F, Andriamihaja M, Kong X. Fate of undigested proteins in the pig large intestine: what impact on the colon epithelium? Anim Nutr. 2021a;9:110–8.
5. Yin L, Yang H, Li J, Ding X, Wu G, Yin Y. Pig models on intestinal development and therapeutics. Amino Acids. 2017;49(12):2099–106.
6. Chalvon-Demersay T, Blachier F, Tomé D, Blais A. Animal models for the study of the relationships between diet and obesity: a focus on dietary protein and estrogen deficiency. Front Nutr. 2017;4:5.

7. Mudd AT, Dilger RN. Early-life nutrition and neurodevelopment: use of the piglet as a translational model. Adv Nutr. 2017;8(1):92–104.

8. Patterson JK, Len XG, Miller DD. The pig as an experimental model for elucidating the mechanisms governing dietary influence on mineral absorption. Exp Biol Med (Maywood). 2008;233(6):651–64.

9. Blachier F, M'Rabet H, Posho L, Darcy-Vrillon B, Duée PH. Intestinal arginine metabolism during development. Evidence for de novo synthesis of L-arginine in newborn pig enterocytes. Eur J Biochem. 1993;216(1):109–17.

10. Kararli TT. Comparison of the gastrointestinal anatomy, physiology, and biochemistry of humans and commonly used laboratory animals. Biopharm Drug Dispo. 1995;16(5):351–80.

11. Xiao L, Estellé J, Kiilerich P, Ramavo-Caldas Y, Xia Z, Feng Q, Liang S, Pedersen AO, Jorgen Kjeldsen N, Liu C, Maguin E, Doré J, Pons N, Le Chatelier E, Prifti E, Li J, Jia H, Liu X, Xu X, Ehrlich SD, Madsen L, Kristiansen K, Rogel-Gaillard C, Wang J. A reference gene catalogue of the pig gut microbiome. Nat Microbiol. 2016;1:16161.

12. Gaudichon C, Bos C, Morens C, Petzke KJ, Mariotti F, Everwand J, Benamouzig R, Daré S, Tomé D, Metges CC. Ileal losses of nitrogen and amino acids in humans and their importance to the assessment of amino acid requirement. Gastroenterology. 2002;123(1):50–9.

13. Davila AM, Blachier F, Gotteland M, Andriamihaja M, Benetti PH, Sanz Y, Tomé D. Intestinal luminal nitrogen metabolism: role of the gut microbiota and consequences for the host. Pharmacol Res. 2013;68(1):95–107.

14. Baglieri A, Mahe S, Zidi S, Huneau JF, Thuillier F, Marteau P, Tomé D. Gastro-jejunal digestion of soya-bean-milk protein in humans. Br J Nutr. 1994;72(4):519–32.

15. Bos C, Juillet B, Fouillet H, Turlan L, Daré S, Luengo C, N'tounda R, Benamouzig R, Gausserès N, Tomé D, Gaudichon C. Postprandial metabolic utilization of wheat protein in humans. Am J Clin Nutr. 2005;81(1):87–94.

16. Gaudichon C, Mahé S, Benamouzig R, Luengo C, Fouillet H, Daré S, Van Oycke M, Ferrière F, Rautureau J, Tomé D. Net postprandial utilization of 15N-labeled milk protein nitrogen is influenced by diet composition in humans. J Nutr. 1999;129(4):890–5.

17. Gausseres N, Mahé S, Benamouzig R, Luengo C, Drouet H, Rautureau J, tomé D. The gastro-ileal digestion of 15N-labelled pea nitrogen in adult humans. Br J Nutr. 1996;76(1):75–85.

18. Mariotti F, Pueyno ME, Tomé D, Berot S, Benamouzig R, Mahé S. The influence of the albumin fraction on the bioavailability and postprandial utilization of pea protein given selectively to humans. J Nutr. 2001;131(6):1706–13.

19. Blachier F, Mariotti F, Huneau JF, Tomé D. Effects of amino acid-derived luminal metabolites on the colonic epithelium and physiopathological consequences. Amino Acids. 2007a;33(4): 547–62.

20. Yao C, Muir JG, Gibson PR. Review article: insights into colonic protein fermentation, its modulation and potential health implications. Aliment Pharmacol Ther. 2016;43(2):181–96.

21. Gibson JA, Sladen GE, Dawson AM. Protein absorption and ammonia production: the effects of dietary protein and removal of the colon. Br J Nutr. 1976;35(1):61–5.

22. Kramer P. The effect of varying sodium loads on the ileal excreta of human ileostomized subjects. J Clin Invest. 1966;45(11):1710–8.

23. Smiddy FG, Gregory SD, Smith IB, Goligher JC. Fecal loss of fluid, electrolytes, and nitrogen in colitis before and after ileostomy. Lancet. 1960;1(7114):14–9.

24. Dubuisson C, Lioret S, Touvier M, Dufour A, Calamassi-Tran G, Volatier JL, Lafay L. Trends in food and nutritional intakes of French adults from 1999 to 2007: results from the INCA surveys. Br J Nutr. 2010;103(7):1035–8.

25. Pasiakos SM, Agarwal S, Lieberman HR, Fulgoni VL 3rd. Sources and amounts of animal, dairy, and plant protein intake of US adults in 2007-2010. Nutrients. 2015;7(8):7058–69.

26. Kaman WE, Hays JP, Endtz HP, Bikker FJ. Bacterial proteases: targets for diagnosis and therapy. Eur J Clin Microbiol Infect Dis. 2014;33(7):1081–7.

27. van der Wielen N, Moughan PJ, Mensink M. Amino acid absorption in the large intestine of human and porcine models. J Nutr. 2017;147(8):1493–8.

28. James PS, Smith MV. Methionine transport by pig colonic mucosa measured during early post-natal development. J Physiol. 1976;262(1):151–68.
29. Just A, Jorgensen H, Fernandez JA. The digestive capacity of the caecum-colon and the value of the nitrogen absorbed from the hind gut for protein synthesis in pigs. Br J Nutr. 1981;46(1): 209–19.
30. Sepulveda FV, Smith MW. Different mechanisms for neutral amino acid uptake by new-born colon. J Physiol. 1979;286:479–90.
31. Nyangale EP, Mottram DS, Gibson GR. Gut microbial activity, implications for health and disease: the potential role of metabolite analysis. J Proteome Res. 2012;11(12):5573–85.
32. Metges CC. Contribution of microbial amino acids to amino acid homeostasis. J Nutr. 2000;130(7):1857S–64S.
33. Gietzen DW, Rogers QR. Nutritional homeostasis and indispensable amino acid sensing: a new solution to an old puzzle. Trends Neurosci. 2006;29(2):91–9.
34. Reeds PJ. Dispensable and indispensable amino acids for humans. J Nutr. 2000;130(7): 1835S–40S.
35. Blachier F, Blais A, Elango R, Saito K, Shimomura Y, Kadowaki M, Matsumoto H. Tolerable amounts of amino acids for human supplementation: summary and lessons from published peer-reviewed studies. Amino Acids. 2021c;53(9):1313–28.
36. Fürst P, Stehle P. What are the essential elements needed for the determination of amino acid requirement in humans? J Nutr. 2004;134(6):1558S–65S.
37. Torrallardona D, Harris CI, Coates ME, Fuller MF. Microbial amino acid synthesis and utilization in rats: incorporation of 15N from 15NH4Cl into lysine in the tissue of germ-free and conventional rats. Br J Nutr. 1996;76(5):689–700.
38. Backes G, Hennig U, Petzke KJ, Elsner A, Junghans P, Nurnberg G, Metges CC. Contribution of intestinal microbiota lysine to lysine homeostasis is reduced in minipigs fed a wheat gluten-based diet. Am J Clin Nutr. 2002;76(6):1317–25.
39. Millward DJ, Forrester T, Ah-Sing E, Yeboah N, Gibson N, Badaloo A, Boyne M, Reade M, Persaud C, Jackson A. The transfer of 15N from urea to lysine in the human infant. Br J Nutr. 2000;83(5):505–12.
40. Metges CC, Petzke KJ, El-Khoury AE, Henneman L, Grant I, Bedry S, Regan MM, Fuller MF, Young VR. Incorporation of urea and ammonia nitrogen into ileal and fecal microbial proteins and plasma free amino acids in normal men and ileostomates. Am J Clin Nutr. 1999;70(6): 1046–58.
41. Abubucker S, Segata N, Goll J, Schubert AM, Izard J, Cantarel BL, Rodriguez-Mueller B, Zucker J, Thiagarajan M, Henrissiat B, White O, Kelley ST, Methé B, Schloss PD, Gevers D, Mitreva M, Huttenhower C. Metabolic reconstruction of metagenomic data and its application to the human microbiome. PLoS Comput Biol. 2012;8(6):e1002358.
42. Bergen WG. Small-intestinal or colonic microbiota as a potential amino acid source in animals. Amino Acids. 2015;47(2):251–8.
43. Chen L, Li P, Wang J, Li X, Gao H, Yin Y, Hou Y, Wu G. Catabolism of nutritionally essential amino acids in developing porcine enterocytes. Amino Acids. 2009;37(1):143–52.
44. Stoll B, Henry J, Reeds PJ, Yu H, Jahoor F, Burrin DG. Catabolism dominates the first-pass intestinal metabolism of dietary essential amino acids in milk protein-fed piglets (1998). J Nutr. 1998;128(3):606–14.
45. Bergen WG, Wu G. Intestinal nitrogen recycling and utilization in health and disease. J Nutr. 2009;139(5):821–5.
46. Grohmann U, Bronte V. Control of immune response by amino acid metabolism. Immunol Rev. 2010;236:243–64.
47. Riedijk MA, Stoll B, Chacko S, Schierbeek H, Sunehag AL, van Goudoever JB, Burrin DG. Methionine transmethylation and transsulfuration in the piglet gastrointestinal tract. Proc Natl Acad Sci U S A. 2007;104(9):3408–13.
48. Shoveller AK, Brunton JA, Pencharz PB, Ball RO. The methionine requirement is lower in neonatal piglets fed parenterally than in those fed enterally. J Nutr. 2003;133(5):1387–90.

49. Portune KJ, Beaumont M, Davila AM, Tomé D, Blachier F, Sanz Y. Gut microbiota role in dietary protein metabolism and health-related outcomes: the two sides of the coin. Trends Food Sci Technol. 2016;57:213–32.
50. Shimizu T, Ohtani K, Hirakawa H, Ohshima K, Yamashita A, Shiba T, Ogasawara N, Hattori M, Kuhara S, Hayashi S. Complete genome sequence of Clostridium perfringens, an anaerobic flesh-eater. Proc Natl Acad Sci U S A. 2002;99(2):996–1001.
51. Pridmore RD, Berger B, Desiere F, Vilanova D, Barretto C, Pittet AC, Zwallen MC, Rouvet M, Altermann E, Barrangou R, Mollet B, Mercenier A, Klaenhammer T, Arigoni F, Schell MA. The genome sequence of the probiotic intestinal bacterium lactobacillus johnsonii NCC 533. Proc Natl Acad Sci U S A. 2004;101(8):2512–7.
52. Yu XJ, Walker DH, Liu Y, Zhang L. Amino acid biosynthesis deficiency in bacteria associated with human and animal hosts. Infect Genet Evol. 2009;9(4):514–7.
53. Nolling J, Breton G, Omeichenko MV, Marakova KS, Zeng Q, Gibson R, Lee HM, Dubois D, Qiu D, Hitti J, Wolf YI, Tatusov LR, Sabathe L, Doucette-Stam P, Soucaille P, Daly MJ, Bennett GM, Koonin EJ, Smith DR. Genome sequence and comparative analysis of the solvent-producing bacterium clostridium acetobutylicum. J Bacteriol. 2001;183(16):4823–38.
54. Bolotin A, Wincker P, Mauger S, Jaillon O, Malarme K, Weissenbach J, Ehrlich SD, Sorokin A. The complete genome sequence of the lactic acid bacterium Lactococcus lactis spp. lactis IL1403. Genome Res. 2011;11(5):731–53.
55. Godon JJ, Delorme C, Bardowski J, Chopin MC, Ehrlich SD, Renault P. Gene inactivation in Lactococcus lactis: branched-chain amino acid biosynthesis. J Bacteriol. 1993;175(14): 4383–90.
56. Fichman Y, Gerdes SY, Kovacs H, Szabados L, Zilbersein A, Csonka LN. Evolution of proline biosynthesis: Enzymology, bioinformatics, genetics, and transcriptional regulation. Biol Rev Camb Philos Soc. 2015;90(4):1065–99.
57. Fischbach MA, Sonnenburg JL. Eating for two: how metabolism establishes interspecies interactions in the gut. Cell Host Microbe. 2011;10(4):336–47.
58. Xu Y, Labedan B, Glansdorff N. Surprising arginine biosynthesis: a reappraisal of the enzymology and evolution of the pathway in microorganisms. Microbiol Mol Biol Rev. 2007;71(1):36–47.
59. Peters-Wendisch P, Stolz M, Etterich H, Kennerknecht N, Sahm H, Eggeling L. Metabolic engineering of Corynebacterium glutamicum for L-serine production. Appl Environ Microbiol. 2005;71(11):7139–44.
60. Trivedi V, Gupta A, Jala VR, Saravanan P, Rao GS, Rao NA. Crystal structure of binary and ternary complexes of serine hydroxymethyltransferase from Bacillus stearothermophilus: insights into the catalytic mechanism. J Biol Chem. 2002;277(19):17161–9.
61. Min B, Pelaschier JT, Graham DE, Tumbula-Hansen D, Soll D. Transfer RNA-dependent amino acid biosynthesis: an essential route to asparagine formation. Proc Natl Acad Sci U S A. 2002;99(5):2678–83.
62. Rodionov DA, Vitreschak AG, Mironov AA, Gelfand MS. Comparative genomics of the methionine metabolism in gram-positive bacteria: a variety of regulatory systems. Nucleic Acids Res. 2004;32(11):3340–53.
63. Vitreschak AG, Lyubetskaya EV, Shirshin MA, Gelfand MS, Lyubetsky A. Attenuation regulation of amino acid biosynthesis operons in proteobacteria. FEMS Microbiol Lett. 2004;234(2):357–70.
64. Allison MJ, Baetz AL, Wiegel J. Alternative pathways for biosynthesis of leucine and other amino acids in Bacteroides ruminicola and Bacteroides fragilis. Appl Environ Microbiol. 1984;48(6):1111–7.
65. Xi G, Keyhani NO, Bonner CA, Jensen RA. Ancient origin of the tryptophan operon and the dynamics of evolutionary change. Microbiol Mol Biol Rev. 2003;67(3):303–42.
66. Kulis-Horn RK, Persicke M, Kalinowski J. Histidine biosynthesis, its regulation and biotechnological application in Corynebacterium glutamicum. Microb Biotechnol. 2014;7(1):5–25.

67. Macfarlane GT, Macfarlane S. Bacteria, colonic fermentation, and gastrointestinal health. J AOAC Int. 2012;95(1):50–60.
68. Wang X, Gibson GR, Costabile A, Sailer M, Theis S, Rastall RA. Prebiotic supplementation of in vitro fecal fermentations inhibits proteolysis by gut bacteria, and host diet shapes gut bacterial metabolism and response to intervention. Appl Environ Microbiol. 2019;85(9): e02749–18.
69. Pessione E. Lactic acid bacteria contribution to gut microbiota complexity: lights and shadows. Front Cell Infect Microbiol. 2012;2:86.
70. Liu M, Bayjanov JR, Renckens B, Nauta A, Siezen RJ. The proteolytic system of lactic acid bacteria revisited: a genomic comparison. BMC Genomics. 2010;11:36.
71. Steiner HY, Naider F, Becker JM. The PTR family: a new group of peptide transporters. Mol Microbiol. 1995;16(5):825–34.
72. Saier MH Jr. Families of transmembrane transporters selective for amino acids and their derivatives. Microbiology (Reading). 2000;146(8):1775–95.
73. Eggeling L, Sahm H. New ubiquitous translocators: amino acid export by Corynebacterium glutamicum and Escherichia coli. Arch Microbiol. 2003;180(3):155–60.
74. Dai ZL, Li XL, Xi PB, Zhang J, Wu G, Zhu WY. Metabolism of select amino acids in bacteria from the pig small intestine. Amino Acids. 2012;42(5):1597–608.
75. Sengupta C, Ray S, Chowdhury R. Fine tuning of virulence regulatory pathways in enteric bacteria in response to varying bile and oxygen concentrations in the gastrointestinal tract. Gut Pathol. 2014;6:38.
76. Riggottier-Gois L. Dysbiosis in inflammatory bowel diseases: the oxygen hypothesis. ISME J. 2013;7(7):1256–61.
77. Lu Z, Imlay JA. When anaerobes encounter oxygen: mechanisms of oxygen toxicity, tolerance and defence. Nat Rev Microbiol. 2021;19(12):774–85.
78. Kim J, Hetzel M, Boiangiu CD, Buckel W. Dehydration of (R)-2-hydroxyacyl-CoA to enoyl-CoA in the fermentation of alpha-amino acids by anaerobic bacteria. FEMS Microbiol Rev. 2004;28(4):455–68.
79. Fonknechten N, Chaussonnerie S, Tricot S, Lajus A, Andreesen JR, Perchat N, Pelletier E, Gouyvenoux M, Barbe V, Salanoubat M, Le Paslier P, Weissenbach J, Cohen GN, Kreimeyer A. Clostridium stricklandii, a specialist in amino acid degradation: revisiting its metabolism through its genome sequence. BMC Genomics. 2010;11:555.
80. Smith EA, Macfarlane GT. Enumeration of human colonic bacteria producing phenolic and indolic compounds: effects of pH, carbohydrate availability and retention time on dissimilatory aromatic amino acid metabolism. J Appl Bacteriol. 1996;81(3):288–302.
81. Birkett A, Muir J, Phillips J, Jones G, O'Dea K. Resistant starch lowers fecal concentrations of ammonia and phenol in humans. Am J Clin Nutr. 1996;63(5):766–72.
82. Geboes KP, De Hertogh G, De Preter V, Luypaerts A, Bammens B, Evenepoel P, Ghoss Y, Geboes K, Rutgeerts P, Verbeke K. The influence of inulin on the absorption of nitrogen and the production of metabolites of protein fermentation in the colon. Br J Nutr. 2006;96(6): 1078–86.
83. Windey K, De Preter V, Huys G, Broekaert WF, Delcour JA, Louat T, Herman J, Verbeke K. Wheat bran extract alters colonic fermentation and microbial composition but does not affect faecal water toxicity: a randomized controlled trial in healthy subjects. Br J Nutr. 2015;113(2):225–38.
84. Macfarlane GT, Gibson GR, Cummings JH. Comparison of fermentation reactions in different regions of the human colon. J Appl Bacteriol. 1992;72(1):57–64.
85. Macfarlane GT, Cummings JH, Macfarlane S, Gibson GR. Influence of retention time on degradation of pancreatic enzymes by human colonic bacteria grown in a 3-stage continuous system. J Appl Bacteriol. 1989;67(5):520–7.
86. Roager HM, Hansen LBS, Bahl MI, Frandsen HL, Carvalho V, Gobel RJ, Dalgaard MD, Plichta DR, Sparholt MH, Vestergaard H, Hansen T, Sicheritz-Ponten T, Bjorn Nielsen H,

Pedersen O, Lauritzen L, Kristensen M, Gupta R, Licht TR. Colonic transit time is related to bacterial metabolism and mucosal turnover in the gut. Nature Microbiol. 2016;1(9):16093.

87. Sridharan GV, Choi K, Klemashevich C, Wu C, Prabakaran D, Pan LB, Steimeyer S, Mueller C, Yousofshani M, Alaniz RC, Lee K, Jayaraman A. Prediction and quantification of bioactive microbiota metabolites in the mouse. Nat Commun. 2014;5:5492.

88. Rechkemmer G, Rönnau K, von Engelhardt W. Fermentation of polysaccharides and absorption of short-chain fatty acids in the mammalian hindgut. Comp Biochem Physiol A Comp Physiol. 1988;90(4):563–8.

89. Hamer HM, Jonkers D, Venema K, Vanhoutvin S, Troost FJ, Brummer RJ. Review article: the role of butyrate on colonic function. Aliment Pharmacol Ther. 2008;27(2):104–19.

90. Laparra JM, Sanz Y. Interactions of gut microbiota with functional food components and nutraceuticals. Pharm Res. 2010;61(3):219–25.

91. Neis EPJG, Dejong CHC, Rensen SS. The role of microbial amino acid metabolism in host metabolism. Nutrients. 2015;7(4):2930–46.

92. Liu X, Blouin JM, Santacruz A, Lan A, Andriamihaja M, Wilkanowicz S, Benetti PH, Tomé D, Sanz Y, Blachier F, Davila AM. High-protein diet modifies colonic microbiota and luminal environment but not colonocyte metabolism: the increased luminal bulk connection. Am J Phys. 2014;307(4):G459–70.

93. Cummings JH, Macfarlane GT. The control and consequences of bacterial fermentation in the human colon. J Appl Microbiol. 1991;70(6):443–59.

94. Rios-Covian D, Gonzalez S, Nogacka AM, Arboleya S, Salazar N, Gueimonde M, de Los Reyes-Gavilan CG. An overview on fecal branched short-chain fatty acids along human life and as related with body mass index: associated dietary and anthropometric factors. Front Microbiol. 2020;11:973.

95. Trefflich I, Dietrich S, Braune A, Abraham K, Weikert C. Short- and branched-chain fatty acids as fecal markers for microbiota activity in vegans and omnivores. Nutrients. 2021;13(6): 1808.

96. Musch MW, Bookstein C, Xie Y, Sellin JH, Chang EB. SCFA increase intestinal Na absorption by induction of NHE3 in rat colon and human intestinal C2/bbe cells. Am J Phys. 2001;280(4):G687–93.

97. Diener M, Helmle-Kolb C, Murer H, Scharrer E. Effect of short-chain fatty acids on cell volume and intracellular pH in rat distal colon. Pflugers Arch. 1993;424(3–4):216–23.

98. Zaharia V, Varzescu M, Djavadi I, Newman E, Egnor RW, Alexander-Chacko J, Charney AN. Effects of short-chain fatty acids on colonic Na$^+$ absorption and enzyme activity. Comp Biochem Physiol. 2001;128(2):335–47.

99. Charney AN, Giannella RA, Egnor RW. Effect of short-chain fatty acids on cyclic 3′,5′- -guanosine monophosphate-mediated chloride secretion. Comp Biochem Physiol A Mol Integr Physiol. 1999;124(2):169–78.

100. Dagher PC, Egnor RW, Taglietta-Kohlbrecher A, Charney AN. Short-chain fatty acids inhibits cAMP-mediated chloride secretion in rat colon. Am J Phys. 1996;271(6):1853–60.

101. Chu S, Montrose MH. Extracellular pH regulation in microdomains of colonic crypts: effects of short-chain fatty acids. Proc Natl Acad Sci U S A. 1995;92(8):3303–7.

102. Jaskiewicz J, Zhao Y, Hawes JW, Shimomura Y, Crabb DW, Harris RA. Catabolism of isobutyrate by colonocytes. Arch Biochem Biophys. 1996;327(2):265–70.

103. Boudry G, Jamin A, Chatelais L, Gras-Le Guen C, Michel C, Le Huërou-Luron I. Dietary protein excess during neonatal life alters colonic microbiota and mucosal response to inflammatory mediators later in life in female pigs. J Nutr. 2013;143(8):1225–32.

104. Petry N, Egli N, Chassard C, Lacroix C, Hurrell R. Inulin modifies the bifidobacteria population, fecal lactate concentration, and fecal pH but does not influence iron absorption in women with low iron status. Am J Clin Nutr. 2017;96(2):325–31.

105. Belenguer A, Duncan SH, Holtrop G, Anderson SE, Lobley GE, Flint HJ. Impact of pH on lactate formation and utilization by human fecal microbial communities. Appl Environ Microbiol. 2007;73(20):6526–33.

106. De Vadder F, Kovatcheva-Datchary P, Zitoun C, Duchampt A, Bäcked F, Mithieux G. Microbiota-produced succinate improves glucose homeostasis via intestinal gluconeogenesis. Cell Metab. 2016;24(1):151–7.

107. Miller TL, Wolin MJ. Fermentations by saccharolytic intestinal bacteria. Am J Clin Nutr. 1979;32(1):164–72.

108. Serena C, Ceperuelo-Mallafré V, Keiran N, Queipo-Ortuno MI, Bernal R, Gomez-Huelgas R, Urpi-Sarda M, Sabater M, Perez-Brocal V, Andrés-Lacueva C, Moya A, Tinahones F, Fernandez-Real JM, Vendrell J, Fernandez-Veledo S. Elevated circulating levels of succinate in human obesity are linked to specific gut microbiota. ISME J. 2018;12(7):1642–57.

109. Koestler BJ, Fisher CR, Payne SM. Formate promotes shigella intercellular spread and virulence gene expression. MBio. 2018;9(5):e01777–18.

110. Shaulov Y, Shimokawa C, Trebicz-Geffren M, Nagaraja S, Methling K, Lalk M, Weiss-Cerem L, Lamm AT, Hisaeda H, Ankri S. Escherichia coli mediated resistance of Entamoeba histolytica to oxidative stress is triggered by oxaloacetate. PLoS Pathog. 2018;14(10): e1007295.

111. Stanley SL, Reed SL. Microbes and microbial toxins: paradigms for microbial-mucosal interactions. VI. Entamoeba histolytica: parasite-host interactions. Am J Phys. 2001;280(6): G1049–54.

112. Felmlee MA, Jones RS, Rodriguez-Cruz V, Follman KE, Morris ME. Monocarboxylate transporters (SLC16): function, regulation, and role in health and disease. Pharmacol Res. 2020;72(2):466–85.

113. Darcy-Vrillon B, Morel MT, Cherbuy C, Bernard F, Posho L, Blachier F, Meslin JC, Duée PH. Metabolic characteristics of pig colonocytes after adaptation to a high fiber diet. J Nutr. 1993;123(2):234–43.

114. Nadjsombati MS, McGinty JW, Lyons-Cohen MR, Jaffe JB, DiPeso L, Schneider C, Miller CN, Pollack JN, Nagana Gowda GA, Fontana MF, Erle DJ, Anderson MS, Locksley RM, Raftery D, van Moltke J. Detection of succinate by intestinal tuft cells triggers a type 2 innate immune circuit. Immunity. 2018;49(1):33–41.e7.

115. Banerjee A, Herring CA, Chen B, Kim H, Simmons AJ, Southard-Smith AN, Allaman MM, White JR, Macedonia MC, Mckinley ET, Ramirez-Solano MA, Scoville EA, Liu Q, Wilson KT, Coffey RJ, Washington MK, Goettel JA, Lau KS. Succinate produced by intestinal microbes promotes specification of tuft cells to suppress ileal inflammation. Gastroenterology. 2020;159(6):2101–15.

116. Mouillé B, Robert V, Blachier F. Adaptative increase of ornithine production and decrease of ammonia metabolism in rat colonocytes following hyperproteic diet ingestion. Am J Phys. 2004;287(2):G344–51.

117. Wrong O, Metcalfgibson A. The electrolyte content faeces. Proc R Soc Med. 1965;58(12): 1007–9.

118. Warren KS, Newton WL. Portal and peripheral blood ammonia concentrations in germ-free and conventional Guinea pigs. Am J Phys. 1959;197:717–20.

119. Mora D, Arioli S. Microbial urease in health and disease. PLoS Pathog. 2014;10(12): e1004472.

120. Ni J, Shen TCD, Chen EZ, Bittinger K, Bailey A, Roggiani M, Sirota-Madi A, Friedman ES, Chau L, Lin A, Nissim I, Scott J, Lauder A, Hoffmann C, Rivas G, Albenberg L, Baldassano RN, Braun J, Xavier RJ, Clish CB, Yudkoff M, Li H, Goulian M, Bushman FD, Lewis JD, Wu GD. A role of bacterial urease in gut dysbiosis and Crohn's disease. Sci Transl Med. 2017;9 (416):eaah6888.

121. Ryvchin R, Dubinsky V, Rabinowitz K, Wasserberg N, Dotan I, Gophna U. Alteration in urease-producing bacteria in the gut microbiomes of patients with inflammatory bowel diseases. J Crohns Colitis. 2021;15(12):2066–77.

122. Codina J, Pressley TA, Dubose TD Jr. The colonic H+,K+ ATPase functions as a Na+−dependent K+(NH4+)-ATPase in apical membranes from rat distal colon. J Biol Chem. 1999;274(28):19693–8.

123. Eklou-Lawson M, Bernard F, Neveux N, Chaumontet C, Bos C, Davila-Gay AM, Tomé D, Cynober L, Blachier F. Colonic luminal ammonia and portal blood L-glutamine and L-arginine concentrations: a possible link between colon mucosa and liver ureagenesis. Amino Acids. 2009;37(4):751–60.

124. Hall MC, Koch MO, McDougal WS. Mechanisms of ammonium transport by intestinal segments following urinary diversion: evidence for ionized NH4+ transport via K(+)-pathways. J Urol. 1992;148(1):453–7.

125. Handlogten ME, Hong SP, Zhang L, Vander AW, Steinbaum ML, Campbell-Thompson M, Weiner ID. Expression of the ammonia transporter proteins rh B glycoprotein and rh C glycoprotein in the intestinal tract. Am J Phys. 2005;288(5):G1036–47.

126. McDougal WS, Stampfer DS, Kirley S, Bennett PM, Lin CW. Intestinal ammonium transport by ammonium and hydrogen exchange. J Am Coll Surg. 1995;181(3):241–8.

127. Singh SK, Binder HJ, Geibel JP, Boron WF. An apical permeability barrier to NH3/NH4+ in isolated, perfused colonic crypts. Proc Natl Acad Sci U S A. 1995;92(25):11573–7.

128. Summerskill and Wolpert. Ammonia metabolism in gut. Am J Clin Nutr. 1970;23(5):633–9.

129. Stewart GS, Fenton RA, Thévenod F, Smith CP. Urea movement across mouse colonic plasma membranes is mediated by UT-A urea transporters. Gastroenterology. 2004;126(3):765–73.

130. Andriamihaja M, Davila AM, Eklou-Lawson M, Petit N, Delpal S, Allek F, Blais A, Delteil C, Tomé D, Blachier F. Colon luminal content and epithelial cell morphology are markedly modified in rats fed with a high-protein diet. Am J Phys. 2010;299(5):G1030–7.

131. Cremin JD Jr, Fitch MD, Fleming SE. Glucose alleviates ammonia-induced inhibition of short-chain fatty acid metabolism in rat colonic epithelial cells. Am J Phys. 2003;285(1):G105–14.

132. Darcy-Vrillon B, Cherbuy C, Morel MT, Durand M, Duée PH. Short-chain fatty acid and glucose metabolism in isolated pig colonocytes. Mol Cell Biochem. 1996;156(2):145–51.

133. Mouillé B, Morel E, Robert V, Guihot-Joubrel G, Blachier F. Metabolic capacity for L-citrulline synthesis from ammonia in rat isolated colonocytes. Biochim Biophys Acta. 1999;1427(3):401–7.

134. Roediger WE. The colonic epithelium in ulcerative colitis: an energy-deficiency disease? Lancet. 1980;2(8197):712–5.

135. Hughes R, Kurth MJ, McGilligan V, McGlynn H, Rowland I. Effect of colonic bacterial metabolites on Caco-2 cell paracellular permeability in vitro. Nutr Cancer. 2008;60(2):259–66.

136. Macfarlane GT, Cummings JH. The colonic flora, fermentation, and large bowel digestive function. In: Phillips SF, Pemberton JH, Shorter RG, editors. The large intestine: physiology, pathophysiology, and disease. New York: Raven Press; 1991.

137. Bone E, Tamm A, Hill M. The production of urinary phenols by gut bacteria and their possible role in the causation of large bowel cancer. Am J Clin Nutr. 1976;29(12):1448–54.

138. Hughes R, Magee EA, Bingham S. Protein degradation in the large intestine: relevance to colorectal cancer. Curr Issues Intest Microbiol. 2000;1(2):51–8.

139. Pedersen G, Brynskov J, Saermark T. Phenol toxicity and conjugation in human colonic epithelial cells. Scand J Gastroenterol. 2002;37(1):74–9.

140. McCall IC, Betanzos A, Weber DA, Nava P, Miller GW, Parkos CA. Effects of phenol on barrier function of a human intestinal epithelial cell line correlate with altered tight junction protein localization. Toxicol Appl Pharmacol. 2009;241(1):61–70.

141. Gostner A, Blaut M, Schäffer V, Kozianowski G, Theis S, Klingeberg M, Drombrowski Y, Martin D, Erhrardt S, Taras D, Schwiertz A, Kleesen B, Lürhs H, Schauber J, Dorbath D, Menzel T, Scheppach W. Effect of isomalt consumption on faecal microflora and colonic metabolism in healthy volunteers. Br J Nutr. 2006;95(1):40–50.

142. Gryp T, De Paepe K, Vanholder R, Kerckhof FM, Van Biesen W, Van de Wiele T, Verbeke F, Speeckaert M, Joossens M, Couttenye MM, Vaneechoutte M, Glorieux G. Gut microbiota generation of protein-bound uremic toxins and related metabolites is not altered at different stages of chronic kidney disease. Kidney Int. 2020;97(6):1230–42.

143. King RA, May BL, Davies DA, Bird AR. Measurement of phenol and p-cresol in urine and feces using vacuum microdistillation and high-performance liquid chromatography. Anal Biochem. 2009;384(1):27–33.

144. Windey K, François I, Broekaert W, De Preter V, Delcour JA, Louat T, Herman J, Verbeke K. High dose of probiotics reduces fecal water cytotoxicity in healthy subjects. Mol Nutr Food Res. 2014;58(11):2206–18.

145. Geypens B, Claus D, Evenepoel P, Hiele M, Maes B, Peeters M, Rutgeers P, Ghoos Y. Influence of dietary protein supplements on the formation of bacterial metabolites in the colon. Gut. 1997;41(1):70–6.

146. Toden S, Bird AR, Topping DL, Conlon MA. Resistant starch attenuates colonic DNA damage induced by higher dietary protein in rats. Nutr Cancer. 2005;51(1):45–51.

147. Paturi G, Nyanhanda T, Butts CA, Herath TD, Monro JA, Ansell J. Effects of potato fiber and potato-resistant starch on biomarkers of colonic health in rats fed diets containing red meat. J Food Sci. 2012;77(10):H216–23.

148. Taciak M, Barszcz M, Swiech E, Tusnio A, Bachanek I. Interactive effects of protein and carbohydrates on production of microbial metabolites in the large intestine of growing pigs. Arch Anim Nutr. 2017;71(3):192–209.

149. Saito Y, Sato T, Nomoto K, Tsuji H. Identification of phenol- and p-cresol-producing intestinal bacteria by using media supplemented with tyrosine and its metabolites. FEMS Microbiol Ecol. 2018;94(9):fiy125.

150. Ji M, Du H, Xu Y. Structural and metabolic performance of p-cresol producing bacteria in different carbon sources. Food Res Int. 2020;132:109049.

151. Passmore IJ, Letertre MPM, Preston MD, Bianconi I, Harrison MA, Nasher F, Kaur H, Hong HA, Baines HD, Cutting SM, Swann JR, Wren BW, Dawson LF. Para-cresol production by Clostridium difficile affects microbial diversity and membrane integrity of gram-negative bacteria. PLoS Pathog. 2018;14(9):e1007191.

152. Abt MC, McKenney PT, Pamer EG. Clostridium difficile colitis: pathogenesis and host defence. Nat Rev Microbiol. 2016;14(10):609–20.

153. Hafiz S, Oakley CL. Clostridium difficile: isolation and characteristics. J Med Microbiol. 1976;9(2):129–36.

154. Dawson LF, Donahue EH, Cartman ST, Barton RH, Bundy J, McNerney R, Minton NP, Wren BW. The analysis of para-cresol production and tolerance in Clostridium difficile 027 and 012 strains. BMC Microbiol. 2011;11:86.

155. Patel M, Fowler D, Sizer J, Walton C. Fecal volatile biomarkers of Clostridium difficile infection. PLoS One. 2019;14(4):e0215256.

156. Mercer KE, Ten Have GAM, Pack L, Lan R, Deutz NEP, Adams SH, Piccolo BD. Net release and uptake of xenometabolites across intestinal, hepatic, muscle, and renal tissue beds in healthy conscious pigs. Am J Phys. 2020;319(2):G133–41.

157. Ramakrishna BS, Gee D, Weiss A, Pannall P, Roberts-Thomson IC, Roediger WE. Estimation of phenolic conjugation by colonic mucosa. J Clin Pathol. 1989;42(6):620–3.

158. Schepers E, Glorieux G, Vanholder R. The gut: the forgotten organ in uremia. Blood Purif. 2010;29(2):130–6.

159. Aronov PA, Luo FJG, Plummer NS, Quan Z, Holmes S, Hostetter TH, Meyer TW. Colonic contribution to uremic solutes. J Am Soc Nephrol. 2011;22(9):1769–76.

160. Zeng Y, Lin Y, Li L, Zhang X, Wang M, Chen Y, Luo L, Lu B, Xie Z, Liao Q. Targeted metabolomics for the quantitative measurement of 9 gut microbiota-host co-metabolites in rat serum, urine and feces by liquid chromatography-tandem mass spectrometry. J Chromatogr B Analyt Technol Biomed Life Sci. 2019;1110-1111:133–43.

161. Andriamihaja M, Lan A, Beaumont M, Audebert M, Wong X, Yamada K, Yin Y, Tomé D, Carrasco-Pozo C, Gotteland M, Kong X, Blachier F. The deleterious metabolic and genotoxic effects of the bacterial metabolite p-cresol on colonic epithelial cells. Free Radic Biol Med. 2015;85:219–27.

162. Wong X, Carrasco-Pozo C, Escobar E, Navarrete P, Blachier F, Andriamihaja M, Lan A, Tomé D, Cires MJ, Pastene E, Gotteland M. Deleterious effect of p-cresol on human colonic epithelial cells prevented by proanthocyanidin-containing polyphenol extracts from fruits and proanthocyanidin bacterial metabolites. J Agric Food Chem. 2016;64(18):3574–83.
163. Gao J, Xu K, Liu H, Liu G, Bai M, Peng C, Li T, Yin Y. Impact of the gut microbiota on intestinal immunity mediated by tryptophan metabolism. Front Cell Infect Microbiol. 2018;8: 13.
164. Lee JY, Wood TK, Lee J. Roles of indole as an interspecies and interkingdom signaling molecule. Trends Microbiol. 2015;23(11):707–18.
165. Keszthelyi D, Troost FJ, Masclee AA. Understanding the role of tryptophan and serotonin metabolism in gastrointestinal function. Neurogastroenterol Motil. 2009;21(12):1239–49.
166. Roager HM, Licht TR. Microbial tryptophan catabolites in health and disease. Nat Commun. 2018;9(1):3294.
167. Dong F, Perdew GH. The aryl hydrocarbon receptor as a mediator of host-microbiota interplay. Gut Microbes. 2020;12(1):1859812.
168. Hyland NP, Cavanaugh CR, Hornby PJ. Emerging effects of tryptophan pathway metabolites and intestinal microbiota on metabolism and intestinal function. Amino Acids. 2022;54(1): 57–70.
169. Jensen MT, Cox RP, Jensen BB. 3-methylindole (skatole) and indole production by mixed populations of pig fecal bacteria. Appl Environ Microbiol. 1995;61(8):3180–4.
170. Oliphant K, Allen-Vercoe E. Macronutrient metabolism by the human gut microbiome major fermentation by-products and their impact on host health. Microbiome. 2019;7(1):91.
171. Taleb S. Tryptophan dietary impacts gut barrier and metabolic diseases. Front Immunol. 2019;10:2113.
172. Dodd D, Spitzer MH, Van Treuren W, Merrill BD, Hryckowian AJ, Higginbottom SK, Le A, Cowan TM, Nolan GP, Fischbach MA, Sonnenburg JL. A gut bacterial pathway metabolizes aromatic amino acids into nine circulating metabolites. Nature. 2017;551(7682):648–52.
173. Wikoff WR, Anfora AT, Liu J, Schultz PG, Lesley SA, Peters EC, Siuzdak G. Metabolomics analysis reveals large effects of gut microbiota on mammalian blood metabolites. Proc Natl Acad Sci U S A. 2009;106(10):3698–703.
174. Chappell CL, Darkoh C, Shimmin L, Farhana N, Kim DK, Okhuysen PC, Hixson J. Fecale indole as a biomarker of susceptibility to cryptosporidium infection. Infect Immun. 2016;84 (8):2299–306.
175. Darkoh C, Chappell C, Gonzales C, Okhuysen P. A rapid and specific method for the detection of indole in complex biological samples. Appl Environ Microbiol. 2015;81(23):8093–7.
176. Leong SC, Sirich TL. Indoxyl sulfate. Review of toxicity and therapeutic strategies. Toxins (Basel). 2016;8(12):358.
177. Huc T, Konop M, Onyszkiewicz M, Podsadni P, Szczepanska A, Turlo J, Ufnal M. Colonic indole, gut bacteria metabolite of tryptophan, increases portal blood pressure in rats. Am J Phys. 2018;315(4):R646–55.
178. Whitt DD, Demoss RD. Effect of microflora on the free amino acid distribution in various regions of the mouse gastrointestinal tract. Appl Microbiol. 1975;30(4):609–15.
179. Kim J, Park W. Indole: a signaling molecule or a mere metabolic byproduct that alters bacterial physiology at a high concentration? J Microbiol. 2015;53(7):421–8.
180. Lee JH, Lee J. Indole as an intercellular signal in microbial communities. FEMS Microbiol Rev. 2010;34(4):426–44.
181. Vega NM, Allison KR, Samuels AN, Klempner MS, Collins JJ. Salmonella typhimurium intercepts Escherichia coli signaling to enhance antibiotic resistance. Proc Natl Acad Sci U S A. 2013;110(35):14420–5.
182. Rattanaphan P, Mittraparp-Arthorn P, Srinoun K, Vuddhakul V, Tansila N. Indole signaling decreases biofilm formation and related virulence of listeria monocytogenes. FEMS Microbiol Lett. 2020;367(14):fnaa116.

183. Li G, Young KD. Indole production by the tryptophanase TnaA in Escherichia coli is determined by the amount of exogenous tryptophan. Microbiology (Reading). 2013;159(2): 402–10.
184. Bansal T, Englert D, Lee J, Hedge M, Wood TK, Jayaraman A. Differential effects of epinephrine, norepinephrine, and indole on Escherichia coli O157:H7 chemotaxis, colonization, and gene expression. Infect Immun. 2007;75(9):4597–607.
185. Nowak A, Libudzisz Z. Influence of phenol, p-cresol and indole on growth and survival of intestinal lactic acid bacteria. Anaerobe. 2006;12(2):80–4.
186. Ledala N, Malik M, Rezaul K, Paveglio S, Provatas A, Kiel A, Caimano M, Zhou Y, Lindgren J, Krasulova K, Illes P, Dvorak Z, Kortagere S, Kienesberger S, Cosic A, Pöltl L, Zechner EL, Ghosh S, Mani S, Radolf JD, Matson AP. Bacterial indole as a multifunctional regulator of Klebsiella oxytoca complex enterotoxicity. MBio. 2022;13(1):e0375221.
187. Agus A, Planchais J, Sokol H. Gut microbiota regulation of tryptophan metabolism in health and disease. Cell Host Microbe. 2018;23(6):716–24.
188. Permonian L, Duarte-Silva M, de Barros R, Cardoso C. The aryl hydrocarbon receptor (AHR) as a potential target for the control of intestinal inflammation: insights from an immune and bacteria sensor receptor. Clin Rev Allergy Immunol. 2020;59(3):382–90.
189. Hubbard TD, Murray IA, Perdew GH. Indole and tryptophan metabolism: endogenous and dietary routes to Ah receptor activation. Drug Metab Dispos. 2015;43(10):1522–35.
190. Lamas B, Richard ML, Leducq V, Pham HP, Michel ML, Da Costa G, Bridonneau C, Jegou S, Hoffmann TW, Natividad JM, Brot L, Taleb S, Couturier-Maillard A, Nion-Larmurier L, Merabtene F, Selsik P, Bourrier A, Cosnes J, Ryffel B, Beaugerie L, Launay JM, Langella P, Xavier RJ, Sokol H. CARD9 impacts colitis by altering gut microbiota metabolism of tryptophan into aryl hydrocarbon receptor ligands. Nat Med. 2016;22(6):598–605.
191. Wlodarska M, Luo C, Kolde R, d'Hennezel E, Annand JW, Heim CE, Krastel P, Schmitt EK, Omar AS, Creasey EA, Garner AL, Mohammadi S, O'Connell DJ, Abubucker S, Arthur TD, Franzosa EA, Huttenhower C, Murphy LO, Haiser HJ, Vlamakis H, Porter JA, Xavier RJ. Indoleacrylic acid produced by commensal Peptostreptococcus species suppresses inflammation. Cell Host Microbe. 2017;22(1):25–37.
192. Zelante T, Iannitis RG, Cunha C, De Luca A, Giovannini G, Pieraccini G, Zecchi R, D'Angelo C, Massi-Benedetti C, Fallarino F, Carvalho A, Puccetti P, Romani L. Tryptophan catabolites from microbiota engage hydrocarbon receptor and balance mucosal reactivity via interleukin-22. Immunity. 2013;39(2):372–85.
193. Natividad JM, Agus A, Planchais JA, Lamas B, Jarry AC, Martin R, Michel ML, Chong-Nguyen C, Roussel R, Straube M, Jegou S, McQuitty C, Le Gall M, da Costa G, Lecornet G, Michaudel C, Modoux M, Glodt J, Bridonneau C, Sovran B, Dupraz L, Bado A, Richard ML, Langella P, Hansel B, Launay JM, Xavier RJ, Duboc H, Sokol H. Impaired aryl hydrocarbon receptor ligand production by the gut microbiota is a key factor in metabolic syndrome. Cell Metab. 2018;28(5):737–749.e4.
194. Holst JJ. The physiology of glucagon-like peptide 1. Physiol Rev. 2007;87(4):1409–39.
195. Bansal T, Alaniz RC, Wood TK, Jayaraman A. The bacterial signal indole increases epithelial-cell tight-junction resistance and attenuates indicators of inflammation. Proc Natl Acad Sci U S A. 2010;107(1):228–33.
196. Shimada Y, Kinoshita M, Harada K, Mizutani M, Masahata K, Kayama H, Takeda K. Commensal bacteria-dependent indole production enhances epithelial barrier function in the colon. PLoS One. 2013;8(11):e80604.
197. Armand L, Fofana M, Couturier-Becavin K, Andriamihaja M, Blachier F. Dual effects of the tryptophan-derived bacterial metabolite indole on colonic epithelial cell metabolism and physiology: comparison with its co-metabolite indoxyl sulfate. Amino Acids. 2022;54 (10):1371–82.
198. Chimerel C, Emery E, Summers DK, Keyser U, Gribble FM, Reimann F. Bacterial metabolite indole modulates incretin secretion from intestinal enteroendocrine L cells. Cell Rep. 2014;9 (4):1202–8.

199. Karlin DA, Mastromarino AJ, Jones RD, Stroehlein JR, Lorentz O. Fecal skatole and indole and breath methane and hydrogen in patients with large bowel polyps or cancer. J Cancer Res Clin Oncol. 1985;109(2):135–41.
200. Yokoyama MT, Carlson JR. Microbial metabolites of tryptophan in the intestinal tract with special reference to skatole. Am J Clin Nutr. 1979;32(1):173–8.
201. Cook KL, Rothrock MJ Jr, Loughrin JH, Doerner KC. Characterization of skatole-producing microbial populations in enriched swine lagoon slurry. FEMS Microbiol Ecol. 2007;60(2): 329–40.
202. Whitehead TR, Price NP, Drake HL, Cotta MA. Catabolic pathway for the production of skatole and indoleacetic acid by the acetogen Clostridium drakei, clostridium scatologenes, and swine manure. Appl Environ Microbiol. 2008;74(6):1950–3.
203. Choi SH, Kim Y, Oh S, Oh S, Chun T, Kim T, Kim SH. Inhibitory effect of skatole (3-methylindole) on enterohemorrhagic Escherichia coli O157:H7 ATCC 43894 biofilm formation mediated by elevated endogenous oxidative stress. Lett Appl Microbiol. 2014;58 (5):454–61.
204. Deng Z, Luo XM, Liu J, Wang H. Quorum sensing, biofilm, and intestinal mucosal barrier: involvement the role of probiotic. Front Cell Infect Microbiol. 2020;10:538077.
205. Probert HM, Gibson GR. Bacterial biofilms in the human gastrointestinal tract. Curr Issues Intest Microbiol. 2002;3(2):23–7.
206. Witte F, Pajic A, Menger F, Tomasevic I, Schubert DC, Visscher C, Terjung N. Preliminary test of the reduction capacity for the intestinal absorption of skatole and indole in weaning piglets by pure and coated charcoal. Animals (Basel). 2021;11(9):2720.
207. Claus R, Raab S. Influence on skatole formation from tryptophan in the pig colon. Adv Exp Med Biol. 1999;467:679–84.
208. Kurata K, Kawahara H, Nishimura K, Jisaka M, Yokota K, Shimizu H. Skatole regulates intestinal epithelial cellular functions through activating aryl hydrocarbon receptors and p38. Biochem Biophys Res Commun. 2019;510(4):649–55.
209. Blachier F, Davila AM, Mimoun S, Benetti PH, Atanasiu C, Andriamihaja M, Benamouzig R, Bouillaud F, Tomé D. Luminal sulfide and large intestine mucosa: friend or foe? Amino Acids. 2010;39(2):335–47.
210. Barton LL, Ritz NL, Fauque GD, Lin HC. Sulfur cycling and the intestinal microbiome. Dig Dis Sci. 2017;62(9):2241–57.
211. Basic A, Blomqvist M, Dahlen G, Svensäter G. The proteins of fusobacterium spp. involved in hydrogen sulfide production from L-cysteine. BMC Microbiol. 2017;17(1):61.
212. Linden DR. Hydrogen sulfide signaling in the gastrointestinal tract. Antioxid Redox Signal. 2014;20(5):818–30.
213. Tittsler RP. The effects of temperature upon the production of hydrogen sulphide by salmonella pullorum. J Bacteriol. 1931;21(2):111–8.
214. Ohge H, Furne JK, Springfield J, Sueda T, Madoff RD, Levitt MD. The effects of antibiotics and bismuth on fecal hydrogen sulfide and sulfate-reducing bacteria in the rat. FEMS Microbiol Lett. 2003;228(1):137–42.
215. Rowan FE, Docherty NG, Coffey JC, O'Connell PR. Sulphate-reducing bacteria and hydrogen sulphide in the aetiology of ulcerative colitis. Br J Surg. 2009;96(2):151–8.
216. Florin T, Neale G, Gibson GR, Christl SU, Cummings JH. Metabolism of dietary sulphate: absorption and excretion in humans. Gut. 1991;32(7):766–73.
217. Lewis S, Cochrane S. Alteration of sulfate and hydrogen metabolism in the human colon by changing intestinal transit rate. Am J Gastroenterol. 2007;102(3):624–33.
218. Pal VK, Bandyopadhyay P, Singh A. Hydrogen sulfide in physiology and pathogenesis of bacteria and viruses. IUBMB Life. 2018;70(5):393–410.
219. Shatalin K, Shatalina E, Mironov A, Nudler E. H2S: a universal defense against antibiotics in bacteria. Science. 2011;334(6058):986–90.

220. Forte E, Borisov VB, Falabella M, Colaço HG, Tinajero-Trejo M, Poole RK, Vicente JB, Sarti P, Giuffrè A. The terminal oxidase cytochrome bd promotes sulfide-resistant bacterial respiration and growth. Sci Rep. 2016;6:23788.
221. Andriamihaja M, Lan A, Beaumont M, Grauso M, Gotteland M, Pastene E, Cires MJ, Carrasco-Pozo C, Tomé D, Blachier F. Proanthocyanidin-containing polyphenol extracts from fruits prevent the inhibitory effect of hydrogen sulfide on human colonocyte oxygen consumption. Amino Acids. 2018;50(6):755–63.
222. Gasaly N, Gotteland M. Interference of dietary polyphenols with potentially toxic amino acid metabolites derived from the colonic microbiota. Amino Acids. 2022;54(3):311–24.
223. Mimoun S, Andriamihaja M, Chaumontet C, Atanasiu C, Benamouzig R, Blouin JM, Tomé D, Bouillaud F, Blachier F. Detoxification of H2S by differentiated colonic epithelial cells: implication of the sulfide oxidizing unit and of the cell respiratory capacity. Antioxid Redox Signal. 2012;17(1):1–10.
224. Lönnerdal B. Dietary factors influencing zinc absorption. J Nutr. 2000;130(5):1378S–83S.
225. Manach C, Williamson G, Morand C, Scalbert A, Rémésy C. Bioavailability and bioefficacy of polyphenols in humans. I. Review of 97 bioavailability studies. Am J Clin Nutr. 2005;81(1): 230S–42S.
226. Shen X, Carlström M, Borniquel S, Jädert C, Kevil CG, Lundberg JO. Microbial regulation of host hydrogen sulfide bioavailability and metabolism. Free Radic Biol Med. 2013;60:195–200.
227. Blachier F, Beaumont M, Kim E. Cysteine-derived hydrogen sulfide and gut health: a matter of endogenous or bacterial origin. Curr Opin Clin Nutr Metab Care. 2019;22(1):68–75.
228. Guo FF, Yu TC, Hong J, Fang JY. Emerging roles of hydrogen sulfide in inflammatory and neoplastic colonic diseases. Front Physiol. 2016;7:156.
229. Beaumont M, Andriamihaja M, Lan A, Khodorova N, Audebert M, Blouin JM, Grauso M, Lancha L, Benetti PH, Benamouzig R, Tomé D, Bouillaud F, Davila AM, Blachier F. Detrimental effects for colonocytes of an increased exposure to luminal hydrogen sulfide: the adaptive response. Free Radic Biol Med. 2016;93:155–64.
230. Goubern M, Andriamihaja M, Nübel T, Blachier F, Bouillaud F. Sulfide, the first inorganic substrate for human cells. FASEB J. 2007;21(8):1699–706.
231. Bouillaud F, Blachier F. Mitochondria and sulfide: a very old story of poisoning, feeding, and signaling? Antioxid Redox Signal. 2011;15(2):379–91.
232. Grieshaber MK, Völkel S. Animal adaptations for tolerance and exploitation of poisonous sulfide. Annu Rev Physiol. 1998;60:33–53.
233. Giuffrè A, Vicente JB. Hydrogen sulfide biochemistry and interplay with other gaseous mediators in mammalian physiology. Oxid Med Cell Longev. 2018;2018:6290931.
234. Lagoutte E, Mimoun S, Andriamihaja M, Chaumontet C, Blachier F, Bouillaud F. Oxidation of hydrogen sulfide remains a priority in mammalian cells and causes reverse electron transfer in colonocytes. Biochim Biophys Acta. 2010;1797(8):1500–11.
235. Leschelle X, Goubern M, Andriamihaja M, Blottière HM, Couplan E, Gonzalez-Barroso MDM, Petit C, Pagniez A, Chaumontet C, Mignotte B, Bouillaud F, Blachier F. Adaptative metabolic response of human colonic epithelial cells to the adverse effects of the luminal compound sulfide. Biochim Biophys Acta. 2005;1725(2):201–12.
236. Pichette J, Fynn-Sackey N, Gagnon J. Hydrogen sulfide and sulfate prebiotic stimulate the secretion of GLP-1 and improve glycemia in male mice. Endocrinology. 2017;158(10): 3416–25.
237. Gilbert MP, Pratley RE. GLP-1 analogs and DPP-4 inhibitors in type 2 diabetes therapy: review of head-to-head clinical trials. Front Endocrinol (Lausanne). 2020;11:178.
238. Di Martino ML, Campilongo R, Casalino M, Micheli G, Colonna B, Prosseda G. Polyamines: emerging players in bacteria-host interactions. Int J Med Microbiol. 2013;303(8):484–91.
239. Blachier F, Davila AM, Benamouzig R, Tomé D. Chanelling of arginine in NO and polyamine pathways in colonocytes and consequences. Front Biosci (Landmark Ed). 2011;16(4): 1331–43.

240. Pegg AE. Mammalian polyamine metabolism and function. IUBMB Life. 2009;61(9):880–94.
241. Thomas TJ, Tajmir-Riahi HA, Thomas T. Polyamine-DNA interactions and development of gene delivery vehicles. Amino Acids. 2016;48(10):2423–31.
242. Williams K. Interactions of polyamines with ion channels. Biochem J. 1997;325(2):289–97.
243. Igarashi K, Kashiwagi K. Modulation of cellular function by polyamines. Int J Biochem Cell Biol. 2010;42(1):39–51.
244. Ramos-Molina B, Queipo-Ortuno MI, Lambertos A, Tinahones FJ, Penafiel R. Dietary and gut microbiota polyamines in obesity- and age-related diseases. Front Nutr. 2019;6:24.
245. Shah P, Swiatlo E. A multifaceted role for polyamines in bacterial pathogens. Mol Microbiol. 2008;68(1):4–16.
246. Tabor CW, Tabor H. Polyamines in microorganisms. Microbiol Rev. 1985;49(1):81–99.
247. Allison C, Macfarlane GT. Influence of pH, nutrient availability, and growth rate on amine production by Bacteroides fragilis and Clostridium perfringens. Appl Environ Microbiol. 1989;55(11):2894–8.
248. Landete JM, Arena ME, Pardo I, Manca de Nadra MC, Ferrer S. Comparative survey of putrescine production from agmatine deamination in different bacteria. Food Microbiol. 2008a;25(7):882–7.
249. Nakamura A, Ooga T, Matsumoto M. Intestinal luminal putrescine is produced by collective biosynthetic pathways of the commensal microbiome. Gut Microbes. 2019;10(2):159–71.
250. Kitada Y, Muramatsu K, Toju H, Kibe R, Benno Y, Kurihara S, Matsumoto M. Bioactive polyamine production by a novel hybrid system comprising multiple indigenous gut bacterial strategies. Sci Adv. 2018;4(6):eaat0062.
251. Noack J, Dongowski G, Hartmann L, Blaut M. The human gut bacteria Bacteroides thetaiomicron and fusobacterium varium produce putrescine and spermidine in cecum of pectin-fed gnotobiotic rats. J Nutr. 2000;130(5):1225–31.
252. Michael AJ. Polyamine function in archaea and bacteria. J Biol Chem. 2018;293(48): 18693–701.
253. Griswold AR, Jameson-Lee M, Burne RA. Regulation and physiological significance of the agmatine deiminase system of Streptococcus mutans UA 159. J Bacteriol. 2006;188(3): 834–41.
254. Higashi K, Ishigure H, Demizu R, Uemura T, Nishino K, Yamaguchi A, Kashiwagi K, Igarashi K. Identification of spermidine excretion protein complex (MdtJI) in Escherichia coli. J Bacteriol. 2008;190(3):872–8.
255. Osborne DL, Seidel ER. Gastrointestinal luminal polyamines: cellular accumulation and enterohepatic circulation. Am J Phys. 1990;258(4):G576–84.
256. Bardocz S. The role of dietary polyamines. Eur J Clin Nutr. 1993;47(10):683–90.
257. Bamba T, Vaja S, Murphy GM, Dowling RH. Effect of fasting and feeding on polyamines and related enzymes along the villus: crypt axis. Digestion. 1990;46(S2):424–9.
258. Blachier F, Darcy-Vrillon B, Sener A, Duée PH, Malaisse WJ. Arginine metabolism in rat enterocytes. Biochim Biophys Acta. 1991;1092(3):304–10.
259. Fitzpatrick LR, Wang P, Eikenburg BE, Haddox MK, Johnson LR. Effect of refeeding on polyamine biosynthesis in isolated enterocytes. Am J Phys. 1986;250(5):G709–13.
260. Porter CW, Dworaczyk D, Ganis B, Weiser MM. Polyamines and biosynthetic enzymes in the rat intestinal mucosa and the influence of methylglyoxal-bis(guanylhydrazone). Cancer Res. 1980;40(7):2330–5.
261. Bekebrede AF, Keijer J, Gerrits WJJ, de Boer VCJ. The molecular and physiological effects of protein-derived polyamines in the intestine. Nutrients. 2020;12(1):197.
262. Blachier F, M'Rabet-Touil H, Posho L, Morel MT, Bernard F, Darcy-Vrillon B, Duée PH. Polyamine metabolism in enterocytes isolated from newborn pigs. Biochim Biophys Acta. 1992;1175(1):21–6.
263. Madeo F, Eisenberg T, Pietrocola F, Kroemer G. Spermidine in health and disease. Science. 2018;359(6374):eaan2788.

264. Bartos F, Bartos D, Grettie DP, Campbell RA. Polyamine levels in normal human serum. Comparison of analytical methods. Biochem Biophys Res Commun. 1977;75(4):915–9.
265. Elitsur Y, Gesell M, Luk GD. ODC activity and polyamine levels in isolated human colonocytes. Life Sci. 1993;53(11):945–52.
266. Uda K, Tsujikawa T, Fujiyama Y, Bamba T. Rapid absorption of luminal polyamines in a rat small intestine ex vivo model. J Gastroenterol Hepatol. 2003;18(5):554–9.
267. Matsumoto M, Kibe R, Ooga T, Aiba Y, Kurihara S, Sawaki E, Koga Y, Benno Y. Impact of intestinal microbiota on intestinal luminal metabolome. Sci Rep. 2012;2:233.
268. Cheng SX, Geibel JP, Hebert SC. Extracellular polyamines regulate fluid secretion in rat colonic crypts via the extracellular calcium-sensing receptor. Gastroenterology. 2004;126(1):148–58.
269. Fioramonti J, Fargeas MJ, Bertrand V, Pradayrol L, Buéno L. Induction of postprandial intestinal motility and release of cholecystokinin by polyamines in rats. Am J Phys. 1994;267(6):G960–5.
270. Ma L, Ni Y, Wang Z, Tu W, Ni L, Zhuge F, Zheng A, Hu L, Zhao Y, Zheng L, Fu Z. Spermidine improves gut barrier integrity and gut microbiota function in diet-induced obese mice. Gut Microbes. 2020;12(1):1–19.
271. Ginty DD, Osborne DL, Seidel ER. Putrescine stimulates DNA synthesis in intestinal epithelial cells. Am J Phys. 1989;257(1):G145–50.
272. McCormack SA, Viar MJ, Johnson LR. Polyamines are necessary for cell migration by a small intestinal crypt cell line. Am J Phys. 1993;264(2):G367–74.
273. Rao JN, Xiao L, Wang JY. Polyamines in gut epithelial renewal and barrier function. Physiology (Bethesda). 2020;35(5):328–37.
274. Timmons J, Chang ET, Wang JY, Rao JN. Polyamines and gut mucosal homeostasis. J Gastrointest Dig Syst. 2012;2(7):001.
275. Wang JY. Polyamines and mRNA stability in regulation of intestinal mucosal growth. Amino Acids. 2007;33(2):241–52.
276. Löser C, Eisel A, Harms D, Fölsch UR. Dietary polyamines are essential luminal growth factors for small intestinal and colonic mucosal growth and development. Gut. 1999;44(1):12–6.
277. Nakamura A, Kurihara S, Takahashi D, Ohashi W, Nakamura Y, Kimura S, Onuki M, Kume A, Sasazawa Y, Furusawa Y, Obata Y, Fukuda S, Saiki S, Matsumoto M, Hase K. Symbiotic polyamine metabolism regulates epithelial proliferation and macrophage differentiation in the colon. Nat Commun. 2021;12(1):2105.
278. Gamet L, Cazenave Y, Trocheris V, Denis-Pouxviel C, Murat JC. Involvement of ornithine decarboxylase in the control of proliferation of the HT29 human colon cancer cell line. Effect of vasoactive intestinal peptide on enzyme activity. Int J Cancer. 1991;47(4):633–8.
279. Mayeur C, Veuillet G, Michaud M, Raul F, Blottière HM, Blachier F. Effects of agmatine accumulation in human colon carcinoma cells on polyamine metabolism, DNA synthesis and the cell cycle. Biochim Biophys Acta. 2005;1745(1):111–23.
280. Grosheva I, Zheng D, Levy M, Polansky O, Lichtenstein A, Golani O, Dori-Bachash M, Moresi C, Shapiro H, Del Mare-Roumani S, Valdes-Mas R, He Y, Karbi H, Chen M, Harmelin A, Straussman R, Yissachar N, Elinav E, Geiger B. High-throughput screen identifies host and microbiota regulators of intestinal barrier function. Gastroenterology. 2020;159(5):1807–23.
281. Mantziari A, Mannila E, Collado MC, Salminen S, Gomez-Gallego C. Exogenous polyamines influence in vitro microbial adhesion to human mucus according to the age of mucus donor. Microorganisms. 2021;9(6):1239.
282. Chu H, Duan Y, Yang L, Schnabl B. Small metabolites, possible big changes: a microbiota-centered view of non-alcoholic fatty liver disease. Gut. 2019;68(2):359–70.
283. Cope K, Risby T, Diehl AM. Increased gastrointestinal ethanol production in obese mice: implications for fatty liver disease pathogenesis. Gastroenterology. 2000;119(5):1340–7.

284. Zhu L, Baker SS, Gill C, Liu W, Alkhouri R, Baker RD, Gill SR. Characterization of gut microbiomes in nonalcoholic steatohepatitis (NASH) patients: a connection between endogenous alcohol and NASH. Hepatology. 2013;57(2):601–9.
285. Baraona E, Julkunen R, Tannenbaum L, Lieber CS. Role of intestinal bacterial overgrowth in ethanol production and metabolism in rats. Gastroenterology. 1986;90(1):103–10.
286. Salaspuro MP. Acetaldehyde, microbes, and cancer of the digestive tract. Crit Rev Clin Lab Sci. 2003;40(2):183–208.
287. Yokoyama S, Takeuchi K, Shibata Y, Kageyama S, Matsumi R, Takeshita T, Yamashita Y. Characterization of oral microbiota and acetaldehyde production. J Oral Microbiol. 2018;10 (1):1492316.
288. Seitz HK, Simanovski UA, Garzon FT, Rideout JM, Peters TJ, Koch A, Berger MR, Einecke H, Maiwald M. Possible role of acetaldehyde in ethanol-related rectal cocarcinogenesis in the rat. Gastroenterology. 1990;98(2):406–13.
289. Jokelainen K, Nosova T, Koivisto T, Väkeväinen S, Jousiemies-Somer H, Heine R, Salaspuro M. Inhibition of bacteriocolonic pathway for ethanol oxidation by ciprofloxacin in rats. Life Sci. 1997;61(18):1755–62.
290. Tillonen J, Homann N, Rautio M, Jousimies-Somer H, Salaspuro M. Ciprofloxacin decreases the rate of ethanol elimination in humans. Gut. 1999;44(3):347–52.
291. Jokelainen K, Matysiak-Budnik T, Mäkisolo H, Höckerstedt K, Salaspuro M. High intracolonic acetaldehyde values produced by a bacteriocolonic pathway for ethanol oxidation in piglets. Gut. 1996;39(1):100–4.
292. Xie G, Zhong W, Zheng X, Li Q, Qiu Y, Li H, Chen Y, Zhou Z, Jia W. Chronic ethanol consumption alters mammalian gastrointestinal content metabolites. J Proteome Res. 2013;12 (7):3297–306.
293. Couch RD, Dailey A, Zaidi F, Navarro K, Forsyth CB, Mutlu E, Engen PA, Keshavarzian A. Alcohol induced alterations to the human fecal VOC metabolome. PLoS One. 2015;10(3): e0119362.
294. Wang Y, Tong J, Chang B, Wang B, Zhang B, Wang B. Effects of alcohol on intestinal epithelial barrier permeability and expression of tight junction-associated proteins. Mol Med Rep. 2014;9(6):2352–6.
295. Elamin E, Masclee A, Troost F, Pieters HJ, Keszthelyi D, Aleksa K, Dekker J, Jonkers D. Ethanol impairs intestinal barrier function in humans through mitogen activated protein kinase signaling: a combined in vivo and in vitro approach. PLoS One. 2014b;9(9):e107421.
296. Elamin E, Masclee A, Troost F, Dekker J, Jonkers D. Cytotoxicity and metabolic stress induced by acetaldehyde in human intestinal LS174T goblet-like cells. Am J Phys. 2014a;307(3):G286–94.
297. Basuroy S, Sheth P, Mansbach CM, Rao RK. Acetaldehyde disrupts tight junctions and adherens junctions in human colonic mucosa: protection by EGF and L-glutamine. Am J Phys. 2005;289(2):G367–75.
298. Margolis KG, Cryan JF, Mayer EA. The microbiota-gut-brain axis: from motility to mood. Gastroenterology. 2021;160(5):1486–501.
299. Strandwitz P. Neurotransmitter modulation by the gut microbiota. Brain Res. 2018;1693 (B):128–33.
300. Wall R, Cryan JF, Ross RP, Fitzgerald GF, Dinan TG, Stanton C. Bacterial neuroactive compounds produced by psychobiotics. Adv Exp Med Biol. 2014;817:221–39.
301. Ngo DH, Vo TS. An updated review on pharmaceutical properties of gamma-aminobutyric acid. Molecules. 2019;24(15):2678.
302. Hyland NP, Cryan JF. A gut feeling about GABA: focus on GABA(B) receptors. Front Pharmacol. 2010;1:124.
303. Barrett E, Ross RP, O'Toole PW, Fitzgerald GF, Stanton C. Υ-aminobutyric acid production by culturable bacteria from the human intestine. J Appl Microbiol. 2012;113(2):411–7.

304. Nomura M, Nakajima I, Fujita Y, Kobayashi M, Kimoto H, Suzuki I, Aso H. Lactococcus lactis contains only one glutamate decarboxylase gene. Microbiology (Reading). 1999;145(6): 1375–80.
305. Siragusa S, De Angelis M, Di Cagno R, Rizzello CG, Coda R, Gobbetti M. Synthesis of gamma-aminobutyric acid by lactic acid bacteria isolated from a variety of Italian cheeses. Appl Environ Microbiol. 2007;73(22):7283–90.
306. Feehily C, Karatzas KAG. Role of glutamate metabolism in bacterial responses towards acid and other stresses. J Appl Microbiol. 2013;114(1):11–24.
307. Karatzas KA, Brennan O, Heavin S, Morrissey J, O'Byrne CP. Intracellular accumulation of high levels of gamma-aminobutyrate by listeria monocytogenes 10403S in response to low pH: uncoupling of gamma-aminobutyrate synthesis from efflux in a chemically defined medium. Appl Environ Microbiol. 2010;76(11):3529–37.
308. Matsumoto M, Kibe R, Ooga T, Aiba Y, Sawaki E, Koga Y, Benno Y. Cerebral low-molecular metabolites influenced by intestinal microbiota. Front Syst Neurosci. 2013;7:9.
309. Aston-Jones G, Waterhouse B. Locus coeruleus: from global projection system to adaptive regulation and behavior. Brain Res. 2016;1645:75–8.
310. Gregersen H, Dall FH, Jorgensen CS, Jensen SL, Ahren JB. Effects of noradrenaline and galanin on duodenal motility in the isolated perfused porcine pancreatico-duodenal block. Regul Pept. 1992;39(2–3):157–67.
311. Kurahashi M, Kito Y, Hara M, Takeyama H, Sanders KM, Hashitani H. Norepinephrine has dual effects on human colonic contractions through distinct subtypes of alpha 1 adrenoceptors. Cell Mol Gastroenterol Hepatol. 2020;10(3):658–71.
312. Boyanova L. Stress hormone epinephrine (adrenaline) and norepinephrine (noradrenaline) effects on the anaerobic bacteria. Anaerobe. 2017;44:13–9.
313. Lustri BC, Sperandio V, Moreira CG. Bacterial chat: intestinal metabolites and signals in host-microbiota-pathogen interactions. Infect Immun. 2017;85(12):e00476–17.
314. O'Donnell PM, Aviles H, Lyte M, Sonnenfeld G. Enhancement of in vitro growth of pathogenic bacteria by norepinephrine: importance of inoculum density and role of transferrin. Appl Environ Microbiol. 2006;72(7):5097–9.
315. Asano Y, Hiramoto T, Nishino R, Aiba Y, Kimura T, Yoshihara K, Koga Y, Sudo N. Critical role of gut microbiota in the production of biologically active, free catecholamines in the gut lumen of mice. Am J Phys. 2012;303(11):G1288–95.
316. Klein MO, Battagello DS, Cardoso AR, Hauser DN, Bittencourt JC, Correa RG. Dopamine: functions, signaling, and association with neurological diseases. Cell Mol Neurobiol. 2019;39 (1):31–59.
317. Claus H, Decker H. Bacterial tyrosinases. Syst Appl Microbiol. 2006;29(1):3–14.
318. Li Y, Zhang Y, Zhang XL, Feng XY, Liu CZ, Zhang XN, Quan ZS, Yan JT, Zhu JX. Dopamine promotes colonic mucus secretion through dopamine D5 receptor in rats. Am J Phys. 2019;316(3):C393–403.
319. Baronio D, Gonchoroski T, Castro K, Zanatta G, Gottfried C, Riesgo R. Histaminergic system in brain disorders: lessons from the translational approach and future perspectives. Ann General Psychiatry. 2014;13(1):34.
320. Maintz L, Novak N. Histamine and histamine intolerance. Am J Clin Nutr. 2007;85(5): 1185–96.
321. Guihot G, Blachier F. Histidine and histamine metabolism in rat enterocytes. Mol Cell Biochem. 1997;175(1–2):143–8.
322. Landete JM, De las Rivas B, Marcobal A, Munoz R. Updated molecular knowledge about histamine biosynthesis by bacteria. Crit Rev Food Sci Nutr. 2008b;48(8):697–714.
323. Molenaar D, Bosscher JS, ten Brink B, Driessen AJ, Konings WN. Generation of a proton motive force by histidine decarboxylation and electrogenic histidine/histamine antiport in lactobacillus buchneri. J Bacteriol. 1993;175(10):2864–70.

324. Bjornsdottir-Butler K, Green DP, Bolton GE, McClellan-Green PD. Control of histamine-producing bacteria and histamine formation in fish muscle by trisodium phosphate. J Food Sci. 2015;80(6):M1253–8.
325. Kovacova-Hanuskova E, Buday T, Gavliakova S, Plevkova J. Histamine, histamine intoxication and intolerance. Allergol Immunopathol (Madr). 2015;43(5):498–506.
326. El-Merahbi R, Löffler M, Mayer A, Sumara G. The roles of peripheral serotonin in metabolic homeostasis. FEBS Lett. 2015;589(15):1728–34.
327. Gershon MD, Tack J. The serotonin signaling system: from basic understanding to drug development for functional GI disorders. Gastroenterology. 2007;132(1):397–414.
328. Yano JM, Yu K, Donaldson GP, Shastri GS, Ann P, Ma L, Nagler CR, Ismagilof RF, Mazmanian SK, Hsiao EY. Indigenous bacteria from the gut microbiota regulate host serotonin biosynthesis. Cell. 2015;161(2):264–76.
329. Williams BB, Van Benschoten AH, Cimermancic P, Donia MS, Zimmermann M, Taketani M, Ishihara A, Kashyap PC, Fraser JS, Fischbar MA. Discovery and characterization of gut microbiota decarboxylases that can produce the neurotransmitter tryptamine. Cell Host Microbe. 2014;16(4):495–503.
330. Bhattarai Y, Williams BB, Battaglioli EJ, Whitaker WR, Till L, Grover M, Linden DR, Akiba Y, Kandimalla KK, Zachos NC, Kaunitz JD, Sonnenburg JL, Fischbach MA, Farrugia G, Kashyap PC. Gut microbiota-produced tryptamine activates an epithelial G-protein-coupled receptor to increase colonic secretion. Cell Host Microbe. 2018;23(6):775–85.
331. Thornburn AN, Macia L, Mackay CR. Diet, metabolites, and "western-lifestyle" inflammatory diseases. Immunity. 2014;40(6):833–42.
332. Verbeke KA, Boobis AR, Chiodini A, Edwards CA, Franck A, Kleerebezem M, Nauta A, Raes J, van Tol EA, Tuohy KM. Towards microbial fermentation metabolites as markers for health benefits of prebiotics. Nutr Res Rev. 2015;28(1):42–66.
333. Schaafsma G. The protein digestibility-corrected amino acid score. J Nutr. 2000;130(7):1865S–7S.
334. Boutry C, Fouillet H, Mariotti F, Blachier F, Tomé D, Bos C. Rapeseed and milk protein exhibit a similar nutritional value but marked difference in postprandial regional nitrogen utilization in rats. Nutr Metab (Lond). 2011;8(1):52.
335. Rand WM, Pellett PL, Young VR. Meta-analysis of nitrogen balance studies for estimating protein requirements in healthy adults. Am J Clin Nutr. 2003;77(1):109–27.
336. Schönfeldt HC, Gibson Hall N. Dietary protein quality and malnutrition in Africa. Br J Nutr. 2012;108(S2):S69–76.
337. Thom G, Lean M. Is there an optimal diet for weight management and metabolic health? Gastroenterology. 2017;152(7):1739–51.
338. Hruby A, Hu FB. The epidemiology of obesity: a big picture. PharmacoEconomics. 2015;33(7):673–89.
339. Pesta DH, Samuel VT. A high-protein diet for reducing body fat: mechanisms and possible caveats. Nutr Metab (Lond). 2014;11(1):53.
340. Tipton KD. Nutrition for acute exercise-induced injuries. Annu Rev Metab. 2010;57(S2):43–53.
341. Phillips SM. A brief review of higher dietary protein diets in weight loss: a focus on athletes. Sports Med. 2014;44(S2):S149–53.
342. Santesso N, Akl EA, Bianchi M, Mente A, Mustafa R, Heels-Ansdell D, Schünemann HJ. Effects of higher-versus lower- protein diets on health outcomes: a systematic review and meta-analysis. Eur J Clin Nutr. 2012;66(7):780–8.
343. Barba CV, Cabrera MI. Recommended energy and nutrient intakes for Filipinos 2002. Asia Pac J Clin Nutr. 2008;17(S2):399–404.
344. Wali JA, Solon-Biet SM, Freire T, Brandon AE. Macronutriment determinants of obesity, insulin resistance and metabolic health. Biology (Basel). 2021;10(4):336.

345. Leidy HJ, Clifton PM, Astrup A, Wycherley TP, Westertep-Plantenga MS, Luscombe-Marsh ND, Woods SC, Mattes RD. The role of protein in weight loss and maintenance. Am J Clin Nutr. 2015;101(6):1320S–9S.

346. Wilkinson SB, Phillips SM, Atherton PJ, Patel R, Yarasheski KE, Tarnopolsky MA, Rennie MJ. Differential effects of resistance and endurance exercise in the fed state on signalling molecule phosphorylation and protein synthesis in human muscle. J Physiol. 2008;586(15): 3701–17.

347. Silvester KR, Cummings JH. Does digestibility of meat protein help explain large bowel risk? Nutr Cancer. 1995;24(3):279–88.

348. Flint HJ. The impact of nutrition on the human microbiome. Nutr Rev. 2012;70(S1):S10–3.

349. Schmidt NS, Lorentz A. Dietary restrictions modulate the gut microbiota: implications for health and disease. Nutr Res. 2021;89:10–22.

350. Zheng X, Wang S, Jia W. Calorie restriction and its impact on gut microbial composition and global metabolism. Front Med. 2018;12(6):634–44.

351. Beaumont M, Portune KJ, Steuer N, Lan A, Cerrudo V, Audebert M, Dumont F, Mancano G, Khodorova N, Andriamihaja M, Airinei G, Tomé D, Benamouzig R, Davila AM, Claus SP, Sanz Y, Blachier F. Quantity and source of dietary protein influence metabolite production by gut microbiota and rectal mucosa gene expression: a randomized, parallel, double-blind trial in overweight humans. Am J Clin Nutr. 2017a;106(4):1005–19.

352. Windey K, De Preter V, Louat T, Schuit F, Herman J, Vansant G, Verbeke K. Modulation of protein fermentation does not affect fecal water toxicity: a randomized cross-over study in healthy subjects. PLoS One. 2012a;7(12):e52387.

353. Cummings JH, Hill MJ, Bone ES, Branch WJ, Jenkins DJ. The effect of meat protein and dietary fiber on colonic function and metabolism. II. Bacterial metabolites in feces and urine. Am J Clin Nutr. 1979;32(10):2094–101.

354. Russell WR, Gratz SW, Duncan SH, Holtrop G, Ince J, Scobbie L, Duncan G, Johnstone AM, Lobley GE, Wallace RJ, Duthie GG, Flint HJ. High-protein, reduced-carbohydrate weight-loss diets promote metabolic profiles likely to be detrimental to colonic health. Am J Clin Nutr. 2011;93(5):1062–72.

355. Louis P, Hold GL, Flint HJ. The gut microbiota, bacterial metabolites and colorectal cancer. Nat Rev Microbiol. 2014;12(10):661–72.

356. David LA, Maurice CF, Carmody RN, Gootenberg DB, Button JE, Wolfe BE, Ling AV, Sloan Devline A, Varma Y, Fischbach MA, Biddinger SB, Dutton RJ, Turnbaugh RJ. Diet rapidly and reproducibly alters the human gut microbiome. Nature. 2014;505(7484):559–63.

357. Duncan SH, Belenguer A, Holtrop G, Johnstone AM, Flint HJ, Lobley GE. Reduced dietary intake of carbohydrates by obese subjects results in decreased concentrations of butyrate and butyrate-producing bacteria in feces. Appl Environ Microbiol. 2007;73(4):1073–8.

358. Thibault R, Blachier F, Darcy-Vrillon B, de Coppet P, Bourreille A, Segain JP. Butyrate utilization by the colonic mucosa in inflammatory bowel diseases: a transport deficiency. Inflamm Bowel Dis. 2010;16(4):684–95.

359. Magee EA, Richardson CJ, Hughes R, Cummings JH. Contribution of dietary protein to sulfide production in the large intestine: an in vitro and a controlled feeding in humans. Am J Clin Nutr. 2000;72(6):1488–94.

360. Windey K, De Preter V, Verbeke K. Relevance of protein fermentation to gut health. Mol Nutr Food Res. 2012b;56(1):184–96.

361. Gryp T, Vanholder R, Vaneechoutte M, Glorieux G. p-cresyl sulfate. Toxins (Basel). 2017;9 (2):52.

362. Tennoune N, Andriamihaja M, Blachier F. Production of indole and indole-related compounds by the intestinal microbiota and consequences for the host: the good, the bad, and the ugly. Microorganisms. 2022;10(5):930.

363. Pearson JR, Gill CIR, Rowland IR. Diet, fecal water, and colon cancer: development of a biomarker. Nutr Rev. 2009;67(9):509–26.

364. Benassi-Evans B, Clifton P, Noakes M, Fenech M. High-protein/high red meat and high-carbohydrate weight-loss diets do not differ in their effects on fecal water genotoxicity tested by use of the WIL2-NS cell line and with other biomarkers of bowel health. Mutat Res. 2010;703(2):130–6.

365. Mu C, Yang Y, Luo Z, Guan L, Zhu W. The colonic microbiome and epithelial transcriptome are altered in rats fed a high-protein diet compared with a normal-protein diet. J Nutr. 2016;146(3):474–83.

366. Beaumont M, Andriamihaja M, Armand L, Grauso M, Jaffrézic F, Lanoë D, Moroldo M, Davila AM, Tomé D, Blachier F, Lan A. Epithelial response to a high-protein diet in rat colon. BMC Genomics. 2017b;18(1):116.

367. Fiorentino M, Landais E, Bastard G, Carriquiry A, Wierenga RT, Berger J. Nutrient intake is insufficient among Segenalese urban school children and adolescents: results from two 24 h recalls in state primary schools in Dakar. Nutrients. 2016;8(10):650.

368. Hautvast JL, van der Heijden LJ, Luneta AK, van Staveren WA, Tolboom JJ, van Gastel SM. Food consumption of young stunted and non-stunted children in rural Zambia. Eur J Clin Nutr. 1999;53(1):50–9.

369. Motadi SA, Matsea Z, Mogane PH, Masidwali P, Makwarela M, Mushapi L. Assessment of nutritional status and dietary intake of pregnant women in rural area of Vhembe district. Limpopo province Ecol Food Nutr. 2020;59(3):229–42.

370. Smith MI, Yatsunenko T, Manary MJ, Trehan I, Mkakosya R, Cheng J, Kau AL, Rich SS, Concannon P, Mychaleckyj JC, Liu J, Houpt E, Li JV, Holmes E, Nicholson J, Knights D, Ursell LK, Knight R, Gordon JI. Gut microbiomes of Malawian twin pairs discordant for kwashiorkor. Science. 2013;339(6119):548–54.

371. Subramanian S, Huq S, Yatsunenko T, Haque R, Mahfuz M, Alam MA, Benezra A, DeStefano J, Meier MF, Muegge BD, Barratt MJ, VanArendonk LG, Zhang Q, Province MA, Petri WA Jr, Ahmed T, Gordon JI. Persistent gut microbiota immaturity in malnourished Bangladeshi children. Nature. 2014;510(7505):417–21.

372. Spring S, Premathilake H, DeSilva U, Shili C, Carter S, Pezeshki A. Low protein-high carbohydrate diets alter energy balance, gut microbiota composition and blood metabolomics profile in young pigs. Sci Rep. 2020;10(1):3318.

373. Matsuoka H, Suda W, Tomitsuka E, Shindo C, Takayasu L, Horwood P, Greenhill AR, Hattori M, Umezaki M, Hirayama K. The influence of low protein diet on the intestinal microbiota in mice. Sci Rep. 2020;10(1):17077.

374. Wright EM, Martin MG, Turk E. Intestinal absorption in health and disease: sugars. Best Pract Res Clin Gastroenterol. 2003;17(6):943–56.

375. Mudgil D, Barak S. Composition, properties and health benefits of indigestible carbohydrate polymers as dietary fiber: a review. Int J Biol Macromol. 2013;61:1–6.

376. DeMartino P, Cockburn DW. Resistant starch: impact on the gut microbiome and health. Curr Opin Biotechnol. 2020;61:66–71.

377. Kumar V, Sinha AK, Makkar HPS, De Boeck G, Becker K. Dietary roles of non-starch polysaccharides in human nutrition: a review. Crit Rev Food Sci Nutr. 2012;52(10):899–935.

378. Macfarlane GT, Macfarlane S. Human colonic microbiota: ecology, physiology and metabolic potential of intestinal bacteria. Scand J Gastroenterol. 1997;222:3–9.

379. Louis P, Flint HJ. Formation of propionate and butyrate by the human colonic microbiota. Environ Microbiol. 2017;19(1):29–41.

380. Haenen D, Zhang J, Souza da Silva C, Bosch G, van der Meer IM, van Arkel J, van den Borne JJ, Pérez Guttiérez O, Smidt H, Kemp B, Müller M, Hooiveld GJ. A diet high in resistant starch modulates microbiota composition, SCFA concentrations, and gene expression in pig intestine. J Nutr. 2013;143(3):274–83.

381. Jenkins DJ, Vuksan V, Kendall CW, Würst P, Jeffcoat R, Waring S, Mehling CC, Vidgen E, Augustin LS, Wong E. Physiological effects of resistant starches on fecal bulk, short-chain fatty acids, blood lipids and glycemic index. J Am Coll Nutr. 1998;17(6):609–16.

382. Le Blay G, Michel C, Blottière HM, Cherbut C. Enhancement of butyrate production in the rat caecocolonic tract by long-term ingestion of potato starch. Br J Nutr. 1999;82(5):419–26.

383. Schwiertz A, Lehmann U, Jacobasch G, Blaut M. Influence of resistant starch on the SCFA production and cell counts of butyrate-producing eubacterium spp. in the human intestine. J Appl Microbiol. 2002;93(1):157–62.

384. Simpson HL, Campbell BJ. Review article: dietary fiber-microbiota interactions. Aliment Pharmacol Ther. 2015;42(2):158–79.

385. Jang C, Hui S, Lu W, Cowan AJ, Morscher RJ, Lee G, Liu W, Tesz GJ, Birnbaum MJ, Rabinowitz JD. The small intestine converts dietary fructose into glucose and organic acids. Cell Metab. 2018;27(2):351–361.e3.

386. Barcenilla A, Pryde SE, Martin JC, Duncan SH, Stewart CS, Henderson C, Flint HJ. Phylogenic relationships of butyrate-producing bacteria from the human gut. Appl Environ Microbiol. 2000;66(4):1654–61.

387. Duncan SH, Hold GL, Harmsen HJM, Stewart CS, Flint HJ. Growth requirements and fermentation products of fusobacterium prausnitzii, and a proposal to reclassify it as Faecalibacterium prausnitzii gen. Nov; comb.nov. Int J Syst Evol Microbiol. 2002;52(6): 2141–6.

388. Lopez-Siles M, Khan TM, Duncan SH, Harmsen HJM, Garcia-Gil J, Flint HJ. Cultured representatives of two major phylogroups of human colonic Faecalibacterium prausnitzii can utilize pectin, uronic acids, and host-derived substrates for growth. Appl Environ Microbiol. 2012;78(2):420–8.

389. Louis P, Flint HJ. Diversity, metabolism and microbial ecology of butyrate-producing bacteria from the human large intestine. FEMS Microbiol Lett. 2009;294(1):1–8.

390. Cummings JH, Pomare EW, Branch WJ, Naylor CP, Macfarlane GT. Short chain fatty acids in human large intestine, portal, hepatic and venous blood. Gut. 1987;28(10):1221–7.

391. Hallert C, Björck I, Nyman M, Pousette A, Grännö C, Svensson H. Increasing fecal butyrate in ulcerative colitis patients by diet: controlled pilot study. Inflamm Bowel Dis. 2003;9(2): 116–21.

392. Topping DL, Clifton PM. Short-chain fatty acids and human colonic function: roles of resistant starch and nonstrach polysaccharides. Physiol Rev. 2001;81(3):1031–64.

393. Weaver GA, Tangel CT, Krause JA, Parfitt MM, Jenkins PL, Rader JM, Lewis BA, Miller TL, Wolin MJ. Acarbose enhances human colonic butyrate production. J Nutr. 1997;127(5): 717–23.

394. Nugent SG, Kumar D, Rampton DS, Evans DF. Intestinal luminal pH in inflammatory bowel disease: possible determinants and implications for therapy with aminosalicylates and other drugs. Gut. 2001;48(4):571–7.

395. Cuff MA, Shirazi-Beechey SP. The human monocarboxylate transporter, MCT1: genomic organization and promoter analysis. Biochem Biophys Res Commun. 2002;292(4):1048–56.

396. Sivaprakasam S, Bhutia YD, Yang S, Ganapathy V. Short-chain fatty acid transporters: role in colonic homeostasis. Compr Physiol. 2017;8(1):299–314.

397. Kawamata K, Hayashi H, Suzuki Y. Propionate absorption associated with bicarbonate secretion in vitro in the mouse cecum. Pflugers Arch. 2007;454(2):253–62.

398. Velasquez OC, Lederer HM, Rombeau JL. Butyrate and the colonocyte. Production, absorption, metabolism, and therapeutic implications. Adv Exp Med Biol. 1997;427:123–34.

399. Hadjiagapiou C, Schmidt L, Dudeja PK, Layden PJ, Ramaswamy K. Mechanism(s) of butyrate transport in Caco-2 cells: role of monocarboxylate transporter 1. Am J Phys. 2000;279(4):G775–80.

400. Ritzhaupt A, Wood IS, Ellis A, Hosie KB, Shirazi-Beechey SP. Identification and characterization of a monocarboxylate transporter (MCT1) in pig and human colon: its potential to transport L-lactate as well as butyrate. J Physiol. 1998;513(3):719–32.

401. Clausen MR, Mortensen PB. Kinetic studies on the metabolism of short-chain fatty acids and glucose by isolated rat colonocytes. Gastroenterology. 1994;106(2):423–32.

402. Jorgensen JR, Clausen MR, Mortensen PB. Oxidation of short and medium chain C2-C8 fatty acids in Sprague-Dawley rat colonocytes. Gut. 1997;40(3):400–5.
403. Roediger WE. Utilization of nutrients by isolated epithelial cells of the rat colon. Gastroenterology. 1982;83(2):424–9.
404. Bergman EN. Energy contribution of volatile fatty acids from the gastrointestinal tract in various species. Physiol Rev. 1990;70(2):567–90.
405. Hall KD, Guo J. Obesity energetics: body weight regulation and the effects of diet composition. Gastroenterology. 2017;152(7):1718–27.
406. Hill JO, Peters JC. Environmental contributions to the obesity epidemic. Science. 1998;280 (5368):1371–4.
407. Bogardus C, Lillioja S, Ravussin E, Abbott W, Zawadzki JK, Young A, Knowler WC, Jacobowitz R, Moll PP. Familial dependence of the resting metabolic rate. N Engl J Med. 1986;315(2):96–100.
408. Bouchard C. Human variation in body mass: evidence for a role of the genes. J Nutr. 1997;55 (1):S21–7.
409. Prentice AM, Jebb SA. Obesity in Britain: gluttony or sloth. BMJ. 1995;311(7002):437–9.
410. Cox AJ, West NP, Cripps AW. Obesity, inflammation, and the gut microbiota. Lancet Diabetes Endocrinol. 2015;3(3):207–15.
411. Gomes AC, Hoffmann C, Mota JF. The human gut microbiota: metabolism and perspective in obesity. Gut Microbes. 2018;9(4):308–25.
412. Binder HJ, Mehta P. Short-chain fatty acids stimulate active sodium and chloride absorption in vitro in the rat distal colon. Gastroenterology. 1989;96(4):989–96.
413. Inagaki A, Hayashi M, Andharia N, Matsuda H. Involvement of butyrate in electrogenic K+ secretion in rat rectal colon. Pflugers Arch. 2019;471(2):313–27.
414. Peng L, Li ZR, Green RS, Holzman IR, Lin J. Butyrate enhances the intestinal barrier by facilitating tight junction assembly via activation of AMP-activated protein kinase in Caco-2 cell monolayers. J Nutr. 2009;139(9):1619–25.
415. Wang HB, Wang PY, Wang X, Wan YL, Liu YC. Butyrate enhances intestinal epithelial barrier function via up-regulation of tight junction protein Claudin-1 transcription. Dig Dis Sci. 2012;57(12):3126–35.
416. Sauer J, Richter KK, Pool-Zobel BL. Products formed during fermentation of the prebiotic inulin with human gut flora enhance expression of biotransformation genes in human primary colon cells. Br J Nutr. 2007;97(5):928–37.
417. Hinnebusch BF, Meng S, Wu JT, Archer SY, Hodin RA. The effects of short-chain fatty acids on human colon cancer cell phenotype are associated with histone hyperacetylation. J Nutr. 2002;132(5):1012–7.
418. Leschelle X, Delpal S, Goubern M, Blottière HM, Blachier F. Butyrate metabolism upstream and downstream acetyl CoA synthesis and growth control of human colon carcinoma cells. Eur J Biochem. 2000;267(21):6435–42.
419. Archer SY, Meng S, Shei A, Hodin RA. p21 (WAF1) is required for butyrate-mediated growth inhibition of human colon cancer cells. Proc Natl Acad Sci U S A. 1998;95(12):6791–6.
420. Nakano K, Mizuno K, Sowa Y, Orita T, Yoshino T, Okuyama Y, Fujita T, Othani-Fujita N, Matsukawa Y, Tokino T, Yamagishi H, Oka T, Nomura H, Sakai T. Butyrate activates the WAF1/Cip1 gene promoter through Sp1 sites in a p53-negative human colon cancer cell line. J Biol Chem. 1997;272(35):22199–206.
421. Siavoshian S, Blottière HM, Cherbut C, Galmiche JP. Butyrate stimulates cyclin D1 and p21 and inhibits cyclin-dependent kinase 2 expression in HT-29 colonic epithelial cells. Biochem Biophys Res Commun. 1997;232(1):169–72.
422. Andriamihaja M, Chaumontet C, Tome D, Blachier F. Butyrate metabolism in human colon carcinoma cells: implications concerning its growth-inhibitory effect. J Cell Physiol. 2009;218 (1):58–65.

423. Donohoe DR, Collins LB, Wali A, Bigler R, Sun W, Bultman SJ. The Warburg effect dictates the mechanism of butyrate-mediated histone acetylation and cell proliferation. Mol Cell. 2012;48(4):612–26.
424. Buttgereit F, Brand MD. A hierarchy of ATP-consuming processes in mammalian cells. Biochem J. 1995;312(1):163–7.
425. Archer SY, Johnson J, Kim HJ, Ma Q, Mou H, Daesety V, Meng S, Hodin RA. The histone deacetylase inhibitor butyrate downregulates cyclin B1 gene expression via a p21/WAF-1 dependent mechanism in human colon cancer cells. Am J Phys. 2005;289(4):G696–703.
426. Zhang Y, Sun Z, Jia J, Du T, Zhang N, Tang Y, Fang Y, Fang D. Overview of histone modification. Adv Exp Med Biol. 2021;1283:1–16.
427. Kaiko GE, Ryu SH, Koues OI, Collins PL, Solnica-Krezel L, Pearce EJ, Pearce EL, Oltz EM, Stappenbeck TS. The colonic crypt protects stem cells from microbiota-derived metabolites. Cell. 2016;165(7):1708–20.
428. Burger-van Paassen N, Vincent A, Puiman PJ, van der Sluis M, Bouma J, Boehm G, van Goudoever JB, van Seuningen I, Renes IB. The regulation of intestinal mucin MUC2 expression by short-chain fatty acids: implications for epithelial protection. Biochem J. 2009;420(2):211–9.
429. Barcelo A, Claustre J, Moro F, Chayvialle JA, Cuber JC, Plaisancié P. Mucin secretion is modulated by luminal factors in the isolated vascularly perfused rat colon. Gut. 2000;46(2): 218–24.
430. Finnie IA, Dwarakanath AD, Taylor BA, Rhodes JM. Colonic mucin synthesis is increased by sodium butyrate. Gut. 1995;36(1):93–9.
431. Karaki S, Tazoe H, Hayashi H, Kashiwabara H, Tooyama K, Suzuki Y, Kuwahara A. Expression of the short-chain fatty acid receptor, GPR43, in the human colon. J Mol Histol. 2008;39(2):135–42.
432. Nohr MK, Pedersen MH, Gille A, Egerod KL, Engelstoft MS, Husted AS, Sichlau RM, Grunddal KV, Poulsen SS, Han S, Jones RM, Offermanns S, Schwartz TW. GPR41/FFAR3 and GPR43/FFAR2 as cosensors for short-chain fatty acids in enteroendocrine cells vs FFAR3 in enteric neurons and FFAR2 in enteric leukocytes. Endocrinology. 2013;154(10):3552–64.
433. Lu VB, Gribble FM, Reimann F. Free fatty acid receptors in enteroendocrine cells. Endocrinology. 2018;159(7):2826–35.
434. Larraufie P, Martin-Gallausiaux C, Lapaque N, Dore J, Gribble FM, Reimann F, Blottiere HM. SCFAs strongly stimulate PYY production in human enteroendocrine cells. Sci Rep. 2018;8(1):74.
435. Chambers ES, Viardot A, Psichas A, Morrison DJ, Murphy KG, Zac-Varghese SE, MacDougall K, Preston T, Tedford C, Finlayson GS, Blundell JE, Bell JD, Thomas EL, Mt-Isa S, Ashby D, Gibson GR, Kolida S, Dhillo WS, Bloom SR, Morley W, Clegg S, Frost G. Effects of targeted delivery of propionate to the human colon on appetite regulation, body weight maintenance and adiposity in overweight adults. Gut. 2015;64(11):1744–54.
436. Van der Beek CM, Canfora EE, Lenaerts K, Troost FJ, Olde Damink SWM, Holst JJ, Masclee AAM, Dejong CHC, Blaak EE. Distal, not proximal, colonic acetate infusions promote fat oxidation and improve metabolic markers in overweight/obese men. Clin Sci (Lond). 2016;130(22):2073–82.
437. Ballantyne GH. Peptide YY(1-36) and peptide YY(3-36): part I. distribution, release and actions. Obes Surg. 2006;16(5):651–8.
438. Everard A, Cani PD. Gut microbiota and GLP-1. Rev Endocr Metab Disord. 2014;15(3): 189–96.
439. Veronese N, Solmi M, Caruso MG, Giannelli G, Osella AR, Evangelou E, Maggi S, Fontana L, Stubbs B, Tzoulaki I. Dietary fiber and health outcomes: an umbrella review of systematic reviews and meta-analyses. Am J Clin Nutr. 2018;107(3):436–44.
440. Krishnamurthy VM, Wei G, Baird BC, Murtaugh M, Chonchol MB, Raphael KL, Greene T, Beddhu S. High dietary fiber intake is associated with decreased inflammation and all-cause mortality in patients with chronic kidney disease. Kidney Int. 2012;81(3):300–6.

441. North CJ, Venter CS, Jerling JC. The effects of dietary fibre on C-reactive protein, an inflammation marker predicting cardiovascular disease. Eur J Clin Nutr. 2009;63(8):921–33.
442. Slavin JL, Lloyd B. Health benefits of fruits and vegetables. Adv Nutr. 2012;3(4):506–16.
443. Kaluza J, Orsini N, Levitan EB, Brzozowska A, Roszkowski W, Wolk A. Dietary calcium and magnesium intake and mortality: a prospective study of men. Am J Epidemiol. 2010;171(7): 801–7.
444. Slavin JL, Martini MC, Jacobs DR Jr, Marquart L. Plausible mechanisms for the protectiveness of whole grains. Am J Clin Nutr. 1999;70(3):459S–63S.
445. Gill SK, Rossi M, Bajka B, Whelan K. Dietary fibre in gastrointestinal health and disease. Nat Rev Gastroenterol Hepatol. 2021;18(2):101–16.
446. Müller M, Canfora EE, Blaak EE. Gastrointestinal transit time, glucose homeostasis and metabolic health: modulation by dietary fibers. Nutrients. 2018;10(3):275.
447. Sasaki D, Sasaki K, Ikuta N, Yasuda T, Fukuda I, Kondo A, Osawa R. Low amounts of dietary fibre increase in vitro production of short-chain fatty acids without changing human colonic microbiota structure. Sci Rep. 2018;8(1):435.
448. Holscher HD. Dietary fiber and prebiotics and the gastrointestinal microbiota. Gut Microbes. 2017;8(2):172–84.
449. Valdes AM, Walter J, Segal E, Spector TD. Role of the gut microbiota in nutrition and health. BMJ. 2018;361:k2179.
450. Belobrajdic DP, Bird AR, Conlon MA, Williams BA, Kang S, McSweeney S, Zhang D, Bryden WL, Gidley MJ, Topping DL. An arabinoxylan-rich fraction from wheat enhances caecal fermentation and protects colonocyte DNA against diet-induced damage in pigs. Br J Nutr. 2012;107(9):1274–82.
451. Khosroshahi HT, Abedi B, Ghojazadeh M, Samadi A, Jouyban A. Effects of fermentable high fiber diet supplementation on gut derived and conventional nitrogenous product in patients on maintenance hemodialysis: a randomized controlled trial. Nutr Metab (Lond). 2019;16:18.
452. Pieper R, Boudry C, Bindelle J, Vahjen W, Zentek J. Interaction between dietary protein content and the source of carbohydrates along the gastrointestinal tract of weaned piglets. Arch Anim Nutr. 2014;68(4):263–80.
453. Williams BA, Zhang D, Lisle AT, Mikkelsen D, McSweeney S, Kang S, Bryden WL, Gidley MJ. Soluble arabinoxylan enhances large intestinal microbial health biomarkers in pigs fed a red meat-containing diet. Nutrition. 2016;32(4):491–7.
454. Zhou XL, Kong XF, Lian GQ, Blachier F, Geng MM, Yin YL. Dietary supplementation with soybean oligosaccharides increases short-chain fatty acids but decreases protein-derived catabolites in the intestinal content of weaned Huanjiang mini-piglets. Nutr Res. 2014;34(9): 780–8.
455. Paone P, Cani PD. Mucus barrier, mucins and gut microbiota: the expected slimy partners? Gut. 2020;69(12):2232–43.
456. Martens EC, Neumann M, Desai MS. Interactions of commensal and pathogenic microorganisms with the intestinal mucosal barrier. Nat Rev Microbiol. 2018;16(8):457–70.
457. Johansson MEV, Hansson GC. Immunological aspects of intestinal mucus and mucins. Nat Rev Immunol. 2016;16(10):639–49.
458. Sonnenburg ED, Sonnenburg JL. Starving out microbial self: the deleterious consequences of a diet deficient in microbiota-accessible carbohydrates. Cell Metab. 2014;20(5):779–86.
459. Sonnenburg JL, Xu J, Leip DD, Chen CH, Westover BP, Weatherford J, Buhler JD, Gordon JI. Glycan foraging in vivo by an intestine-adapted bacterial symbiont. Science. 2005;307 (5717):1955–9.
460. Shon DJ, Kuo A, Ferracane MJ, Malaker SA. Classification, structural biology, and applications of mucin domain-targeting proteases. Biochem J. 2021;478(8):1585–603.
461. Celli JP, Turner BS, Afdhal NH, Keates S, Ghiran I, Kelly CP, Ewoldt RH, McKinley GH, So P, Erramilli S, Bansil R. Helicobacter pylori moves through mucus by reducing mucin viscoelasticity. Proc Natl Acad Sci U S A. 2009;106(34):14321–6.

462. Van der Lugt B, van Beek AA, Aalvink S, Meijer B, Sovran B, Vermeij WP, Brandt MRC, de Vos WM, Savelkoul HFJ, Steegenga WT, Belzer C. Akkermansia muciniphila ameliorates the age-related decline in colonic mucus thickness and attenuates immune activation in accelerated aging Ercc1−/Δ7 mice. Immun Ageing. 2019;16:6.

463. Morales P, Fujio S, Navarrete P, Ugalde JA, Magne F, Carrasco-Pozo C, Tralma K, Quezada M, Hurtado C, Covarrubias N, Brignardello J, Henriquez D, Gotteland M. Impact of dietary lipids on colonic function and microbiota: an experimental approach involving orlistat-induced fat malabsorption in human volunteers. Clin Transl Gastroenterol. 2016;7(4): e161.

464. Jaeger KE, Ransac S, Dijkstra BW, Colson C, van Heuvel C, Misset O. Bacterial lipases. FEMS Microbiol Rev. 1994;15(1):29–63.

465. Hofmann AF, Hagey LR. Key discoveries in bile acid chemistry and biology and their clinical applications: history of the last eight decades. J Lipid Res. 2014;55(8):1553–95.

466. Asare PT, Zurfluh K, Greppi A, Lynch D, Scwab C, Stephan R, Lacroix C. Reuterin demonstrates potent antimicrobial activity against a broad panel of human and poultry meat campylobacter spp. Isolates Microorganisms. 2020;8(1):78.

467. Engels C, Scwab C, Zhang J, Stevens MJA, Bieri C, Ebert MO, McNeill K, Sturla SJ, Lacroix C. Acrolein contributes strongly to antimicrobial and heterocyclic amine transformation activities of reuterin. Sci Rep. 2016;6:36246.

468. Liu F, Yu B. Efficient production of reuterin from glycerol by magnetically immobilized lactobacillus reuteri. Appl Microbiol Technol. 2015;99(11):4659–66.

469. Cleusix V, Lacroix C, Vollenweider S, Duboux M, Le Blay G. Inhibitory activity spectrum of reuterin produced by lactobacillus reuteri against intestinal bacteria. BMC Microbiol. 2007;7: 101.

470. Engevik MA, Danhof HA, Shrestha R, Chang-Graham AL, Hyser JM, Haag AM, Mohammad MA, Britton MA, Versalovic J, Sorg JA, Spinler JK. Reuterin disrupts Clostridioides difficile metabolism and pathogenicity through reactive oxygen species generation. Gut Microbes. 2020;12(1):1788898.

471. Chen WY, Wang M, Zhang J, Barve SS, McClain CJ, Joshi-Barve S. Acrolein disrupts tight junction proteins and causes endoplasmic reticulum stress-mediated epithelial cell death leading to intestinal barrier dysfunction and permeability. Am J Pathol. 2017;187(12): 2686–97.

472. Wilson Tang WH, Wang Z, Levison BS, Koeth RA, Britt EB, Fu X, Wu Y, Hazen SL. Intestinal microbial metabolism of phosphatidylcholine and cardiovascular risk. N Engl J Med. 2013;368(17):1575–84.

473. Rath S, Heidrich B, Pieper DH, Vital M. Uncovering the trimethylamine-producing bacteria of the human gut microbiota. Microbiome. 2017;5(1):54.

474. Romano KA, Vivas EI, Amador-Noguez D, Rey FE. Intestinal microbiota composition modulates choline bioavailability from diet and accumulation of the proatherogenic metabolite trimethylamine-N-oxide. MBio. 2015;6(2):e02481.

475. Zeisel SH, Warrier M. Trimethylamine N-oxide, the microbiome, and heart and kidney disease. Annu Rev Nutr. 2017;37:157–81.

476. Candido FG, Valente FX, Grzeskowiak LM, Boroni Moreira AP, Rocha DMUP, de Cassia Gonçalves Alfenas R. Impact of dietary fat on gut microbiota and low-grade systemic inflammation: mechanisms and clinical implications on obesity. Int J Food Sci Nutr. 2018;69(2):125–43.

477. Desbois AP, Smith VJ. Antibacterial free fatty acids: activities, mechanisms of action and biotechnological potential. Appl Microbiol Biotechnol. 2010;85(6):1629–42.

478. Hinzman MJ, Novotny C, Ullah A, Shamsuddin AM. Fecal mutagen fecapentaene-12 damages mammalian colon epithelial DNA. Carcinogenesis. 1987;8(10):1475–9.

479. Van Tassell RL, Kingston DG, Wilkins TD. Metabolism of dietary genotoxins by the human colonic microflora; the fecapentaenes and heterocyclic amines. Mutat Res. 1990;238(3): 209–21.

480. Tomas J, Mulet C, Saffarian A, Cavin JB, Ducroc R, Regnault B, Tan CK, Duska K, Burcelin R, Wahli W, Sansonetti PJ, Pédron T. High-fat diet modifies the PPAR-Υ pathway leading to disruption of microbial and physiological ecosystem in murine small intestine. Proc Natl Acad Sci U S A. 2016;113(40):E5934–43.
481. Kankaanpää PE, Salminen SJ, Isolauri E, Lee YK. The influence of polyunsaturated fatty acids on probiotic growth and adhesion. FEMS Microbiol Lett. 2001;194(2):149–53.
482. Araujo JR, Tomas J, Brenner C, Sansonetti PJ. Impact of high-fat diet on the intestinal microbiota and small intestinal physiology before and after the onset of obesity. Biochimie. 2017;141:97–106.
483. Cani PD, Bibiloni R, Knauf C, Waget A, Neyrinck AM, Delzenne NM, Burcelin R. Changes in gut microbiota control metabolic endotoxemia-induced inflammation in high-fat diet-induced obesity and diabetes in mice. Diabetes. 2008;57(6):1470–81.
484. Ding S, Chi MM, Scull BP, Rigby R, Schwerbrock NMJ, Magness S, Jobin C, Lund PK. High-fat diet: bacteria interactions promote intestinal inflammation which precedes and correlates with obesity and insulin resistance in mouse. PLoS One. 2010;5(8):e12191.
485. Kim KA, Gu W, Lee IA, Joh EH, Kim DH. High fat diet-induced gut microbiota exacerbates inflammation and obesity in mice via the TLR4 signaling pathway. PLoS One. 2012;7(10): e47713.
486. Rohr MW, Narasimhulu CA, Rudeski-Rohr TA, Parthasarathy S. Negative effects of a high-fat diet on intestinal permeability: a review. Adv Nutr. 2020;11(1):77–91.
487. Russell DW. The enzymes, regulation, and genetics of bile acid synthesis. Annu Rev Biochem. 2003;72:137–74.
488. Chiang JYL. Bile acid metabolism and signaling. Compr Physiol. 2013;3(3):1191–212.
489. Ticho AL, Malhotra P, Dudeja PK, Gill RK, Alrefai WA. Intestinal absorption of bile acids in health and disease. Compr Physiol. 2019;10(1):21–56.
490. Wahlström A, Sayin SI, Marschall HU, Bäcked F. Intestinal crosstalk between bile acids and microbiota and its impact on host metabolism. Cell Metab. 2016;24(1):41–50.
491. Connor WE, Witiak DT, Stone DB, Armstrong ML. Cholesterol balance and fecal neutral steroid and bile acid excretion in normal men fed dietary fats of different fatty acid composition. J Clin Invest. 1969;48(8):1363–75.
492. Stenman LK, Holma R, Korpela R. High-fat-induced intestinal permeability dysfunction associated with altered fecal bile acids. Word J Gastroenterol. 2012;18(9):923–9.
493. Ridlon JM, Kang DJ, Hylemon PB, Bajaj JS. Bile acids and the gut microbiome. Curr Opin Gastroenterol. 2014;30(3):332–8.
494. Setchell KD, Lawson AM, Tanida N, Sjövall J. General methods for the analysis of metabolic profiles of bile acids and related compounds in feces. J Lipid Res. 1983;24(8):1085–100.
495. Buffie CG, Bucci V, Stein RR, McKenney PT, Ling L, Gobourne A, No D, Liu H, Kinnebrew M, Viale A, Littmann E, van den Brink MRM, Jenq RR, Taur Y, Sander C, Cross JR, Toussaint NC, Xavier JB, Pamer EG. Precision microbiome reconstitution restores bile acid mediated resistance to clostridium difficile. Nature. 2015;517(7533):205–8.
496. Ridlon JM, Harris SC, Bhowmik S, Kang DJ, Hylemon PB. Consequences of bile salt biotransformations by intestinal bacteria. Gut Microbes. 2016;7(1):22–39.
497. Kuehne SA, Cartman ST, Heap JT, Kelly ML, Cockayne A, Minton NP. The role of toxin a and toxin B in Clostridium difficile infection. Nature. 2010;467(7316):711–3.
498. Rupnik M, Wilcox MH, Gerding DN. Clostridium difficile infection: new developments in epidemiology and pathogenesis. Nat Rev Microbiol. 2009;7(7):526–36.
499. Fry RJ, Staffeldt E. Effect of a diet containing sodium deoxycholate on the intestinal mucosa of the mouse. Nature. 1964;203:1396–8.
500. Leschelle X, Robert V, Delpal S, Mouillé B, Mayeur C, Martel P, Blachier F. Isolation of pig colonic crypts for cytotoxic assay of luminal compounds: effects of hydrogen sulfide, ammonia and deoxycholic acid. Cell Biol Toxicol. 2002;18(3):193–203.
501. Münch A, Ström M, Söderholm JD. Dihydroxy bile acids increase mucosal permeability and bacterial uptake in human colon biopsies. Scand J Gastroenterol. 2007;42(10):1167–74.

502. Stenman LK, Holma R, Eggert A, Korpela R. A novel mechanism for gut barrier dysfunction by dietary fat: epithelial disruption by hydrophobic bile acids. Am J Phys. 2013;304(3): G227–34.
503. Mars RAT, Yang Y, Ward T, Houtti M, Priya S, Lekatz HR, Tang X, Sun Z, Kalari KR, Korem T, Bhattarai Y, Zheng T, Bar N, Frost G, Johnson AJ, van Treuren W, Han S, Ordog T, Grover M, Sonnenburg J, D'Amato M, Camilleri M, Elinav E, Segal E, Blekhman R, Farrugia G, Swann JR, Knights D, Kashyap PC. Longitudinal multi-omics reveals subset-specific mechanisms underlying irritable bowel syndrome. Cell. 2020;182(6):1460–1473.e17.
504. Lee JS, Wang RX, Alexeev EE, Lanis JM, Battista KD, Glover LE, Colgan SP. Hypoxanthine is a checkpoint stress metabolite in colonic epithelial energy modulation and barrier function. J Biol Chem. 2018;293(16):6039–51.
505. Tsao R. Chemistry and biochemistry of dietary polyphenols. Nutrients. 2010;2(12):1231–46.
506. Hussain T, Tan B, Yin Y, Blachier F, Tossou MCB, Rahu N. Oxidative stress and inflammation: what polyphenols can do for us? Oxidative Med Cell Longev. 2016;2016:7432797.
507. Pérez-Jiménez J, Fezeu L, Touvier M, Arnault N, Manach C, Hercberg S, Galan P, Scalbert A. Dietary intake of 337 polyphenols in Franch adults. Am J Clin Nutr. 2011;93(6):1220–8.
508. Cladis DP, Weaver CM, Ferruzzi MG. (poly)phenol toxicity in vivo following oral administration: a targeted narrative review of (poly)phenols from green tea, grape, and anthocyanin-rich extracts. Phytother Res. 2022;36(1):323–35.
509. Guasch-Ferré M, Merino J, Sun Q, Fito M, Salas-Salvado J. Dietary polyphenols, Mediterranean diet, prediabetes, and type 2 diabetes: a narrative review of the evidence. Oxidative Med Cell Longev. 2017;2017:6723931.
510. Yahfoufi N, Alsadi N, Jambi M, Matar C. The immunomodulatory and anti-inflammatory role of polyphenols. Nutrients. 2018;10(11):1618.
511. Weaver SR, Rendeiro C, McGettrick HM, Philp A, Lucas SJE. Fine wine or sour grapes? A systematic review and meta-analysis of the impact of red wine polyphenols on vascular health. Eur J Nutr. 2021;60(1):1–28.
512. Rauf A, Imran M, Abu-Izneid T, Ul-Haq I, Patel S, Pan X, Naz S, Sanches Silva A, Saeed F, Rasul Suleria HA. Proanthocyanidins: a comprehensive review. Biomed Pharmacol. 2019;116:108999.
513. Rowland I, Gibson G, Heinken A, Scott K, Swann J, Thiele I, Tuohy K. Gut microbiota functions: metabolism of nutrients and other food components. Eur J Nutr. 2018;57(1):1–24.
514. Kawabata K, Yoshioka Y, Terao J. Role of intestinal microbiota in the bioavailability and physiological functions of dietary polyphenols. Molecules. 2019;24(2):370.
515. Murota K, Terao J. Antioxidative flavonoid quercetin: implication of its intestinal absorption and metabolism. Arch Biochem Biophys. 2003;417(1):12–7.
516. Manach C, Scalbert A, Morand C, Rémézy C, Jiménez L. Polyphenols: food sources and bioavailability. Am J Clin Nutr. 2004;79(5):727–47.
517. Cassidy A, Minihane AM. The role of metabolism (and the microbiome) in defining the clinical efficacy of dietary flavonoids. Am J Clin Nutr. 2017;105(1):10–22.
518. Clifford MN, van der Hooft JJJ, Crozier A. Human studies on the absorption, distribution, metabolism, and excretion of tea polyphenols. Am J Clin Nutr. 2013;98(6):1619S–30S.
519. Duda-Chodak A, Tarko T, Satora P, Sroka P. Interactions of dietary compounds, especially polyphenols, with the intestinal microbiota: a review. Eur J Nutr. 2015;54(3):325–41.
520. Russell WR, Scobbie L, Chesson A, Richardson AJ, Stewart CS, Duncan SH, Drew JE, Duthie GG. Anti-inflammatory implications of the microbial transformation of dietary phenolic compounds. Nutr Cancer. 2008;60(5):636–42.
521. Braune A, Engst W, Blaut M. Identification and functional expression of genes encoding flavonoid O- and C- glycosidases in intestinal bacteria. Environ Microbiol. 2016;18(7): 2117–29.
522. Aura AM, O'Leary KA, Williamson G, Ojala M, Bailey M, Puupponen-Pimiä R, Nuutila AM, Oksman-Caldentey KM, Poutanen K. Quercetin derivatives are deconjugated and converted to

hydroxyphenylacetic acids but not methylated by human fecal flora in vitro. J Agric Food Chem. 2002;50(6):1725–30.
523. Bode LM, Bunzel D, Huch M, Cho GS, Ruhland D, Bunzel M, Bub A, Franz CMAP, Kulling SE. In vivo and in vitro metabolism of trans-resveratrol by human gut microbiota. Am J Clin Nutr. 2013;97(2):295–309.
524. Braune A, Blaut M. Bacterial species involved in the conversion of dietary flavonoids in the human gut. Gut Microbes. 2016;7(3):216–34.
525. Hein EM, Rose K, van't Slot G, Friedrich AW, Humpf HU. Deconjugation and degradation of flavonol glycosides by pig cecal microbiota characterized by fluorescence in situ hybridization (FISH). J Agric Food Chem. 2008;56(6):2281–90.
526. Jaganath IB, Mullen W, Lean MEJ, Edwards CA, Crozier A. In vitro catabolism of rutin by human fecal bacteria and the antioxidant capacity of its catabolites. Free Radic Biol Med. 2009;47(8):1180–9.
527. Jiménez-Giron A, Ibanez C, Cifuentes A, Simo C, Munoz-Gonzalez I, Martin-Alvarez PJ, Bartolome B, Moreno-Arribas MV. Faecal metabolomic fingerprint after moderate consumption of red wine by healthy subjects. J Proteome Res. 2015;14(2):897–905.
528. Peng X, Zhang Z, Zhang N, Liu L, Li S, Wei H. In vitro catabolism of quercitin by human fecal bacteria and the antioxidant capacity of its catabolites. Food Nutr Res. 2014;58:23406. https://doi.org/10.3402/fnr.v58.23406.
529. Rechner AR, Smith MA, Kuhnle G, Gibson GR, Debnam ES, Srai SKS, Moore KP, Rice-Evans CA. Colonic metabolism of dietary polyphenols: influence of structure on microbial fermentation products. Free Radic Biol Med. 2004;36(2):212–25.
530. Schneider H, Simmering R, Hartmann L, Pforte H, Blaut M. Degradation of quercetin-3-glucoside in gnotobiotic rats associated with human intestinal bacteria. J Appl Microbiol. 2000;89(6):1027–37.
531. Gill CIR, McDougall GJ, Glidewell S, Stewart D, Shen Q, Tuohy K, Dobbin A, Boyd A, Brown E, Haldar S, Rowland IR. Profiling of phenols in human fecal water after raspberry supplementation. J Agric Food Chem. 2010;58(19):10389–95.
532. Quartieri A, Garcia-Villalba R, Amaretti A, Raimondi S, Leonardi A, Rossi M, Tomàs-Barberàn F. Detection of novel metabolites of flaxseed lignans in vitro and in vivo. Mol Nutr Food Res. 2016;60(7):1590–601.
533. Tomas-Barberan F, Garcia-Villalba R, Quartieri A, Raimondi S, Amaretti A, Leonardi A, Rossi M. In vitro transformation of chlorogenic acid by human gut microbiota. Mol Nutr Food Res. 2014;58(5):1122–31.
534. Fraga CG, Croft KD, Kennedy DO, Tomàs-Barberàn FA. The effects of polyphenols and other bioactives on human health. Food Funct. 2019;10(2):514–28.
535. Setchell KDR, Clerici C, Lephart ED, Cole SJ, Heenan C, Castellani D, Wolfe BE, Nechemias-Zimmer L, Brown ND, Lund TD, Handa RJ, Heubi JE. S-equol, a potent ligand for estrogen receptor beta, is the exclusive enantiomeric form of the soy isoflavone metabolite produced by human intestinal bacterial flora. Am J Clin Nutr. 2005;81(5):1072–9.
536. Tang Y, Nakashima S, Saiki S, Myoi Y, Abe N, Kuwazuru S, Zhu B, Ashida H, Murata Y, Nakamura Y. 3,4-dihydroxyphenylacetic acid is a predominant biologically-active catabolite of quercetin glycosides. Food Res Int. 2016;89(1):716–23.
537. Monogas M, Khan N, Andrés-Lacueva C, Urpi-Sardà M, Vàsquez-Agell M, Lamuela-Raventos RM, Estruch R. Dihydroxylated phenolic acids derived from microbial metabolism reduce lipopolysaccharide-stimulated cytokine secretion by human peripheral blood mononuclear cells. Br J Nutr. 2009;102(2):201–6.
538. Crespo I, San-Miguel B, Mauriz JL, Ortiz de Urbina JJ, Almar M, Tunon MJ, Gonzàlez-Gallego J. Protective effect of protocatechuic acid on TNBS-induced colitis in mice is associated with modulation of the SphK/S1P signaling pathway. Nutrients. 2017;9(3):288.
539. Hu R, He Z, Liu M, Tan J, Zhang H, Hou DX, He J, Wu S. Dietary protocatechuic acid ameliorates inflammation and up-regulates intestinal tight junction proteins by modulating gut microbiota in LPS-challenged piglets. J Anim Sci Biotechnol. 2020;11:92.

540. Farombi EO, Adedara IA, Awoyemi OV, Njoku CR, Micah GO, Esogwa CU, Owumi SE, Olopade JO. Dietary protocatechuic acid ameliorates dextran sulphate sodium-induced ulcerative colitis and hepatotoxicity in rats. Food Funct. 2016;7(2):913–21.

541. Boughton-Smith NK, Evans SM, Hawkey CJ, Cole AT, Balsitis M, Whittle BJ, Moncada S. Nitric oxide synthase activity in ulcerative colitis and Crohn's disease. Lancet. 1993;342 (8867):338–40.

542. Guihot G, Guimbaud R, Bertrand V, Narcy-Lambare B, Couturier D, Duée PH, Chaussade S, Blachier F. Inducible nitric oxide synthase activity in colon biopsies from inflammatory areas: correlation with inflammation intensity in patients with ulcerative colitis but not with Crohn's disease. Amino Acids. 2000;18(3):229–37.

543. Singer II, Kawka DW, Scott S, Weidner JR, Mumford RA, Riehl TE, Stenson WF. Expression of inducible nitric oxide synthase and nitrotyrosine in colonic epithelium in inflammatory bowel disease. Gastroenterology. 1996;111(4):871–85.

544. Beckman JS, Koppenol WH. Nitric oxide, superoxide, and peroxynitrite: the good, the bad, and ugly. Am J Phys. 1996;271(5):C1424–37.

545. Jenner AM, Rafter J, Halliwell B. Human fecal water content of phenolics: the extent of colonic exposure to aromatic compounds. Free Radic Biol Med. 2005;38(6):763–72.

546. Russell DW, Duncan SH, Scobbie L, Duncan G, Cantlay L, Graham Calder A, Anderson SE, Flint HJ. Major phenylpropanoid-derived metabolites in the human gut can arise from microbial fermentation of protein. Mol Nutr Food Res. 2013;57(3):523–35.

547. Liu Y, Shi C, Zhang G, Zhan H, Liu B, Li C, Wang L, Wang H, Wang J. Antimicrobial mechanism of 4-hydroxyphenylacetic acid on listeria monocytogenes membrane and virulence. Biochem Biophys Res Commun. 2021;572:145–50.

548. Cires MJ, Navarrete P, Pastene E, Carrasco-Pozo C, Valenzuela R, Medina DA, Andriamihaja M, Beaumont M, Blachier F, Gotteland M. Protective effect of avocado peel polyphenolic compounds rich in proanthocyanidins on the alterations of colonic homeostasis induced by a high-protein diet. J Agric Food Chem. 2019a;67(42):11616–26.

549. Cires MJ, Navarrete P, Pastene E, Carrasco-Pozo C, Valenzuela R, Medina DA, Andriamihaja M, Beaumont M, Blachier F, Gotteland M. Effects of proanthocyanidin-rich polyphenol extract from avocado on the production of amino acid-derived bacterial metabolites and the microbiota composition in rats fed a high-protein diet. Food Funct. 2019b;10(7):4022–35.

550. Costongs GM, Bos LP, Engels LG, Janson PC. A new method for chemical analysis of feces. Clin Chim Acta. 1985;150(3):197–203.

551. Schilli R, Breuer RI, Klein F, Dunn K, Gnaedinger A, Bernstein J, Paige M, Kaufman M. Comparison of the composition of faecal fluid in Crohn's disease and ulcerative colitis. Gut. 1982;23(4):326–32.

552. Madara JL. Increases in Guinea pig small intestinal transepithelial resistance induced by osmotic loads are accompanied by rapid alterations in absorptive-cell tight-junction structure. J Cell Biol. 1983;97(1):125–36.

553. Samak G, Suzuki T, Bhargava A, Rao RK. C-Jun NH2-terminal kinase-2 mediates osmotic stress-induced tight junction disruption in the intestinal epithelium. Am J Phys. 2010;229(3): G572–284.

554. Hubert A, Cauliez B, Chedeville A, Husson A, Lavoinne A. Osmotic stress, a proinflammatory signal in Caco-2 cells. Biochimie. 2004;86(8):533–41.

555. Schwartz L, Guais A, Pooya M, Abolhassani M. Is inflammation a consequence of extracellular hyperosmolarity? J Inflamma (Lond). 2009;6:21.

556. Grauso M, Lan A, Andriamihaja M, Bouillaud F, Blachier F. Hyperosmolar environment and intestinal epithelial cells: impact on mitochondrial oxygen consumption, proliferation, and barrier function in vitro. Sci Rep. 2019;9(1):11360.

557. McDougall CJ, Wong R, Scudera P, Lesser M, DeCosse JJ. Colonic mucosal pH in humans. Dig Dis Sci. 1993;38(3):542–5.

558. Reiffenstein RJ, Hulbert WC, Roth SH. Toxicology of hydrogen sulfide. Annu Rev Toxicol. 1992;32:109–34.
559. Cuff MA, Shirazi-Beechey SP. The importance of butyrate transport to the regulation of gene expression in the colonic epithelium. Biochem Soc Trans. 2004;32(6):1100–2.
560. Cohen RM, Stephenson RL, Feldman GM. Bicarbonate secretion modulates ammonium absorption in rat distal colon in vivo. Am J Phys. 1988;245(5):F657–67.
561. Christl SU, Bartram HP, Paul A, Kelber E, Scheppach W, Kasper H. Bile acid metabolism by colonic bacteria in continuous culture: effects of starch and pH. Ann Nutr Metab. 1997;41(1): 45–51.
562. De Preter V, Hamer HM, Windey K, Verbeke K. The impact of pre- and/or probiotics on human colonic metabolism: does it affect human health? Mol Nutr Food Res. 2011;55(1): 46–57.
563. Mowat AM, Agace WW. Regional specialization within the intestinal immune system. Nat Rev Immunol. 2014;14(10):667–85.
564. Mu C, Yang Y, Zhu W. Crosstalk between the immune receptors and gut microbiota. Curr Protein Pept Sci. 2015;16(7):622–31.
565. Zhou B, Yuan Y, Zhang S, Guo C, Li X, Li G, Xiong W, Zeng Z. Intestinal flora and disease mutually shape the regional immune system in the intestinal tract. Front Immunol. 2020;11: 575.
566. Peterson LW, Artis D. Intestinal epithelial cells: regulators of barrier function and immune homeostasis. Nat Rev Immunol. 2014;14(3):141–53.
567. Hooper LV, Littman DR, Macpherson AJ. Interactions between the microbiota and the immune system. Science. 2012;336(6086):1268–73.
568. Gonçalves P, Araujo JR, Di Santo JP. A crosstalk between microbiota-derived short-chain fatty acids and the host mucosal immune system regulates intestinal homeostasis and inflammatory bowel disease. Inflamm Bowel Dis. 2018;24(3):558–72.
569. Kelsall BL, Strober W. Distinct populations of dendritic cells are present in the subepithelial dome and T cell regions of the murine Peyer's patch. J Exp Med. 1996;183(1):237–47.
570. Reboldi A, Cyster JG. Peyer's patches: organizing B-cell responses at the intestinal frontier. Immunol Rev. 2016;271(1):230–45.
571. Cheroutre H, Madakamutil L. Acquired and natural memory T cells join forces at the mucosal front line. Nat Rev Immunol. 2004;4(4):290–300.
572. Olivares-Villagomez D, van Kaer L. Intestinal intraepithelial lymphocytes: sentinels of the mucosal barrier. Trends Immunol. 2018;39(4):264–75.
573. Stagg AJ. Intestinal dendritic cells in health and gut inflammation. Front Immunol. 2018;9: 2883.
574. Yadav M, Stephan S, Bluestone JA. Peripherally induced tregs: role in immune homeostasis and autoimmunity. Front Immunol. 2013;4:232.
575. Reséndiz-Albor AA, Esquivel R, Lopez-Revilla R, Verdin L, Moreno-Fierros L. Striking phenotypic and functional differences in lamina propria lymphocytes from the large and small intestine of mice. Life Sci. 2005;76(24):2783–803.
576. Gaboriau-Routhiau V, Rakotobe S, Lécuyer S, Mulder I, Lan A, Bridonneau C, Rochet V, Pisi A, De Paepe M, Brandi G, Eberl G, Snel J, Kelly D, Cerf-Bensussan N. The key role of segmented filamentous bacteria in the coordinated maturation of gut helper T cell responses. Immunity. 2009;31(4):677–89.
577. Colgan SP, Hershberg RM, Furuta GT, Blumberg RS. Ligation of intestinal epithelial CD1d induces bioactive IL-10: critical role of the cytoplasmic tail in autocrine signaling. Proc Natl Acad Sci USA. 1999;96(24):13938–43.
578. Iliev ID, Mileti E, Matteoli G, Chieppa M, Rescigno M. Intestinal epithelial cells promote colitis-protective regulatory T-cell differentiation through dendritic cell conditioning. Mucosal Immunol. 2009a;2(4):340–50.

579. Iliev ID, Spadoni I, Mileti E, Matteoli G, Sonzogni A, Sampietro GM, Foschi D, Caprioli F, Viale G, Rescigno M. Human intestinal epithelial cells promote the differentiation of tolerogenic dendritic cells. Gut. 2009b;58(11):1481–9.
580. Maynard CL, Elson CO, Hatton RD, Weaver CT. Reciprocal interactions of the intestinal microbiota and immune system. Nature. 2012;489(7415):231–41.
581. Thaiss CA, Zmora N, Levy M, Elinav E. The microbiome and innate immunity. Nature. 2016;535(7610):65–74.
582. Kayama H, Okumura R, Takeda K. Interaction between the microbiota, epithelia, and immune cells in the intestine. Annu Rev Immunol. 2020;38:23–48.
583. Rooks MG, Garrett WS. Gut microbiota, metabolites and host immunity. Nat Rev Immunol. 2016;16(6):341–52.
584. Chang PV, Hao L, Offermanns S, Medzhitov R. The microbial metabolite butyrate regulates intestinal macrophage function via histone deacetylase inhibition. Proc Natl Acad Sci U S A. 2014;111(6):2247–52.
585. Schulthess J, Pandey S, Capitani M, Rue-Albrecht KC, Arnold I, Franchini F, Chomka A, Ilott NE, Johnston DGW, Pires E, McCullagh J, Sansom SN, Arancibia-Carmano CV, Uhlig HH, Powrie F. The short-chain fatty acid butyrate imprints an antimicrobial program in macrophages. Immunity. 2019;50(2):432–45.
586. Yang W, Yu T, Huang X, Bilotta AJ, Xu L, Lu Y, Sun J, Pan F, Zhou J, Zhang W, Yao S, Maynard CL, Singh N, Dann SM, Liu Z, Cong Y. Intestinal microbiota-derived short-chain fatty acids regulation of immune cell IL-22 production and gut immunity. Nat Commun. 2020;11(1):4457.
587. Park J, Kim M, Kang SG, Jannasch AH, Cooper B, Patterson J, Kim CH. Short-chain fatty acids induce both effector and regulatory T cells by suppression of histone deacetylases and regulation of the mTOR-S6K pathway. Mucosal Immunol. 2015;8(1):80–93.
588. Kim M, Qie Y, Park J, Kim CH. Gut microbial metabolites fuel host antibody responses. Cell Host Microbe. 2016;20(2):202–14.
589. Vinolo MAR, Rodrigues HG, Hatanaka E, Sato FT, Sampaio SC, Curi R. Suppressive effect of short-chain fatty acids on production of proinflammatory mediators by neutrophils. J Nutr Biochem. 2011;22(9):849–55.
590. Green BT, Brown BR. Interactions between bacteria and the gut mucosa: do enteric neurotransmitters acting on the mucosal epithelium influence intestinal colonization or infection? Adv Exp Med Biol. 2016;874:121–41.
591. Thomas CM, Hong T, van Pijkeren JP, Hemarajata P, Trinh DV, Hu W, Britton RA, Kalkum M, Versalovic J. Histamine derived from probiotic lactobacillus reuteri suppresses TNF via modulation of PKA and ERK signaling. PLoS One. 2012;7(2):e31951.
592. Zhang M, Wang H, Tracey KJ. Regulation of macrophage activation and inflammation by spermine: a new chapter in an old story. Crit Care Med. 2000;28(4):N60–6.
593. Campbell C, McKenney PT, Konstantinovsky D, Isaeva OI, Schizas M, Verter J, Mai C, Jin WB, Guo CJ, Violante S, Ramos RJ, Cross JR, Kadaveru K, Hambor J, Rudensky AY. Bacterial metabolism of bile acids promotes generation of peripheral regulatory T cells. Nature. 2020;581(7809):475–9.
594. Celis AI, Relman DA. Competitors versus collaborators: micronutrient processing by pathogenic and commensal human-associated gut bacteria. Mol Cell. 2020;78(4):570–6.
595. Bielik V, Kolisek M. Bioaccessibility and bioavailability of minerals in relation to a healthy gut microbiome. Int J Mol Sci. 2021;22(13):6803.
596. Bronner F, Pansu D. Nutritional aspects of calcium absorption. J Nutr. 1999;129(1):9–12.
597. Trinidad TP, Wolever TM, Thompson LU. Effect of acetate and propionate on calcium absorption from the rectum and distal colon of humans. Am J Clin Nutr. 1996;63(4):574–8.
598. Whang R, Hampton EM, Whang DD. Magnesium homeostasis and clinical disorders of magnesium deficiency. Ann Pharmacother. 1994;28(2):220–6.
599. Schuchardt J, Hahn A. Intestinal absorption and factors influencing bioavailability of magnesium. An update Curr Nutr Food Sci. 2017;13(4):260–78.

600. Yamasaki D, Funato Y, Miura J, Sato S, Toyosawa S, Furutani K, Kurachi Y, Omori Y, Furakawa T, Tsuda T, Kuwabata S, Mizukami S, Kikuchi K, Miki H. Bacterial Mg2+ extrusion via CNNM4 mediates transcellular Mg2+ transport across epithelia: a mouse model. PLoS Genet. 2013;9(12):e1003983.

601. Pyndt Jorgensen B, Winther G, Kihl P, Nielsen DS, Wegener G, Hansen AK, Sorensen DB. Dietary magnesium deficiency affects gut microbiota and anxiety-like behavior in C57BL/6N mice. Acta Neuropsychiatr. 2015;27(5):307–11.

602. Gommers LMM, Ederveen THA, van der Wijst J, Overmas-Bos C, Kortman GAM, Boekhorst J, Bindels RJM, de Baaij JHF, Hoenderop JGJ. Low gut microbiota diversity and dietary magnesium intake are associated with the development of PPI-induced hypomagnesemia. FASEB J. 2019;33(10):11235–46.

603. Anderson GJ, Frazer DM, McKie AT, Vulpe CD, Smith A. Mechanisms of haem and non-haem iron absorption: lessons from inherited disorders of iron metabolism. Biometals. 2005;18(4):339–48.

604. Fuqua BK, Vulpe CD, Anderson GJ. Intestinal iron absorption. J Trac Elem Med Biol. 2012;26(2–3):115–9.

605. Gulec S, Anderson GJ, Collins JF. Mechanistic and regulatory aspects of intestinal iron absorption. Am J Phys. 2014;307(4):G397–409.

606. Blachier F, Vaugelade P, Robert V, Kibangou B, Canonne-Hergaux F, Delpal S, Bureau F, Blottière H, Boglé D. Comparative capacities of the pig colon and duodenum for luminal iron absorption. Can J Physiol Pharmacol. 2007b;85(2):185–92.

607. Cappellini MD, Musallam KM, Taher AT. Iron deficiency anaemia revisited. J Intern Med. 2020;287(2):153–70.

608. Jaeggi T, Kortman GA, Moretti D, Chassard D, Holding P, Dostal A, Boekhorst J, Timmerman HM, Swinkels DW, Tjalsma H, Njenga J, Mwangi A, Kvalsvig S, Lacroix C, Zimmermann MB. Iron fortification adversely affects the gut microbiome, increases pathogen abundance and induces intestinal inflammation in Kenyan infants. Gut. 2015;64(5):731–42.

609. Paganini D, Zimmermann MB. The effects of iron fortification and supplementation on the gut microbiome and diarrhea in infants and children. Am J Clin Nutr. 2017;106(6):1688S–93S.

610. Salovaara S, Standberg AS, Andlid T. Combined impact of pH and organic acids on iron uptake by Caco-2 cells. J Agric Food Chem. 2003;51(26):7820–4.

611. Yilmaz B, Li H. Gut microbiota and iron: the crucial actors in health and disease. Pharmaceuticals (Basel). 2018;11(4):98.

612. Bouglé D, Vaghefi-Vaezzadeh N, Roland N, Bouvard G, Arhan P, Bureau F, Neuville D, Maubois JL. Influence of short-chain fatty acids on iron absorption by proximal colon. Scand J Gastroenterol. 2002;37(9):1008–11.

613. Chae TU, Kim WJ, Choi S, Park SJ, Lee SY. Metabolic engineering of Escherichia coli for the production of 1,3-diaminopropane, a three-carbon diamine. Sci Rep. 2015;5:13040.

614. Das NK, Schwartz AJ, Barthel G, Inohara N, Liu Q, Sankar A, Hill DR, Ma X, Lamberg O, Schnizlein MK, Arqués JL, Spence JR, Nunez G, Patterson AD, Sun D, Young VB, Shah YM. Microbial metabolite signaling is required for systemic iron homeostasis. Cell Metab. 2020;31(1):115–130.e6.

615. Krall RF, Tzounopoulos T, Aizenman E. The function and regulation of zinc in the brain. Neuroscience. 2021;457:235–58.

616. Prasad AS. Clinical, endocrinological and biochemical effects of zinc deficiency. Clin Endocrinol Metab. 1985;14(3):567–89.

617. Read SA, Obeid S, Ahlenstiel G, Ahlenstiel G. The role of zinc in antiviral immunity. Adv Nutr. 2019;10(4):696–710.

618. Smith JC Jr, McDaniel EG, McBean LD, Doft FS, Halsted JA. Effect of microorganisms upon zinc metabolism using germ-free and conventional rats. J Nutr. 1972;102(6):711–9.

619. Sauer AK, Grabrucker AM. Zinc deficiency during pregnancy leads to altered microbiome and elevated inflammatory markers in mice. Front Neurosci. 2019;13:1295.

620. Zackular JP, Skaar EP. The role of zinc and nutritional immunity in Clostridium difficile infection. Gut Microbes. 2018;9(5):469–76.
621. Guetterman HM, Huey SL, Knight R, Fox AM, Mehta S, Finkelstein JL. Vitamin B-12 and the gastrointestinal microbiome: a systematic review. Adv Nutr. 2021;13(2):530–58.
622. Rossi M, Amaretti A, Raimondi S. Folate production by probiotic bacteria. Nutrients. 2011;3 (1):118–34.
623. Suttie JW. The importance of menaquinones in human nutrition. Annu Rev Nutr. 1995;15: 399–417.
624. Frick PG, Riedler G, Brögli H. Dose response and minimal daily requirement for vitamin K in man. J Appl Physiol. 1967;23(3):387–9.
625. Magnusdottir S, Ravcheev D, de Crécy-Lagard V, Thiele I. Systemic genome assessment of B-vitamin biosynthesis suggests co-operation among gut microbes. Front Genet. 2015;6:148.
626. Radionov DA, Arzamasov AA, Khoroshkin MS, Iablokov SN, Leyn SA, Peterson SN, Novichkov PS, Osterman AL. Micronutrient requirements and shaping capabilities of the human gut microbiome. Front Microbiol. 2019;10:1316.
627. Sharma V, Rodionov DA, Leyn SA, Tran D, Iablokov SN, Ding H, Peterson DA, Osterman AL, Peterson SN. B-vitamin sharing promotes stability of gut microbial communities. Front Microbiol. 2019;10:1485.
628. Jaehme M, Slotboom DJ. Diversity of membrane transport proteins for vitamins in bacteria and archaea. Biochim Biophys Acta. 2015;1850(3):565–76.
629. Putnam EE, Goodman AL. B vitamin acquisition by gut commensal bacteria. PLoS Pathog. 2020;16(1):e1008208.
630. Said HM. Recent advances in transport of water-soluble vitamins in organs of the digestive system: a focus on the colon and the pancreas. Am J Phys. 2013;305(9):G601–10.
631. Tardy AL, Pouteau E, Marquez D, Yilmaz C, Scholey A. Vitamins and minerals for energy, fatigue and cognition: a narrative review of the biochemical and clinical evidence. Nutrients. 2020;12(1):228.
632. Blachier F, Andriamihaja M, Blais A. Sulfur-containing amino acids and lipid metabolism. J Nutr. 2020;150(1):2524S–31S.
633. Clarke G, Sandhu KV, Griffin BT, Dinan TG, Cryan JF, Hyland NP. Gut reactions: breaking down xenobiotic-microbiome interactions. Pharmacol Rev. 2019;71(2):198–224.
634. Croom E. Metabolism of xenobiotics of human environments. Prog Mol Biol Transl Sci. 2012;112:31–88.
635. Koppel N, Maini Rekdal V, Balskus EP. Chemical transformation of xenobiotics by the human gut microbiota. Science. 2017;356(6344):eaag2770.
636. Dietert R, Silbergeld EK. Biomarkers for the 21st century: listening to the microbiome. Toxicol Sci. 2015;144(2):208–16.
637. Jin Y, Wu S, Zeng Z, Fu Z. Effects of environmental pollutants on gut microbiota. Environ Pollut. 2017;222:1–9.
638. Calatayud-Arroyo M, Garcia Barrera T, Callejon Leblic T, Arias Borrego A, Collado MC. A review of the impact of xenobiotics from dietary sources on infant health: early life exposures and the role of the microbiota. Environ Pollut. 2021;269:115994.
639. Krishnan R, Wilkinson I, Joyce L, Rofe AM, Bais R, Conyers RA, Edwards JB. The effect of dietary xylitol on the ability of rat caecal flora to metabolize xylitol. Aust J Exp Biol Med Sci. 1980;58(6):639–52.
640. Renwick AG, Tarka SM. Microbial hydrolysis of steviol glycosides. Food Chem Toxicol. 2008;46(7):S70–4.
641. Renwick AG. The metabolism of intense sweeteners. Xenobiotica. 1986;16(10–11):1057–71.
642. Franco R, Navarro G, Martinez-Pinilla E. Antioxidants versus food antioxidant additives and food preservatives. Antioxidants (Basel). 2019;8(11):542.
643. Vally H, Misso NLA, Madan V. Clinical effects of sulphite additives. Clin Exp Allergy. 2009;39(11):1643–51.

644. Wever J. Appearance of sulphite and S-sulphonates in the plasma of rats after intraduodenal sulphite application. Food Chem Toxicol. 1985;23(10):895–8.
645. Blachier F, Andriamihaja M, Larraufie P, Ahn E, Lan A, Kim E. Production of hydrogen sulfide by the intestinal microbiota and epithelial cells and consequences for the colonic and rectal mucosa. Am J Phys. 2021b;320(2):G125–35.
646. Bedale W, Sindelar JJ, Milkowski AL. Dietary nitrate and nitrite: benefits, risks, and evolving perceptions. Meat Sci. 2016;120:85–92.
647. Smith TJ, Hill KK, Raphael BH. Historical and current perspectives on clostridium botulinum diversity. Res Microbiol. 2015;166(4):290–302.
648. Engemann A, Focke C, Humpf HU. Intestinal formation of N-nitroso compounds in the pig cecum model. J Agric Food Chem. 2013;61(4):998–1005.
649. Carlström M, Moretti CH, Weitzberg E, Lundberg JO. Microbiota, diet and the generation of reactive nitrogen compounds. Free Radic Biol Med. 2020;161:321–5.
650. Barnes JL, Zubair M, John K, Poirier MC, Martin FL. Carcinogens and DNA damage. Biochem Soc Trans. 2018;46(5):1213–24.
651. Duncan SH, Iyer A, Russell WR. Impact of protein on the composition and metabolism of the human gut microbiota and health. Proc Nutr Soc. 2021;80(2):173–85.
652. Thresher A, Foster R, Ponting DJ, Stalford SA, Tennant RE, Thomas R. Are all nitrosamines concerning? A review of mutagenicity and carcinogenicity data. Regul Toxicol Pharmacol. 2020;116:104749.
653. De Pilli T, Alessandrino O. Effects of different food technologies on biopolymers modifications of cereal-based foods: impact of nutritional and quality characteristics review. Crit Rev Food Sci Nutr. 2020;60(4):556–65.
654. Inam-Eroglu E, Ayaz A, Buyuktuncer Z. Formation of advanced glycation endproducts in foods during cooking process and underlying mechanisms: a comprehensive review of experimental studies. Nutr Res Rev. 2020;33(1):77–89.
655. Uribarri J, del Castillo MD, de la Maza MP, Filip R, Gugliucci A, Luevano-Contreras C, Macias-Cervantes MH, Markowicz Bartos D, Medrano A, Menini T, Portero-Otin M, Rojas A, Rodrigues Sampaio G, Wrobel K, Wrobel K, Garay-Sevilla ME. Dietary advanced glycation end products and their role in health and disease. Adv Nutr. 2015;6(4):461–73.
656. Kim E, Coelho D, Blachier F. Review of the association between meat consumption and risk of colorectal cancer. Nutr Res. 2013;33(12):983–94.
657. Kassie F, Rabot S, Kundi M, Chabikovsky M, Qin HM, Knasmüller S. Intestinal microflora plays a crucial role in the genotoxicity of the cooked food mutagen 2-amino-3-methylimidazo (4,5-f) quinoline. Carcinogenesis. 2001;22(10):1721–5.
658. Humblot C, Murkovic M, Rigottier-Gois L, Bensaada M, Bouclet A, Andrieux C, Anba J, Rabot S. Beta-glucuronidase in human intestinal microbiota is necessary for the colonic genotoxicity of the food-borne carcinogen 2-amino-3-methylimidazo (4,5-f) quinoline in rats. Carcinogenesis. 2007;28(11):2419–25.

Alimentation, Bacterial Metabolites, and Host Intestinal Epithelium

4

Abstract

Inflammatory bowel diseases are characterized by chronic inflammation of the intestinal mucosa with alternating relapse and remission episodes. During remission, the inflamed mucosa has the possibility to heal progressively, leading to the partial or complete disappearance of the mucosal damages. Regarding the process of mucosal inflammation, it has been shown in pediatric Crohn's disease patients that diminished capacity for hydrogen sulfide (H_2S) disposal amplifies the pro-inflammatory effect of this bacterial metabolite on the intestinal mucosa when overproduced by the intestinal microbiota, while indole and indole-related compounds exert beneficial immune-regulatory effects in experimental studies. Although butyrate exerts several regulatory effects on the intestinal immune system, results from clinical trials with butyrate fail to support butyrate enemas as an effective treatment for the treatment of ulcerative colitis. Regarding the effects of bacterial metabolites on mucosal healing, some experimental studies indicate beneficial effects of polyamines for mucosal regeneration. During the process of colorectal carcinogenesis, an increased supply of H_2S within adenomatous colonocytes increases their proliferative capacity and their ability to promote tumor formation, suggesting that overproduction of this bacterial metabolite is implicated in colorectal cancer occurrence. H_2S overproduction may also indirectly alter DNA integrity by destabilizing the protective mucous layer, then increasing accessibility of genotoxic compounds to colonic epithelial cells. Spermine is another bacterial metabolite that indirectly alters DNA integrity by increasing reactive oxygen species production in colonocytes when present in excess in the luminal fluid. Other bacterial metabolites like *p*-cresol, fecapentaenes, acetaldehyde, 4-hydroxyphenylacetate, and some specific N-nitrosamines have been shown to alter DNA integrity at excessive concentrations.

The physiological functions of the large intestine mucosa and epithelium can be altered in different pathological situations, including notably inflammatory bowel diseases and colorectal cancer. In these pathologies, apart from the genetic background of the individuals which likely plays a central role in the etiology of these diseases, there are emerging reasons to implicate some of the bacterial metabolites produced from different substrates, including those from alimentary origin, in the development of these diseases. As we will see in this chapter, some of the bacterial metabolites, depending notably on their luminal concentrations, appear to play either beneficial or detrimental roles in the processes involved in the transition from a healthy colorectal mucosa into an inflammatory or cancerous mucosa.

In addition, some among the numerous metabolites produced by the intestinal microbiota have been identified as playing potential roles in the diarrheal processes, notably in case of chronic inflammatory bowel disease.

4.1   Dietary Compounds, Bacterial Metabolites, and Mucosal Inflammation and Subsequent Healing

Inflammation of the Intestinal Mucosa Results Notably from an Inappropriate Immune Response

Inflammatory bowel diseases are characterized by alterations of the intestinal mucosa associated with exacerbated immune functions [1]. Inflammatory bowel diseases, which include ulcerative colitis and Crohn's disease, are characterized by chronic inflammation of the intestinal mucosa resulting from inappropriate mucosal immune responses against luminal components in genetically predisposed individuals, with alternating relapse and remission episodes [2, 3]. Although Crohn's disease and ulcerative colitis have different clinical features [4], the inflammation of the intestinal mucosa is often observed in the distal parts of the intestine for both diseases [5]. The etiology of inflammatory bowel diseases, although still largely elusive, is related to both genetic [6] and environmental factors, including the ones from dietary origin [7]. Although patients suffering from inflammatory bowel disease often ask clinicians for the optimal dietary practices to reduce relapse risk between episodes of inflammatory flare and to correct for nutritional deficiencies, the optimal dietary composition in terms of macronutrients, vitamins, minerals, and micronutrients remain overall largely unknown, notably because extent, duration, severity of inflammation, and disease history are different between patients [8].

Chronic inflammatory bowel diseases often require long-term treatment based on a combination of drugs such as corticosteroids and immunosuppressive compounds to control the disease by reducing the inflammatory process [9–11].

Mucosal Healing After an Episode of Inflammatory Flare Is Associated with Clinical Remission

In Crohn's disease remission, the inflamed mucosa has the possibility to heal progressively, leading to the total disappearance of all mucosal ulcerations; although in clinical practice, this endpoint is difficult to achieve [12–15]. For patients with ulcerative colitis, mucosal healing is described by the absence of friability, blood, erosions, and ulcers in all visualized segments of the gut mucosa [16, 17]. According to available evidence, it seems that an advanced mucosal healing coincides with a sustained clinical remission and improved clinical outcomes, as well as reduced rates of hospitalization and surgical resection. This explains why mucosal healing can be considered a full therapeutic goal and an endpoint for clinical trials for both Crohn's disease and ulcerative colitis [18–24], even if mucosal healing by itself is obviously not sufficient to cure inflammatory bowel diseases [25].

In these diseases, due to a loss of epithelial cells, the intestinal barrier is disrupted, leading to the contact of the laminal propria compartment with compounds and microorganisms present in the intestinal lumen. Normal human colon mucosa is equipped with an inner mucus layer that cannot be penetrated by bacteria. However, the structure of the mucus in the colon of animals that develop colitis and in patients with active ulcerative colitis allows bacteria to penetrate and reach the epithelium [26].

When the mucosal damages are limited, restitution is made through the migration of viable epithelial cells from the local subepithelial stem cell niche in the proximity of the wound edges, then allowing the reestablishment of the intestinal epithelial continuity [27–29]. This process is followed by nearby cell proliferation to repopulate the damaged area. In such restitution process, cells in the proximity of the wound are characterized by a loss of classical absorptive epithelial cell polarization, with the reorganization of the cytoskeleton and loss of microvilli [30]. These cells extend protrusions (called lamellipodia) into the denuded area to allow full wound closure [31]. In case of severe mucosal injury, the regeneration capacity of local stem cells is overwhelmed, so that adequate tissue healing is not possible. In such situations, bone-marrow-derived mesenchymal stem cells migrate into the intestinal wall where they contribute, as differentiated mesenchymal cells such as myofibroblasts, to the mucosa repair process [32–35]. In fact, an increased number of myofibroblasts is observed in the lamina propria from colitis patients when compared with normal lamina propria [36]. Wound healing can be viewed as the succession of steps necessary to restore tissue structure and functions, such as barrier function and water absorption [37], whereas fibrosis can be viewed as the pathophysiological response after intestinal mucosa injury that leads to inappropriate tissue repair [38]. Indeed, fibrosis represents an exaggerated response characterized by the accumulation of collagen-rich extracellular matrix produced by a permanent or transient numerical expansion of mesenchymal cells, including fibroblasts and smooth muscle cells [39]. Fibrogenesis in inflammatory bowel diseases is driven notably by a complex set of interactions between cytokines, adhesion molecules, and growth factors [40].

The challenge for the medical and scientific communities is thus to find ways to accelerate mucosal healing without promoting fibrosis. Of note, preclinical works with rodents indicate that mucosal healing progression can be initiated even in the context of not-fully resolved inflamed mucosa [41, 42].

Several bacterial metabolites in excess increase expression of pro-inflammatory genes in colon

In that overall context, what do we know on the effects of bacterial metabolites toward mucosal inflammatory flare, and healing following an inflammatory episode? Regarding the mucosal inflammatory process by itself, the amino acid and urea-derived bacterial metabolite ammonia, as well as the cysteine and sulfate-derived metabolite hydrogen sulfide (H_2S) have been shown to induce, when present in excess, the expression of pro-inflammatory genes in colonic tissues, like the ones involved in the production of pro-inflammatory cytokines and pro-oxidant mediators [43, 44].

Mucosal Inflammation in the Colon Is Associated with Impaired Mitochondrial Energy Production

Interestingly, ammonia and H_2S, which increase the expression of pro-inflammatory genes in colonic tissues, are known to inhibit mitochondrial oxygen consumption in colonocytes, and thus mitochondrial ATP production, when present above a threshold value [45, 46]. These results can be usefully related to the fact that intestinal inflammation is associated with lower level of mucosal ATP, thus pointing out the potential importance of altered mitochondrial function in the pathophysiology of the disease. Indeed, in the inflamed colon of humans and experimental animals, decreased mitochondrial metabolic activity together with decreased ATP are recorded [47]. In support of a causal link between decreased ATP concentration in inflamed colon mucosa and progression of the disease, an increased mucosal level of ATP was found to protect mice from colitis [48]. In addition, proteomic analysis of colon mucosa obtained from volunteers with active ulcerative colitis has shown decreased expression of three mitochondrial enzymes involved in mitochondrial energy production when compared with normal colon mucosa [49], suggesting here again the implication of mitochondrial dysfunction in the pathogenesis of inflammatory bowel diseases [50]. These results were corroborated by a study showing that the activities of three complexes of the mitochondrial respiratory chain were markedly decreased in the colonic mucosa of patients with ulcerative colitis when compared with the activities measured in the normal colonic mucosa [51].

Hydrogen Sulfide in Excess in the Luminal Fluid Exerts Pro-inflammatory Effect on the Intestinal Mucosa, but Tiny Amounts of This Compound Are Required to Maintain Mucosa in Healthy State

Regarding H_2S, the concept that excessive concentrations of H_2S in the intestinal luminal content may participate in mucosal inflammation has been proposed more than two decades ago [52]. In support of this proposition, a study has shown that a high concentration of H_2S destabilizes the protective mucous layer that covers the intestinal epithelium through the reduction of disulfide bounds linking the mucin-2 network, a process that would increase the interactions between bacteria and the epithelium [53]. This result is of potential importance since it has been elegantly shown that invalidation of the gene coding for one of the main mucins in the intestine leads to the appearance of mucosal inflammation [54].

By studying the microbiota composition of patients with new-onset pediatric Crohn's disease, it has been found that such composition is characterized by a high relative abundance of *Atopobium, Fusobacterium, Veillonella, Prevotella, Streptococcus*, and *Leptotrichia* [55]. Several members of those genera are known to produce H_2S through the catabolism of sulfur-containing amino acids. In this cohort, the abundance of H_2S producers from cysteine was correlated with the severity of mucosal inflammation. To search for a possible causal link between H_2S production and intestinal inflammation, the authors colonized interleukin$10^{-/-}$ mice (mice that are invalidated for the regulatory cytokine interleukin-10), with the H_2S producer *Atopobium parvulum*, and measured a worsening of colonic inflammation [55]. Such worsening was attenuated by the H_2S scavenger bismuth. In addition, in this Crohn's disease pediatric cohort, the colonic mucosa biopsies displayed decreased expression of the mitochondrial enzymes involved in H_2S detoxification. Overall, the results from this study indicate that diminished capacity for sulfide disposal in the colonic mucosa of patients with pediatric Crohn's disease amplifies the pro-inflammatory effect of H_2S overproduction by the intestinal microbiota. Another study is in accordance with this conclusion, as gene expression and the activity of one of the H_2S detoxification enzyme thiosulfur transferase is downregulated in the mucosa of ulcerative colitis patients [56]. Decreased capacity for sulfide detoxification was also identified in the mucosa of patients with Crohn's disease [57], reinforcing the view that H_2S, over the capacity of the intestinal mucosa to detoxify it, exerts pro-inflammatory effects (Fig. 4.1). The study by Jowett et al. also indirectly indicates a pro-inflammatory effect of excessive H_2S. In this study, following for 1-year patients with ulcerative colitis in remission, it was found that patients with the highest dietary protein consumption were associated with threefold increase in the risk of relapse when compared with patients with the lower protein intake [58]. One plausible explanation for such a result is related to the increased amount of H_2S, since, although H_2S was not measured in this study, high sulfate dietary intake was also associated with a higher risk of relapse in volunteers, a result that favors the implication of luminal H_2S for such risk. Indeed, sulfate is one of the substrates that is converted by the intestinal microbiota into H_2S [59]. A study, in a mouse model deficient in the regulatory interleukin-10, showed that a diet high in saturated fat increases, in a process modulated by taurine-containing bile acids, the release of H_2S produced by the intestinal microbiota from sulfite, such event appearing to play a role in the intestinal mucosa inflammation in this experimental model [60] (Fig. 4.1).

However, from experimental works with rodents with chemically induced colitis, in which pharmacological inhibitors of H_2S synthesis or H_2S-releasing compounds were used, it appears that a low minimal amount of sulfide produced endogenously or supplied from the colonic luminal content is likely necessary to limit the extent of colonic mucosal inflammation, thus suggesting that a complete depletion of H_2S supply to colonocytes is counter-productive in inflammatory situation [61]. Indeed, several in vivo studies found that inhibition of H_2S endogenous synthesis favors intestinal inflammation and delays colitis resolution [62, 63]. Endogenous synthesis

exert some antioxidant activity, notably by persulfidation of cysteine residues that protects these residues from oxidative damage [66]. This result is of special interest when considering that mitochondria-derived reactive oxygen species play a role in epithelial barrier dysfunction [67]. A loss of epithelial integrity apparently plays a pivotal role during inflammatory bowel diseases [68]. This action of H_2S may thus contribute to the anti-inflammatory effect of low supply of H_2S in colonocytes, regardless of its endogenous or exogenous origin.

Finally, endogenous production of H_2S appears to contribute to mucus production, thus favoring segregation between luminal bacteria and the intestinal epithelium [69]. Then, the concept that H_2S is a double-edged sword for the intestinal epithelium has been proposed [70]. In that view, the effects of the diffusible bacterial metabolite H_2S on the inflammatory process in the intestinal mucosa depend on both the low endogenous production by the host colonic tissue, that is protective, and on the exogenous production by the intestinal microbiota, that is deleterious when produced in excess.

However, as presented in the following paragraph, it is worth noting that some bacterial metabolites have also been shown to be able, at least in preclinical studies, to diminish the risk of intestinal mucosa inflammation. Overall, the idea is that the severity of mucosal inflammation in the colon and rectum with active inflammatory bowel diseases would be partly related to the ratio of deleterious versus beneficial bacterial metabolite concentrations [59].

Indole and Indole-Related Compounds Exert Protective Effect Against Intestinal Mucosa Inflammation

Beneficial effects of bacterial metabolites in the context of intestinal mucosal inflammation have been reported for the tryptophan-derived metabolite indole and its related compounds which exert immune-regulatory effects [71–73]. Indole, the main tryptophan-derived bacterial metabolite [74], decreases mucosal inflammation and injury in an experimental model of enteropathy [75]. Interestingly, these effects were paralleled by an effect of indole on the expression of different cytokines in a human enterocyte cell line, with decreased expression of the pro-inflammatory IL-8, and increased expression of the regulatory cytokine IL-10 [76] (Fig. 4.2). However, in this study, expression of several genes linked to inflammation were found to be also increased [76], making the indole effects on the intestinal epithelial cells more complicated than it appears at first sight.

Other tryptophan-derived bacterial metabolites have shown some modulatory roles against intestinal mucosal inflammation. In a mice model with colitis, aryl hydrocarbon receptor ligands, such as tryptophan-derived indole and indole-like compounds, that are produced by the intestinal microbiota, were able to attenuate intestinal inflammation [77]. Indole-3-propionate, when given by the oral route, was found to exert beneficial effects on the intestinal barrier function when this latter was experimentally altered in rodent models, such as in the model of radiation injury [78], or in the model of rodents fed with a high-fat diet [79] (Fig. 4.2). In such experimental context, it is of interest to note that circulating indole-3-propionic acid is reduced in patients suffering from ulcerative colitis when compared with their

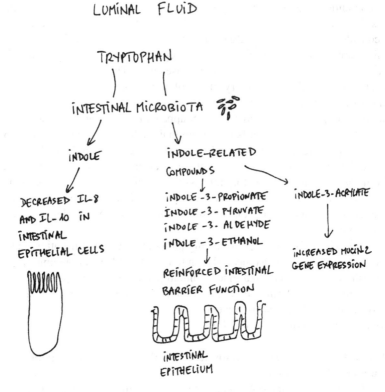

Fig. 4.2 Effects of indole and indole-related compounds on intestinal barrier function and cytokine expression in intestinal epithelial cells. IL-8 is a pro-inflammatory interleukin while IL-10 is a regulatory interleukin

healthy counterparts, while the increased level of this bacterial metabolite is associated with remission [80]. In a model of experimental colitis induced in mice, other indole-related compounds given by the oral route, namely indole-3-pyruvate, indole-3-aldehyde, and indole-3-ethanol, were found to protect against increased intestinal permeability observed in this model [81].

Indole-3-acrylate, produced by *Peptostreptococcus* species, diminishes intestinal inflammation in mice and upregulates Mucin 2 gene expression [82] (Fig. 4.2). Indole-3 propionate, produced by Clostridium species, reduced intestinal permeability and inflammation in a rodent model [72]. Overall, the protective effects of bacterial metabolites derived from tryptophan on intestinal mucosa might contribute to the beneficial effects of tryptophan supplementation observed in animal models of colitis [83, 84].

However, in order to establish the overall beneficial vs. deleterious effects of increased production of indole and indole derivatives by the gut microbiota, it appears important to consider that indoxyl sulfate, a co-metabolite derived from indole produced by the host, contributes to renal disease progression and associated pathologies, as will be developed in Chap. 5 of this book [85–88] (see Sect. 5.3).

Histamine and Spermine Modulate the Inflammasome Signaling

Other amino acid-derived bacterial metabolites such as histamine and spermine have been shown to modulate inflammation in the intestinal epithelium by modulating inflammasome signaling and IL-18 secretion [89]. Limited inflammasome activation serves critical functions in pathogen defense by intervening in the removal of damaged host cells, and by stimulating an adaptive immune response, while inappropriate inflammasome activation is linked to several inflammatory disorders [90–92]. To balance the potentially beneficial modulatory effect of spermine on intestinal inflammation, attention must be given to the otherwise deleterious effect of excessive spermine catabolism on colonocyte DNA integrity in response to enteroxigenic *B. fragilis* ([93], see Sect. 4.2).

Can Butyrate Exert Modulatory Response on the Inflamed Intestinal Mucosa?

Attention has also been paid to the potentially protective effects of the short-chain fatty acid butyrate on intestinal mucosal inflammation, given its regulatory effects on the intestinal immune system in experimental studies as presented in Chap. 3. Regarding the process of intestinal mucosa inflammation, treatment of lipopolysaccharide-treated macrophages with butyrate reduced the production of pro-inflammatory mediators including nitric oxide, interleukin-6, and interleukin-12 [94]. In addition, butyrate promoted the differentiation of colonic regulatory T cells (Treg), which act as an anti-inflammatory effector [95].

Also of interest, butyrate enhances MUC2 production in colonic biopsies obtained from ulcerative colitis patients [96]. In several in vivo experimental works with rodents in which colitis is chemically induced, butyrate displays anti-inflammatory effects [97–99].

Several controlled studies have been performed on the effects of butyrate in patients with inflammatory bowel disease. In ulcerative colitis patients receiving butyrate enemas, thus receiving injection of butyrate solution into the distal bowel by the rectal way, the results were rather heterogeneous depending on the studies considered. One study found worse clinical situation in butyrate treated-patients than in placebo-treated patients [100], two studies found no therapeutic value of short-chain fatty acid enemas because of no significant clinical improvement [101, 102], and four studies found some modest improvement of inflammatory parameters after enemas with buyrate or other short-chain fatty acids [103–106]. These discrepancies are likely due to differences among the clinical status of the volunteers included in each study, the number of patients, the doses of short-chain fatty administered, the duration of treatment, and the parameters measured. Finally, by recapitulating the data obtained in randomized controlled trials, it was concluded that the results obtained do not overall support the application of butyrate enemas for the treatment of ulcerative colitis [107]. Regarding Crohn's disease, there are no reliable data supporting the use of butyrate enemas for the treatment of this disease [107]. Such a conclusion suggests that treatment with butyrate is not able by itself to reduce the inflammatory flare to any valuable extent in ulcerative colitis. These disappointing results are likely due, in part at least, to reduction of butyrate uptake by the inflamed mucosa through downregulation of the monocarboxylate transporter MCT1, thus limiting its action on the colonic epithelium [108].

Bacterial Metabolites and Intestinal Mucosa Healing: A Potential Role of Polyamines

Regarding the effects of bacterial metabolites on intestinal mucosal healing after an inflammatory episode, there is still a paucity of data regarding this important specific area of research [7, 13]. Experiments with animal models and intestinal epithelial cells have provided some information on the roles played by bacteria-derived polyamines on this process. In rats with intestinal mucosal injury induced by chemotherapy, it has been shown that increased polyamine content in mucosa plays a central role in mucosal regeneration [109]. The beneficial effects of polyamines on regeneration seem to be partly related to the effects of these molecules on epithelial cell migration in the mucosal healing processes [110–112]. Regarding short-chain fatty acids produced by the intestinal microbiota, despite promising effects in experimental models of treatment with butyrate or mixture of short-chain fatty acids on mucosal healing observed after anastomosis (connection between two parts of the intestine) [113–115], enemas with butyrate used at high concentration and for long-term treatment (100 mM for 20 days) displays no measurable beneficial effect in patients with ulcerative colitis in remission [116].

Irritable Bowel Syndrome: What Role for the Bacterial Metabolite Profile?

Irritable bowel syndrome is a common functional intestinal disorder in the Western world characterized by recurrent abdominal pain, associated with stool of abnormal form and frequency [117, 118]. The diagnosis of irritable bowel syndrome relies mainly on the identification of characteristic symptoms and on the exclusion of other organic diseases [119]. The pathophysiology of irritable bowel syndrome seems to be related to enhanced intestinal epithelium permeability [120–122], mucosal immune activation [123, 124], visceral hypersensitivity [125], and intestinal dysmotility [126], but the etiology (causes) of this syndrome remains obscure. In patients with irritable bowel syndrome characterized by abdominal pain and frequent defecation or diarrhea, motility index and number of high amplitude propagation contractions were higher when compared to healthy patients, and the colonic transit time was shortened in a way that implies the colonic enteric nervous system [127].

Several authors have proposed gluten-free and low FODMAP (meaning Fermentable Oligo-, Di- and Monosaccharides, And Polyols) diets to reduce irritable bowel symptoms. Gluten proteins, predominantly gliadins in wheat, are resistant to complete degradation by mammalian enzymes, resulting in the production of polypeptides with potential immunogenic sequences in genetically susceptible individuals [128]. Gluten-related disorders are prevalent conditions that encompass diseases triggered by dietary gluten, including notably and most importantly celiac disease [129, 130]. In celiac disease patients, the only treatment is a life-long, strict gluten-free diet [131]. Of note, bacteria isolated from the human small intestine exhibit distinct metabolic patterns in terms of gluten protein degradation, increasing or reducing gluten peptide immunogenicity in laboratory animals [129, 132].

Regarding FODMAP, that is thus a group of highly fermentable oligo-, di-, and monosaccharides, as well as polyols, these latter compounds being poorly absorbed and easily fermented by the gut bacteria with the production of methane, hydrogen, and short-chain fatty acids [133]. According to the available dietary intervention studies, there is insufficient evidence to recommend gluten-free diets to reduce irritable bowel symptoms [134]. Regarding low FODMAP diets, there is some evidence that it may possibly reduce global symptoms in patients, but the evidence obtained are of low quality, asking for additional studies to reach more robust conclusions [134, 135].

The association between usual symptoms of irritable bowel syndrome and modification of the intestinal microbiota composition, if any, remains poorly documented [126]. In a clinical setting, subjects with irritable bowel symptoms and who experienced pain, as assessed by questionnaire during the 7-week duration of the study, were characterized by an over fivefold less bifidobacteria when compared with other subjects with irritable bowel symptoms but without declared pain [136].

Regarding the bacterial metabolites that would be implicated in irritable bowel syndrome, here again, there is a paucity of information. When the feces from patients with irritable bowel syndrome are analyzed and compared to data obtained in healthy volunteers, decreased branched-chain fatty acid concentrations are recorded [137]. This result is of potential importance given that branched-chain fatty acids act as regulators of electrolyte movements across the colonic epithelium [138] and prevent the disruption of the intestinal epithelial barrier in vitro [139]. Analysis of metabolites in feces originating from patients with diarrhea-predominant irritable bowel syndrome and from healthy counterparts, revealed that esters of short-chain fatty acids are associated with irritable bowel syndrome [140], but the effects of these metabolites on the colonic epithelium remain to be determined. Of note, higher concentrations of acetate and propionate are measured in the feces of patients with irritable bowel symptoms than in the feces produced by healthy subjects [141], but the role of these short-chain fatty acids, if any, on the etiology of irritable bowel syndrome remains unknown.

Key Points
- Hydrogen sulfide produced by the gut microbiota in excess contributes to the inflammation of the colonic mucosa.
- Indole and several indole-related compounds produced by the gut microbiota generally exert beneficial effects on the intestinal mucosa in a context of inflammatory flare.
- Histamine and spermine are bacterial metabolites that modulate the inflammasome pathway.
- Although butyrate displays immunomodulatory roles on the intestinal mucosa in a context of inflammatory flare, the capacity of this short-chain fatty acid to reduce inflammation in clinical practice remains doubtful.

- Polyamines represent potential candidates for mucosal healing after an episode of mucosal inflammation.
- Differences in fecal concentrations of short-chain and branched-chain fatty acids have been measured in patients with irritable bowel syndrome.

4.2   Dietary Compounds, Bacterial Metabolites, and Colorectal Cancer

Colorectal cancer is one of the leading global causes of cancer-related death [142]. Colorectal cancer mortality has increased during the last decades in economically emerging countries [143, 144]. Of note, in the high-risk countries for colorectal cancer, the distal large intestine areas (left colon and rectum) are the intestinal segments where majority of neoplasms are observed when compared with proximal parts of the large intestine (right and transverse colon) [145].

A classical model for the genetic and epigenetic changes that occur during colorectal tumorigenesis was proposed in 1990 by Fearon and Volgenstein [146]. In this model, the genetic alterations involved in the development of colorectal cancer were described. Of note, the order of genetic events occurring in the process of colorectal carcinogenesis appears to be crucial for the development of the disease [147].

These genetic and epigenetic alterations have an impact on the normal processes of large intestine epithelium renewal [148]. Briefly, the accumulation of these abnormalities, in a process that takes over a decade to occur, interferes with the process of normal cell division, differentiation, migration, and apoptosis as well as cell signaling in response to extracellular and intracellular regulators [149–152]. The loss of genetic and epigenetic stability has been observed in most neoplastic lesions in the colon, but the exact implication of epigenetic modifications measured in cancerous cells for the development of the disease remains poorly understood [153, 154]. It is believed that such instability accelerates the accumulation of mutations and epigenetic alterations in colonic and rectal epithelial cells, such accumulation driving the malignant transformation of these cells [155, 156].

Colorectal Cancer: One Name for Several Disease Forms

It is obvious that colorectal cancer is not one single disease, but rather gathers several diseases with different genetic and environmental influences. Indeed, both genetic and environmental factors play important roles in the etiology of the different forms of colorectal cancer [157]. The great majority of colorectal cancer are called sporadic in reference to the fact that approximately three-quarters of patients have a negative family history of colorectal cancer. In the classic colorectal cancer formation model, most cancers arise from polyps beginning with aberrant crypt formation, such aberrant crypts possibly evolving into neoplastic precursor lesions within the adenoma-carcinoma pathway which evolves from early adenoma to advanced adenoma, before finally possibly becoming colorectal cancer [158, 159]. Although the histology of the so-called conventional tubular adenomas is homogeneous, the specific biochemical parameters of these polyps are heterogeneous, explaining likely why some adenomas progress up to colorectal cancer (approximately 10% of polyps), while the vast majority do not [155, 160, 161]. Briefly, the biochemical parameters identified in the different colorectal cancer molecular subgroups that are affected by the different mutations observed are related to epithelial cell immune destruction avoidance after transformation, growth suppressor evasion of transformed epithelial cells, inactivation of DNA repair mechanisms in epithelial cells with alteration of DNA integrity, replicative immortality of cancerous cells, resistance of transformed epithelial cells to cell death, dysregulation of cellular energetics by modified fuel utilization in transformed epithelial cells, and tumor-promoting inflammation [157]. Some mutations are common to a vast majority of subgroups, suggesting a central role of these mutations in colorectal cancer in general [162–166].

A minor subgroup of the patient population is formed by those affected by a hereditary form of colorectal cancer, accounting for 5–10% of all patients. The most common syndrome in these categories is the Lynch syndrome which is caused by mutation in one of the DNA-repair genes [167]. Impaired DNA repair during cell replication gives rise to the accumulation of DNA mutations. The second hereditary form of colorectal cancer is familial adenomatous polyposis syndrome [168]. This syndrome is notably caused by mutations in the adenomatous polyposis coli (APC) gene, which controls the activity of the Wnt signaling pathway [169–171].

Regarding non-hereditary forms of colorectal cancer, chronic inflammatory bowel disease is associated with an increased risk of colorectal cancer [172, 173]. However, chronic inflammatory bowel disease explains a very minor part (approximately 1%) of all colorectal cancer in western populations, and several studies indicate that the incidence of colorectal cancer in individuals affected by chronic inflammatory bowel diseases is decreasing in the last decades because of effective anti-inflammatory treatments and improved surveillance [174, 175].

Etiology of Colorectal Cancer Is Related to Both Genetic and Environmental Parameters

The etiology of colorectal cancer includes both genetic and environmental factors. Among colorectal cancer, approximately between 10 and 20% can be attributed to heritable gene variations [176], thus suggesting that the largest fraction of sporadic colorectal cancer cases is linked to environmental and lifestyle causes [177]. Among the environmental factors that are involved in the modulation of colorectal cancer risk, dietary parameters are presumed to play a significant role [178]. The variation of colorectal cancer incidence between countries appears largely associated with differences in dietary habits. However, it is only fair to say that epidemiological studies are markedly dependent on the methodologies used to obtain and analyze dietary and lifestyle information, and this contributes to the variability of the results obtained. Furthermore, western diets contain an enormous number of different compounds, making the identification of compounds, or mixture of compounds, that may increase or decrease the risk of colorectal cancer difficult [179]. In addition, epidemiological studies cannot establish causal relationships between dietary parameters and the endpoint measured which is often colorectal cancer appearance [180]. Lastly, numerous biases related to other non-dietary lifestyle parameters must all be considered to more robustly interpret the epidemiological data obtained in the different studies. With these reservations in mind, a range of environmental and lifestyle parameters have been proposed to modify the risk of developing colorectal cancer [181]. Briefly, the risk appears to increase notably in association with prolonged smoking [182, 183], and high body mass index is associated with increased risk [184], while regular physical activity is associated with decreased risk [158]. Regarding alcohol consumption, high intake is associated with an increased risk of colorectal cancer [185].

Identifying the Dietary Compounds that Are Likely to Modulate the Risk of Colorectal Cancer: A Difficult Task When Considering the Complexity and Changing Nature of Western Diets

Concerning dietary habits, several epidemiological and experimental studies have suggested that consumption of typical western-style diets that are rich in energy, fat, and protein, and poor in fiber and phytochemicals increase colorectal cancer risk [186–189].

More specifically, regarding the consumption of meat, most population studies suggest that high consumption of red and processed meat increases colorectal cancer risk in a dose-dependent manner [190–195]. Not only the amount, but the frequency of red meat intake has been associated with the risk of colorectal carcinogenesis [196]. Of note, a meta-analysis of the prospective cohort studies (studies in which a large number of volunteers are followed and observed over a period of time to record the development of chosen outcomes) found that the relative risk of colon cancer was slightly increased with an odds ratio equal to 1.18 for individuals with the highest consumption of heme iron compared with individuals with the lowest consumption (meaning that the risk of colorectal cancer is increased by 18% for individuals with the highest meat consumption) [197]. However, it must be mentioned that the relationship between red meat consumption and colorectal cancer risk is still the subject of scientific debates. Some reviews of prospective epidemiological studies have concluded that the association between red meat consumption and colorectal cancer risk is generally rather weak [198]. In fact, in a large multiethnic cohort, it was concluded that no role for meat in the etiology of colorectal cancer could be clearly identified [199]. Similarly, another study found no association between colorectal cancer risk and levels of intake of meat [200]. The Nurses' Health Study and Health Professionals Follow-up study showed no strong correlation between consumption of heme iron and incidence of colorectal cancer in an analysis performed during a 22-years period of follow-up [201]. Similarly, Kabat et al. reported no association between intake of iron, heme iron, or iron in meat and colorectal cancer incidence in a large cohort study with Canadian women [202]. These latter results make the association between heme consumption and increased colorectal cancer risk questionable, and suggest a relatively modest role of red meat consumption in the risk of colorectal cancer.

However, the fact that many studies found a positive association between high red and processed meat consumption and risk of colorectal cancer, and even if this increased risk appears modest, has led to recommending a consumption of red meat not above 2 times 230 g red meat per week [203]. Recommendation concerning red and processed meat has indeed to integrate the fact that red meat represents an important source of proteins and of micronutrients, including iron, zinc, and vitamin B12, notably in a context of the large prevalence of anemia worldwide [204]. Indeed, anemia affects roughly one-third of world's population, with half the cases being due to iron deficiency which increases maternal and child mortality, while affecting physical performance. Even in developed countries, iron deficiency still affects a significant proportion of young females [205].

Another indirect evidence of the influence of environment on the risk of sporadic colorectal cancer is related to the fact that occurrence of colorectal cancer has been shown to be significantly increased for instance in Japanese individuals who have migrated to the United States of America and adopted a Western-style diet/-lifestyle compared with those living in Japan [206]. Other studies indicate that the incidence of colorectal cancer has dramatically increased in Japan, and this coincides with an overall increased consumption of the Westernized diet [207, 208].

Several Bacterial Metabolites Affect the DNA Integrity in Colonic Epithelial Cells

In that overall context, this chapter is devoted to the current emerging knowledge on typical luminal environment of colorectal epithelium that may affect epithelium homeostasis by either affecting its coordinated renewal, or by driving the stem and progenitor cells from normal to transformed phenotypes.

Regarding the risk of colorectal cancer, research aiming at identifying bacterial metabolites that may accumulate and participate in the process of pre-neoplasia and neoplasia in the colon and rectum is motivated notably by the fact that if cancer of the small intestine is rare [209], colorectal cancer remains one of the predominant cancers in both males and females [157]. Then, the idea is that the luminal environment of the residing stem cells in the colonic crypts may favor (or limit) the molecular events that, by accumulating in these cells during decades, drive the process of evolution of a normal colorectal epithelium to epithelium with cancerous lesions.

Emphasis will be given to the bacterial metabolites that are known to be genotoxic for colonic epithelial cells when present in excess, either by direct action on DNA, or by indirect effect through modification of cell metabolism, resulting in altered integrity of their nuclear and/or mitochondrial DNA [210–214]. This topic is of particular importance as long-term exposure of colonic crypt cells to DNA-damaging agents is likely to increase the risk of unrepaired DNA lesions in these cells [215]. To make a long and complicated story short, cancer stem cell hypothesis proposes that cells at the origin of most colorectal cancer are stem or stem cell-like cells that reside at the base of the colon crypts [216–218]. However, whether a colorectal cancer stem cell is a transformed descendent of a normal intestinal stem cell, or whether differentiated cells can also acquire a cancer stem cell phenotype upon transformation remains debated [219].

The mechanisms for genetic changes in colorectal cancer and their interactions with environmental risk factors are difficult to unravel. However, recent studies have begun to better clarify how the intestinal microbiota can generate genomic changes in colorectal cancer, for instance through bacterial metabolite synthesis, toxin production, and, in some cases, production of reactive species [220].

Is Excessive Hydrogen Sulfide Genotoxic for Colonic Epithelial Cells, and Is This Bacterial Metabolite Involved in Colorectal Carcinogenesis?

Among the bacterial metabolites with reported genotoxicity, the bacterial metabolite hydrogen sulfide that is produced from cysteine and sulfate was reported to alter genomic DNA integrity in intestinal colonic epithelial cells [221]. However, probably because of different experimental designs, the genotoxicity of H_2S was not confirmed in further experiments either using human colonocytes incubated with the H_2S donor sodium sulfide up to 3 millimolar, or in in vivo experiments using colonic short-term intraluminal instillation with H_2S in rodent [43]. In this latter study, maybe because of the activity of the enzymatic machinery devoted to DNA reparation, no measurable effects of H_2S on DNA integrity were recorded using the sensitive gamma H2AX genotoxicity test. This assay detects DNA double-strand breaks that is one of the events that is occurring during colorectal carcinogenesis [222–224].

Some experimental and clinical studies robustly suggest that both endogenous synthesis of H_2S in colonic epithelial cells, and luminal exogenous H_2S are implicated in the process of colorectal carcinogenesis. Gene and protein expression have revealed that the H_2S-synthesizing enzyme cystathionine β-synthase is increased in colonic tumors when compared with the surrounding mucosa [225], suggesting an increased capacity for H_2S synthesis in the cancerous samples. However, since colonic tumors and surrounding tissues contained numerous different cell phenotypes, it would be important to determine the expression of this enzyme in the colonic epithelial cell and other cell types.

Upregulation of cystathionine β-synthase in human colonic biopsies obtained from precancerous adenomatous polyps has been measured when compared with normal mucosa, and, of great significance, forced upregulation of cystathionine β-synthase in late adenoma-like colonic epithelial cells resulted in differences in the expression of genes involved in colorectal cancer development, notably genes involved in NF-kappaB, KRAS, and p53 signaling [226]. In addition, the cells overexpressing cystathionine β-synthase were characterized by increased proliferative activity and enhanced cellular bioenergetics capacity in terms of respiration and glycolysis. This increased cystathionine β-synthase reinforced cell tumorigenicity in athymic nude mice. On the opposite, genetic ablation of cystathionine β-synthase in mice resulted in a reduction of the number of mutagen-induced aberrant crypt foci [226]. Thus, from these experimental results, it appears that increased endogenous production of H_2S within adenomatous colonocytes increases their proliferative capacity and their ability to promote tumor formation.

Decreased expression of cystathionine β-synthase in HCT116 colonic carcinoma cells reduces cell growth and size of HCT116 xenografts (cell transplant in mice) [225]. In accordance with these results, pharmacological inhibition of cystathionine β-synthase catalytic activity inhibits HCT116 oxygen consumption and glycolysis, thus seriously affecting ATP production in these cells, an effect that was paralleled by an arrest in the cell cycle, resulting in a reduction of cell growth without loss of cell viability [227]. In vivo, the silencing or pharmacological inhibition of cystathionine β-synthase attenuates the growth of colon carcinoma cell xenografts in nude mice, as well as neo-vessel density, suggesting a role of endogenous H_2S in colorectal cancer cell growth and tumor angiogenesis [225]. Thus, in cancerous cells, hydrogen sulfide produced endogenously appears to favor cell proliferation through its utilization as an energy substrate.

Increased supply of H_2S produced by the intestinal microbiota has been also implicated in the process of colorectal carcinogenesis. Indeed, by collecting feces and tissue samples collected on and off the tumor site within the same individuals, and by using multi-omic data and community metabolic models to assess H_2S production in colorectal cancer, Hale and collaborators [228] obtained data indicating increased H_2S production by the gut bacteria in colonic cancer samples. The predicted increased H_2S production at the tumor site was associated with the relative abundance of *Fusobacterium nucleatum*, a known hydrogen sulfide producer largely suspected to promote colorectal cancer [229–233]. By using paired colon tumor tissues and adjacent tissues from volunteers, it has been shown that hydrogen sulfide-producing sulfidogenic *Fusobacterium nucleatum* was enriched in the colon tumors of patients with deficient mismatched nucleotide repair in DNA [234]. In accordance with these latter results, in volunteers, sulfidogenic bacteria abundance is higher in colonic tissue biopsies recovered from patients with colorectal cancer when compared with those from healthy subjects [235], thus reinforcing the view that these bacteria participate, likely in part by producing excessive amounts of hydrogen sulfide, in colorectal cancer development. Lastly, in the same line of thinking, sulfur-containing compounds in the samples of flatus (mixture of gas produced by the large intestine microbiota and evacuated) recovered from patients with colorectal cancer were more abundant when compared with samples recovered from healthy subjects [236].

P-cresol Is One of the Greatest Predictors of Genotoxicity in Colonocytes

Using the measurement of DNA double-strand breaks with the gamma H2AX genotoxicity test, the tyrosine-derived bacterial metabolite *p*-cresol was found to dose-dependently alter the DNA integrity in colonocytes [237]. Interestingly, by using a fermentation system containing fecal samples from volunteers in the presence of various substrates, together with the test of fermentation supernatants on human colonocytes for their genotoxic potential, *p*-cresol was identified among the compounds present in the supernatants as the greatest predictor of genotoxicity against colonocytes [238] (Fig. 4.3). In addition, association has been found between the fecal *p*-cresol concentration and the total DNA adduct formation leading to DNA damage in the distal colon of mice [239, 240]. Although of considerable interest to

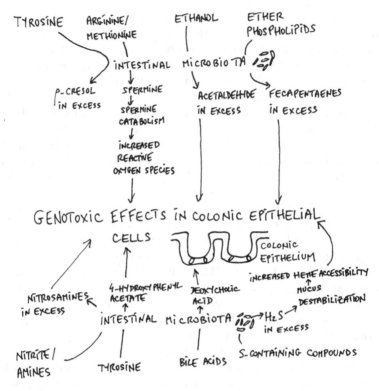

Fig. 4.3 Effects of different bacterial metabolites in excessive concentrations on DNA integrity in colonic epithelial cells

identify the bacterial metabolites that may increase the risk of colorectal neoplasia development, the results of these in vitro studies ask for in vivo studies with relevant experimental models.

Phenol Is a Precursor of Mutagenic Compound

Phenol, which is produced by the intestinal microbiota from tyrosine, after reacting with nitrite, leads to the formation of the compound p-diazoquinone which is mutagenic in tests with bacteria [241]. Unfortunately, no data are available regarding the mutagenic potential of p-diazoquinone on colonic epithelial cells.

Fecapentaenes Are Potent Mutagens for Colonic Epithelial Cells

Fecapentaenes, which are produced by the intestinal microbiota presumably from polyunsaturated ether phospholipids, represent potent mutagens toward colon epithelial cells [242] (Fig. 4.3).

Secondary Bile Acids in Excess Are DNA-Damaging Agent

A wealth of experimental studies has shown that the secondary bile acid deoxycholic acid that is produced by the intestinal microbiota from bile acids can act as a DNA-damaging agent [188, 243, 244] (Fig. 4.3).

Acetaldehyde Is a Potential Carcinogen in the Rectum

Although ethanol is likely not genotoxic by itself for colonic epithelial cells, this compound can be converted to acetaldehyde by the gut microbiota. Acetaldehyde is considered as a potential carcinogenic compound in the rectum [245] (Fig. 4.3). One study reported modest genotoxic effect of micromolar concentration of acetaldehyde on human colonocytes [246].

Excessive Spermine Catabolism Increases Damages to DNA in Colonic Epithelial Cells

Regarding spermine, it appears that, when present in excess, this polyamine which is efficiently taken up by colonic epithelial cells [247], increases reactive oxygen species production and DNA damage in these cells through its increased catabolism by the spermine oxidase activity in response to enteroxigenic *B. fragilis* infection [93] (Fig. 4.3). In regards with the group of polyamines, metabolites derived from these compounds were found to accumulate in microbial biofilms that adhere to human colorectal biopsies when compared to biopsies without biofilms [248]. Microbial biofilms are aggregations of microbial communities encased in a polymeric matrix that are present on the mucosal surface of colon. Biofilms that invade the colonic mucus layers and come into direct contact with epithelial cells generally indicate a pathologic state [249, 250]. The results obtained by Johnson and collaborators [248] are of great interest when considering the role of polyamines on colon cancer cell growth [251] (see also Sect. 3.1).

4-Hydroxyphenylacetic Acid in Excess Decreases Oxygen Consumption and Increases Reactive Oxygen Species Production in Colon Epithelial Cells, in Association with DNA Alteration

The microbial fermentation of aromatic amino acids by bacterial species present in the human colon produces 4-hydroxyphenylacetic acid [252]. This metabolite, when present in excess, decreases the respiration of human colonocytes and increases reactive oxygen species (also called ROS) production. This increased reactive oxygen species production has been associated with DNA alteration in these cells [246] (Fig. 4.3).

Elevated mitochondrial reactive oxygen species production over the capacity of colonocytes to detoxicate them, may damage and produce mutations in mitochondrial DNA which is in proximity with the electron transport chain, and which is not protected by histones [253]. Of interest, mitochondrial DNA mutations observed in colon cancer are mutations in the gene coding for a subunit of cytochrome c oxidase (complex IV of the mitochondrial respiratory chain) with associated proton leaks and thus decreased mitochondrial energy production efficiency, together with perturbations of transport processes, and metabolite trafficking [254].

Heme, by Increasing H$_2$S Production, Destabilizes the Mucus Layer, Thus Increasing Heme Accessibility to Colonic Epithelial Cells, Then Favoring Its Genotoxic Effect

A special interest has been devoted to the possible role of heme, the pigment of red meat, on the process of colorectal carcinogenesis, since as presented above, high red and processed meat consumption is suspected to increase the risk of colorectal cancer. Although iron in heme is much better absorbed than elemental iron [255], dietary heme is relatively poorly absorbed by the enterocytes in the small intestine, and accordingly, part of dietary heme is transferred from the small to the large intestine [256]. Hemin, the reactive ferric protoporphyrin-IX groups of hemoglobin [257], represents the stable form of heme [258]. Hemin has been shown to be genotoxic for colonocytes [259]. In experimental works with rodents, dietary supplementation with hemin results in alteration of the colonic epithelium, with crypt depth being increased by the so-called compensatory hyperproliferation of crypt cells [260], thus leading to hyperplasia. Such hemin-induced hyperplasia is believed to represent an early event that would drive the normal colonic epithelium toward the neoplastic pathway [53]. This process appears to be dependent on the intestinal microbiota and on its metabolic activity. Indeed, by supplementing rodents with heme, a colonic epithelial hyperproliferation was observed that was completely suppressed by treatment with antibiotics [53], thus implying the gut microbiota in such effects. Some evidence suggest that H$_2$S was one of the bacterial metabolites indirectly implicated in the deleterious effect of heme on the colonic epithelium. Indeed, the abundance of the mucin-degrading bacteria *Akkermansia muciniphila* was increased after dietary heme supplementation [53]. Since mucin-degrading bacteria metabolize notably the sulfate-containing mucin, thus producing H$_2$S [261]; and since H$_2$S in excess fragilizes the mucus barrier, the authors have proposed that heme by favoring H$_2$S production, increases the accessibility of heme to colonic epithelial cells, thus reinforcing its genotoxic effect (Fig. 4.3). The identification of the bacterial metabolites that fragilize the mucus layers in the colon is of paramount importance because, as exemplified above for the H$_2$S case, increased accessibility of genotoxic compounds to the colonic epithelial cells is likely a way to adversely increase alteration of DNA integrity. Indeed, elegant studies performed in rodent models that have been invalidated for the production of specific mucins reveal that these animals are more prone to develop colorectal carcinomas than their wild-type counterpart [262].

Of note, the addition of calcium to the heme-enriched diet allows to trap the luminal heme, then counteracting the deleterious effects of heme on the colonic epithelium [263].

Dietary supplementation with heme iron affects the microbial community structure [264]. Numerous bacteria do not possess the capacity to synthesize heme [265] and are therefore equipped with diverse heme transport systems [266]. Bacterial pathogens are notably efficient at capturing heme and utilizing it for their metabolism [267]. Then the overall capacity of the colonic microbiota to take up heme plays a role in fixing the luminal heme concentration within the large intestine, and thus presumably the effects of heme on the stem cells in the colonic crypts.

N-Nitrosamines Are Suspected to Increase the Risk of Colorectal Cancer

N-nitrosamines are a large family of compounds among which several of them have been shown to induce cancer development in various organs in animal models [268]. As introduced in Chap. 3, bacteria are believed to play a central role in nitrosamine synthesis in the human intestine by catalyzing nitrosation of diet-derived amines [269]. The carcinogenic effect of N-nitrosamines in excess are likely mediated via direct interaction with DNA in cells causing chemical structure alterations [270–273] (Fig. 4.3). Of note, the level of dietary intake of the nitrosamine N-nitrosodimethylamine was positively associated with risk of rectal cancer [274], thus pointing out potential implication of this compound, from either dietary source or resulting from the activity of the intestinal microbiota, in rectal carcinoma formation.

Is Butyrate Protective Against Colorectal Cancer Risk?

Regarding the undigestible polysaccharide-derived short-chain fatty acid butyrate, this bacterial metabolite has been considered to display some protective action on the risk of developing colorectal cancer.

The evidence for such a proposition were based notably on experimental works showing in a rodent model of colitis-associated colon cancer that a dietary supplementation with resistant starch decreased tumor multiplicity and adenoma formation, an effect that was associated with an increased short-chain fatty acid production, including butyrate [275]. In addition, numerous studies have shown that butyrate is able in vitro to inhibit the proliferation of colonocytes recovered from human carcinomas. However, even if these experiments bring important new information regarding the mechanisms of action of butyrate on colonic cancerous cells, from a strategic point of view, this information is of little relevance in a cancer prevention perspective. As indicated in previous paragraph (see Chap. 3), the anti-proliferative effect of butyrate on colonic cancerous cells is related to the dose-dependent capacity of this bacterial metabolite to inhibit histone deacetylase activity [276], therefore increasing the acetylation of histones, such a process being linked to the alteration of cell cycle regulatory proteins [277–279]. These alterations result in a marked reduction of cancerous cell growth. This effect appears rather specific for butyrate, as acetate has no effect on the acetylation of histone proteins and on cancerous cell growth [280, 281].

An experimental study has seriously contradicted the initial hypothesis of a role of butyrate for decreasing the risk of colorectal cancer. Indeed, in a model of rodent that spontaneously develops multiple polyps in small intestine and colon, gut microbes stimulated colon cancer development through the production of butyrate that was used as a fuel to sustain hyperproliferation of colon epithelial cells [282]. In addition, butyrate in drinking water given to a model of chemically induced colon cancer in rodents has been shown to enhance the development of colonic neoplasia, an effect that was associated with an increased butyrate content in feces [283].

Then butyrate may be viewed, depending on its luminal concentration in the colon, and on the characteristics (notably in terms of metabolic capacities) of colonic differentiated and undifferentiated epithelial cells (either healthy, pre-neoplasic, or neoplasic); as either an oncometabolite (a metabolite associated with cancer [284]) or as a tumor-suppressive metabolite [285]. This may explain why, although dietary fibers, and notably butyrogenic fibers, are generally believed to decrease colorectal cancer risk, there is a lack of consensus because of conflicting results obtained from prospective-cohort studies [286].

Key Points
- Several bacterial metabolites derived from dietary compounds that are partially transferred from the small to the large intestine are suspected in excessive amounts and after long-term exposure to increase the risk of colorectal cancer.
- These bacterial metabolites, which include p-cresol, phenol, fecapentaenes, secondary bile acids, acetaldehyde, spermine, and the hemin group of hemoglobin, have all been shown to act as DNA-damaging agents on colonic epithelial cells when present in excess.
- Some bacterial metabolites when present in excessive concentrations, like hydrogen sulfide, may indirectly favor DNA alteration in colonic epithelial cells by altering the mucous layer, therefore increasing the accessibility of genotoxic compounds.
- Increased hydrogen sulfide production by adenomatous colonocytes and by intestinal bacteria is likely to favor colorectal carcinogenesis.
- The initial hypothesis that butyrate would reduce the risk of colorectal cancer has been challenged by results obtained from preclinical and prospective-cohort studies.

4.3   Dietary Compounds, Bacterial Metabolites, and Diarrhea

One Main Physiological Function of the Colonic Epithelium Is Water and Electrolyte Absorption
Approximately 1.2 L (including water in food) is consumed every day by an adult individual. In the digestive tract, there are exocrine secretions by salivary gland, stomach, exocrine pancreas, and liver. In addition, 1.0 L of fluid moves per day from blood to the intestinal lumen. Thus, it is estimated that 8 to 9 L of fluid traverse every day the digestive tract [287]. On this amount of fluid, 6.7 L are reabsorbed every day by the small intestine. The remaining water (approximately 1.5 L per day) moves from the small to the large intestine. The large intestine absorbs in such condition 1.4 L of water per day, thus leading to the excretion of approximately 100 ml water per day in feces [288]. In addition to water, the colon, and to a much lower extent, rectum, are the final intestinal segments for recapturing solutes present in the luminal fluid [289].

To maintain electrolyte homeostasis, and to provide electrolytes for different physiological functions, complex interactions between absorptive and secretory processes are necessary. Although the net result of electrolyte transport in the healthy gut is largely orientated toward transfer from the lumen to the blood, there

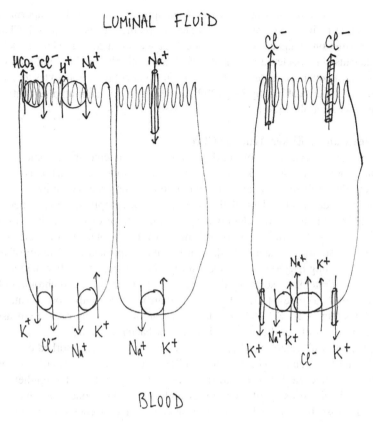

LUMINAL   FLUID

BLOOD

Fig. 4.4 Schematic view of the different transporters and channels in the colonic absorptive epithelial cells

is also secretion of electrolytes, mainly chloride (Cl^-) and bicarbonate ($HCO3^-$) that play a role in digestion and absorption of nutrients [290, 291].

Regarding the large intestine, there are significant qualitative and quantitative differences in electrolyte transport processes between the different segments [292]. Both surface and crypt colonocytes perform secretory and absorptive functions that can happen simultaneously [293, 294]. Regarding the colonic absorption of electrolytes, sodium (Na^+), chloride and potassium (K^+) are transferred from the lumen to the blood by a series of dedicated transporters, co-transporters, exchangers, and channels present in the brush border membranes at the luminal side and in the baso-lateral membranes situated in proximity of the blood capillaries (Fig. 4.4). Incidentally, and importantly, the Na/K ATPase situated in the baso-lateral membranes of colonocytes transfer Na^+ to the blood while transferring K^+ from blood to colonocyte intracellular medium [295]. To make a long story short and more easily understandable, it can be useful to consider that overall, water follows Na^+ during its transcellular journey from the lumen to the blood.

Absorptive colonocytes are in addition equipped with several aquaporin water channels that facilitate the transport of water across these cells [296, 297]. Changes in the distribution of aquaporins in colonocytes are associated in a rodent model with diarrhea caused by bacterial pathogens [298]. Interestingly, in vitro inhibition of the expression of aquaporin 8, one of the aquaporins that equips colonocytes, decreased osmotic water permeability by 38% [299], thus suggesting major involvement of this channel for water absorption.

Diarrhea Can Be Either Acute or Chronic
Classification for evaluation of diarrhea considers both acute and chronic (chronic meaning more than 4 weeks-duration) forms of diarrhea. This classification considers the probability of an infectious etiology for acute conditions, whereas chronic diarrhea is much less likely to be infectious, implying to consider other causes [287]. An alternative classification for the type of diarrhea is based on the appearance of stool either fatty, or inflammatory (associated in some cases with blood in the stool), or watery [300]. Diarrhea may represent a symptom of many diseases including bacterial, viral, and parasitic infections, altered bile acid absorption, carbohydrate malabsorption, dissacharidase insufficiency, and chronic inflammatory bowel diseases [301]. In watery diarrhea, the body loses water and electrolytes (Na^+, Cl^-, K^+, $HCO3^-$) in aqueous stool, and such losses are measured in different types of diarrhea from either viral or bacterial origin.

In patients with watery diarrhea, solutes are not sufficiently absorbed or secreted in excess into the luminal content, or both [287]. The ability of the colonic epithelium to dehydrate the luminal fluid depends on the integrity of the epithelium so preventing back diffusion of electrolytes and other solutes once they have been absorbed across the epithelial layer. For instance, the pathogenesis of inflammatory bowel disease-associated diarrhea is essentially an outcome of mucosal damage caused by persistent inflammation resulting in dysregulated intestinal ion transport and impaired epithelial barrier function [302].

Watery diarrheas are categorized as either osmotic or secretory. In patients with secretory diarrhea, stool osmolarity is almost entirely accounted for by excessive electrolyte secretion (Na^+, K^+, and accompanying anions). In osmotic diarrhea, poorly absorbed compounds draw fluid in the lumen [303]. This is the case for instance for dietary lactose in case of lactase deficiency [304], a deficiency that affects a large part of the population worldwide, notably in Asian countries [305, 306]. Undigested and/or unabsorbed carbohydrates in addition to provoking osmotic retention of fluid can increase bacterial fermentation with the production of various compounds including gases [307, 308]. Clinical studies indicate an increased intestinal motility, either as a cause or consequence, in chronic watery diarrhea, thus limiting the contact time between absorbable solutes and colonic epithelium [309].

Short-Chain Fatty Acids Can Modulate Water and Electrolyte Movements Through the Colonic Epithelium in Case of Diarrhea
Chronic diarrhea in mammalian models and in humans has been associated with an overall shift in the composition of the microbiota and of its metabolic activity

[310, 311]. Several metabolites produced by the intestinal microbiota have been shown to interfere with the process of electrolyte absorption and secretion. Short-chain fatty acids increase colonic fluid and electrolyte absorption in healthy individuals by involving Na^+/H^+, short-chain fatty acid/$HCO3^-$, and Cl^-/short-chain fatty acid exchangers in apical membranes of colonocytes [312]. Among short-chain fatty acids, butyrate upregulates the sodium channel ENaC in human colonocytes as well as sodium absorption across the colonic epithelium in rodent model [313]. The other short-chain fatty acid propionate increases Na^+ absorption in the rabbit colon [314]. This latter short-chain fatty acid also induced chloride secretion across colonic epithelium in in vitro experiments [315], indicating that this bacterial metabolite has an impact on both electrolyte absorption and secretion. Adding resistant starch to oral rehydration solution (that contains glucose and electrolytes [316], and which are used notably to rehydrate children with diarrhea [317]), has been proposed to increase short-chain fatty acid production by the intestinal microbiota, thus further improving the effectiveness of this solution for the treatment of acute diarrhea in children [312]. Some beneficial effects of such addition have been reported in children with acute watery diarrhea [318, 319], while one multicenter study found no effect of the addition of a mixture of non-digestible carbohydrates to oral rehydration therapy for the treatment of children with acute infectious diarrhea [320], suggesting that the effects of non-digestible carbohydrates on pediatric diarrhea depend on their chemical nature and/or the clinical form of diarrhea.

The Bacterial Metabolites Deoxycholate and Succinate Modulate the Growth of the Diarrhea Inducer Clostridium difficile
The secondary bile acids deoxycholate that is produced by the intestinal microbiota promotes germination but markedly reduces the vegetative growth of *Clostridium difficile*, bacteria that represent a major cause of intestinal infection and diarrhea among individuals following antibiotic treatment [321, 322]. Germination occurs only in the distal part of the intestine partly because oxygen concentration at this site is negligible [323]. Ex vivo studies with intestinal extracts from antibiotic-treated and untreated rodents revealed a correlation between the capacity of the extracts to support the growth of *C. difficile* and decreased levels of secondary bile acids in these extracts [324, 325]. These observations suggest that the production of secondary bile acids by the usual bacterial community, and the reduction of the size of this community by antibiotic treatment, and thus of such production, increases the susceptibility to *C. difficile* overgrowth [321]. In other words, deoxycholic acid, which is known to be deleterious for the colonic epithelial cells when present in excess in the luminal fluid (see previous paragraphs), is also, at moderate concentrations, able to limit the growth of *Clostridium difficile*.

   C. difficile may cause colitis through a range of virulence factors, including toxins, as well as adherence and motility factors. In response to limited nutrient availability, *C. difficile* produces toxins that primarily target intestinal epithelial cells, provoking necrosis, and thus leading to a loss of apical membrane integrity, and finally increasing exposure of the host intestinal mucosa to microbes and their bacterial metabolites [321]. Such increased exposure may result in an activation of the host intestinal inflammatory response and associated diarrhea.

In addition to deoxycholate, other metabolites produced by the intestinal microbiota, namely mucin-derived monosaccharides, and succinate, are used by *C. difficile* which takes up these compounds and used them for growth [326, 327]. Succinate is produced by the intestinal microbiota from different substrates including amino acids and carbohydrates [328, 329].

Is Excessive Serotonin Produced by the Intestinal Microbiota Involved in Chronic Diarrhea?

Several compounds secreted by enteroendocrine cells, mast cells, and submucosal neurons are involved in chronic diarrhea. Among these compounds, serotonin is playing an important role, notably since this compound stimulates gut motility [330–332]. Antagonists of the serotonin receptor 5-HT$_3$ are pharmaceuticals used for the treatment of chronic watery diarrhea [287]. As serotonin is also produced by the intestinal microbiota, it would be of major interest to evaluate the role played by this bacterial metabolite regarding the net fluid fluxes across the colonic epithelium.

Key Points

- Short-chain fatty acids increase colonic fluid and electrolyte absorption by colonocytes.
- Sufficient production of secondary bile acids by the intestinal microbiota limits the growth of the diarrhea inducer *Clostridium difficile*.
- Succinate produced by the intestinal microbiota is used by *Clostridium difficile* for growth.

References

1. Fiocchi C. Inflammatory bowel disease: etiology and pathogenesis. Gastroenterology. 1998;115(1):182–205.
2. Kaser A, Zeissig S, Blumberg RS. Inflammatory bowel disease. Annu Rev Immunol. 2010;28: 573–621.
3. Molodecky NA, Soon IS, Rabi DM, Ghali WA, Ferris M, Chernoff G, Benchimol EI, Panaccione R, Ghosh S, Barkema HW, Kaplan GG. Increasing incidence and prevalence of the inflammatory bowel diseases with time, based on systematic review. Gastroenterology. 2012;142(1):46–54.
4. Owczarek D, Rodacki T, Domagala-Rodacka R, Cibor D, Mach T. Diet and nutritional factors in inflammatory bowel diseases. World J Gastroenterol. 2016;22(3):895–905.
5. Gecse KB, Vermeire S. Differential diagnosis of inflammatory bowel disease: imitations and complications. Lancet Gastroenterol Hepatol. 2018;3(9):644–53.
6. Khor B, Gardet A, Xavier RJ. Genetics and pathogenesis of inflammatory bowel disease. Nature. 2011;474(7351):307–17.
7. Vidal-Lletjos S, Beaumont M, Tomé D, Benamouzig R, Blachier F, Lan A. Dietary protein and amino acid supplementation in inflammatory bowel disease course: what impact on the colonic mucosa? Nutrients. 2017;9(3):310.
8. Lucendo AJ, De Rezende LC. Importance of nutrition in inflammatory bowel disease. World J Gastroenterol. 2009;15(17):2081–8.

9. Colombel JF, Mahadevan U. Inflammatory bowel disease 2017: innovations and changing paradigms. Gastroenterology. 2017;152(2):309–12.
10. Jeong DY, Kim S, Son MJ, Son CY, Kim JY, Kronbichler A, Lee KH, Shin JI. Induction and maintenance treatment of inflammatory bowel disease: a comprehensive review. Autoimmun Rev. 2019;18(5):439–54.
11. Leppkes M, Neurath MF. Cytokines in inflammatory bowel diseases. Update 2020. Pharmacol Res. 2020;158:104835.
12. De Cruz P, Kamm MA, Prideaux L, Allen PB, Moore G. Mucosal healing in Crohn's disease: a systematic review. Inflamm Bowel Dis. 2013;19(2):429–44.
13. Lan A, Blachier F, Benamouzig R, Beaumont M, Barrat C, Coehlo C, Lancha A Jr, Kong X, Yin Y, Marie JC, Tomé D. Mucosal healing in inflammatory bowel diseases: is there a place for nutritional supplementation? Inflamm Bowel Dis. 2015;21(1):198–207.
14. Sandborn WJ, Abreu MT, Dubinsky MC. A noninvasive method to assess mucosal healing in patients with Crohn's disease. Gastroenterol Hepatol (NY). 2018;14(5):1–12.
15. Shah SC, Colombel JF, Sands BE, Narula N. Systematic review with meta-analysis: mucosal healing is associated with improved long-term outcomes in Crohn's disease. Aliment Pharmacol Ther. 2016;43(3):317–33.
16. D'Haens G, Sandborn WJ, Feagan BG, Geboes K, Hanauer SB, Irvine EJ, Lémann M, Marteau P, Rutgeerts P, Schölmerich J, Sutherland LR. A review of activity indices and efficacy end points for clinical trials of medical therapy in adults with ulcerative colitis. Gastroenterology. 2007;132(2):763–86.
17. Lichtenstein GR, Rutgeerts P. Importance of mucosal healing in ulcerative colitis. Inflamm Bowel Dis. 2010;16(2):338–46.
18. Bouguen G, Levesque BG, Pola S, Evans E, Sandborn WJ. Endoscopic assessment and treating to target increase the likelihood of mucosal healing in patients with Crohn's disease. Clin Gastroenterol Hepatol. 2014;12(6):978–85.
19. Colombel JF, Rutgeerts P, Reinisch W, Esser D, Wang Y, Lang Y, Marano CW, Strauss R, Oddens BJ, Feagan BJ, Hanauer SB, Lichtenstein GR, Present D, Sands BE, Sandborn WJ. Early mucosal healing with infliximab is associated with improved long-term clinical outcomes in ulcerative colitis. Gastroenterology. 2011;141(4):1194–201.
20. Dave M, Loftus EV Jr. Mucosal healing in inflammatory bowel disease. A true paradigm of success? Gastroenterol Hepatol (NY). 2012;8(1):29–38.
21. Neurath MF, Travis SPL. Mucosal healing in inflammatory bowel diseases: a systematic review. Gut. 2012;61(11):1619–35.
22. Papi C, Fasci-Spurio F, Rogai F, Settesoldi A, Margagnoni G, Annese V. Mucosal healing in inflammatory bowel disease: treatment efficacy and predictive factors. Dig Liver Dis. 2013;45 (12):978–85.
23. Pineton de Chambrun G, Peyrin-Biroulet L, Lémann M, Colombel JF. Clinical implications of mucosal healing for the management of IBD. Nat Rev Gastroenterol Hepatol. 2010;7(1): 15–29.
24. Rieder F, Karrasch T, Ben-Horin S, Schirbel A, Ehehalt R, Wehkamp J, de Haar C, Velin D, Latella G, Scaldaferri F, Rogler G, Higgins P, Sans M. Results of the 2nd scientific workshop of the ECCO (III): basic mechanisms of intestinal healing. J Crohns Colitis. 2012;6(3):373–85.
25. Rutgeerts P, Vermeire S, Van Assche G. Mucosal healing in inflammatory bowel disease: impossible ideal or therapeutic target? Gut. 2007;56(4):453–5.
26. Johansson MEV, Gustafsson JK, Holmén-Larsson J, Jabbar KS, Xia L, Xu H, Ghishan FK, Carvalho FA, Gewirtz AT, Sjövall T, Hansson GC. Bacteria penetrate the normally impenetrable inner colon mucus layer in both murine colitis models and patients with ulcerative colitis. Gut. 2014;63(2):281–91.
27. Feil W, Wenzl E, Vattay P, Starlinger M, Sogukoglu T, Schiessel R. Repair of rabbit duodenal mucosa after acid injury in vivo and in vitro. Gastroenterology. 1987;92(6):1973–86.

28. McKaig BC, Makh SS, Hawkey CJ, Podolsky DK, Mahida YR. Normal human colonic subepithelial myofibroblasts enhance epithelial migration (restitution) via TGF-beta3. Am J Physiol. 1999;276(5):G1087–93.
29. Moore R, Carlson S, Madara JL. Rapid barrier restitution in an in vivo model of intestinal epithelial injury. Lab Invest. 1989;60(2):237–44.
30. Lacy ER. Epithelial restitution in the gastrointestinal tract. J Clin Gastroenterol. 1988;10(1): S72–7.
31. Nusrat A, Delp C, Madara JL. Intestinal epithelial restitution. Characterization of a cell culture model and mapping of cytoskeletal elements in migrating cells. J Clin Invest. 1992;89(5): 1501–11.
32. Brittan M, Hunt T, Jeffery R, Poulsom R, Forbes SJ, Hodivala-Dilke K, Goldman J, Alison MR, Wright NA. Bone marrow derivation of pericryptal myofibroblasts in the mouse and human small intestine and colon. Gut. 2002;50(6):752–7.
33. Sipos F, Galamb O. Epithelial-to-mesenchymal and mesenchymal-to-epithelial transitions in the colon. World J Gastroenterol. 2012;18(7):601–8.
34. Tanaka F, Tominaga K, Ochi M, Tanigawa T, Watanabe T, Fujiwara Y, Ohta K, Oshitani N, Higuchi K, Arakawa T. Exogenous administration of mesenchymal stem cells ameliorates dextran sulfate sodium-induced colitis via anti-inflammatory action in damaged tissue in rats. Life Sci. 2008;83(23–24):771–9.
35. Wei Y, Nie Y, Lai J, Yvonne Wan YJ, Li Y. Comparison of the population capacity of hematopoietic and mesenchymal stem cells in experimental colitis rat model. Transplantation. 2009;88(1):42–8.
36. Hayashi Y, Tsuji S, Tsuji M, Nishida T, Ishii S, Nakamura T, Eguchi H, Kawano S. The transdifferentiation of bone-marrow-derived cells in colonic mucosal regeneration after dextran-sulfate-sodium-induced colitis in mice. Pharmacology. 2007;80(4):193–9.
37. Sturm A, Dignass AU. Epithelial restitution and wound healing in inflammatory bowel disease. World J Gastroenterol. 2008;14(3):348–53.
38. White ES, Mantovani AR. Inflammation, wound repair, and fibrosis: reassessing the spectrum of tissue injury and resolution. J Pathol. 2013;229(2):141–4.
39. Rieder F, Fiocchi C, Rogler G. Mechanisms, management, and treatment of fibrosis in patients with inflammatory bowel diseases. Gastroenterology. 2017;152(2):340–350.e6.
40. Principi M, Giorgio F, Losurdo G, Neve V, Contaldo A, Di Leo A, Ierardi E. Fibrogenesis and fibrosis in inflammatory bowel diseases: Good and bad side of same coin? World J Gastrointest Pathophysiol. 2013;4(4):100–7.
41. Liu X, Beaumont M, Walker F, Chaumontet C, Andriamihaja M, Matsumoto H, Khodorova N, Lan A, Gaudichon C, Benamouzig R, Tomé D, Davila AM, Marie JC, Blachier F. Beneficial effects of an amino acid mixture on colonic mucosal healing in rats. Inflamm Bowel Dis. 2013;19(13):2895–905.
42. Vidal-Lletjos S, Andriamihaja M, Blais A, Grauso M, Lepage P, Davila AM, Gaudichon C, Leclerc M, Blachier F, Lan A. Mucosal healing progression after acute colitis in mice. World J Gastroenterol. 2019;25(27):3572–89.
43. Beaumont M, Andriamihaja M, Lan A, Khodorova N, Audebert M, Blouin JM, Grauso M, Lancha L, Benetti PH, Benamouzig R, Tomé D, Bouillaud F, Davila AM, Blachier F. Detrimental effects for colonocytes of an increased exposure to luminal hydrogen sulfide: The adaptive response. Free Radic Biol Med. 2016;93:155–64.
44. Villodre Tudela C, Boudry C, Stumpff F, Aschenbach J, Vahjen W, Zentek J, Pieper R. Down-regulation of monocarboxylate transporter 1 (MCT1) gene expression in the colon of piglets is linked to bacterial protein fermentation and pro-inflammatory cytokine-mediated signalling. 2015;113(4):610–7.
45. Andriamihaja M, Davila AM, Eklou-Lawson M, Petit N, Delpal S, Allek F, Blais A, Delteil C, Tomé D, Blachier F. Colon luminal content and epithelial cell morphology are markedly modified in rats fed with a high-protein diet. Am J Physiol. 2010;299(5):G1030–7.

46. Mimoun S, Andriamihaja M, Chaumontet C, Atanasiu C, Benamouzig R, Blouin JM, Tomé D, Bouillaud F, Blachier F. Detoxification of H2S by differentiated colonic epithelial cells: implication of the sulfide oxidizing unit and of the cell respiratory capacity. Antioxid Redox Signal. 2012;17(1):1–10.
47. Heller S, Penrose HM, Cable C, Biswas D, Nakhoul H, Baddoo M, Flemington E, Crawford SE, Savkovic SD. Reduced mitochondrial activity in colonocytes facilitates AMPKα2-dependent inflammation. FASEB J. 2017;31(5):2013–25.
48. Bär F, Bochmann W, Widok A, von Medem K, Pagel R, Hirose M, Yu X, Kalies K, König P, Böhm R, Herdegen T, Reinicke AT, Büning J, Lehnert H, Fellermann K, Ibrahim S, Sina C. Mitochondrial gene polymorphisms that protect mice from colitis. Gastroenterology. 2013;145(5):1055–1063.e3.
49. Hsieh SY, Shih TC, Yeh CY, Lin CJ, Chou YY, Lee YS. Comparative proteomic studies on the pathogenesis of human ulcerative colitis. Proteomics. 2006;6(19):5322–31.
50. Novak EA, Mollen KP. Mitochondrial dysfunction in inflammatory bowel disease. Front Cell Dev Biol. 2015;3:62.
51. Sifroni KG, Damiani CR, Stoffel C, Cardoso MR, Ferreira GK, Jeremias IC, Rezin GT, Scaini G, Schuck PF, Dal-Pizzol F, Streck EL. Mitochondrial respiratory chain in the colonic mucosa of patients with ulcerative colitis. Mol Cell Biochem. 2010;342(1–2):111–5.
52. Levine J, Ellis CJ, Furne JK, Springfield J, Levitt MD. Fecal hydrogen sulfide production in ulcerative colitis. Am J Gastroenterol. 1998;93(1):83–7.
53. Ijssennagger N, Belzer C, Hooiveld G, Dekker J, van Mil SWC, Müller M, Kleerebezem M, van der Meer R. Gut microbiota facilitates dietary heme-induced epithelial hyperproliferation by opening the mucus barrier in colon. Proc Natl Acad Sci USA. 2015;112(32):10038–43.
54. Van der Sluis M, De Koning BAE, De Bruijn ACJM, Velcich A, Meijerink JPP, Van Goudoever JB, Büller HA, Dekker J, Van Seuningen I, Renes IB, Einerhand AWC. Muc-2-deficient mice spontaneously develop colitis, indicating that MUC2 is critical for colonic protection. Gastroenterology. 2006;131(1):117–29.
55. Mottawea W, Chiang CK, Mühlbauer M, Starr AE, Butcher J, Abujamel T, Deeke SA, Brandel A, Zhou H, Shokralla S, Hajibabaei M, Singleton R, Benchimol EI, Jobin C, Mack DR, Figeys D, Stintzi A. Altered intestinal microbiota-host mitochondria crosstalk in new onset Crohn's disease. Nat Commun. 2016;7:13419.
56. Ramasamy S, Singh S, Taniere P, Langman MJS, Eggo MC. Sulfide-detoxifying enzymes in the human colon are decreased in cancer and upregulated in differentiation. Am J Physiol. 2006;291(2):G288–96.
57. Arijs I, Vanhove W, Rutgeerts P, Schuit F, Verbeke K, De Preter V. Decreased mucosal sulfide detoxification capacity in patients with Crohn's disease. Inflamm Bowel Dis. 2013;19(5): E70–2.
58. Jowett SL, Seal CJ, Pearce MS, Phillips E, Gregory W, Barton JR, Welfare MR. Influence of dietary factors on the clinical course of ulcerative colitis: a prospective cohort study. Gut. 2004;53(10):1479–84.
59. Blachier F, Andriamihaja M, Larraufie P, Ahn E, Lan A, Kim E. Production of hydrogen sulfide by the intestinal microbiota and epithelial cells and consequences for the colonic and rectal mucosa. Am J Physiol. 2021;320(2):G125–35.
60. Devkota S, Wang Y, Musch MW, Leone V, Fehlner-Peach H, Nadimpalli A, Antonopoulos DA, Jabri B, Chang EB. Dietary-fat-induced taurocholic acid promotes pathobiont expansion and colitis in IL-10$^{-/-}$ mice. Nature. 2012;487(7405):104–8.
61. Guo FF, Yu TC, Hong J, Fang JY. Emerging roles of hydrogen sulfide in inflammatory and neoplastic colonic diseases. Front Physiol. 2016;7:156.
62. Hirata I, Naito Y, Takagi T, Mizushima K, Suzuki T, Omatsu T, Handa O, Ichikawa H, Ueda H, Yoshikawa T. Endogenous hydrogen sulfide is an anti-inflammatory molecule in dextran sodium sulfate-induced colitis in mice. Dig Dis Sci. 2011;56(5):1379–86.
63. Wallace JL, Vong L, McKnight W, Dicay M, Martin GR. Endogenous and exogenous hydrogen sulfide promotes resolution of colitis in rats. Gastroenterology. 2009;137(2):569–78.

64. Flanigan KL, Agbor TA, Motta JP, Ferraz JGP, Wang R, Buret AG, Wallace JL. Proresolution effects of hydrogen sulfide during colitis are mediated through hypoxia-inducible factor-1α. FASEB J. 2015;29(4):1591–602.
65. Qin M, Long F, Wu W, Yang D, Huang M, Xiao C, Chen X, Liu X, Zhu YZ. Hydrogen sulfide protects against DSS-induced colitis by inhibiting NLRP3 inflammasome. Free Radic Biol Med. 2019;137:99–109.
66. Giuffrè A, Vicente JB. Hydrogen sulfide biochemistry and interplay with other gaseous mediators in mammalian physiology. Oxid Med Cell Longev. 2018;2018:6290931.
67. Wang A, Keita AV, Phan V, McKay CM, Schoultz I, Lee J, Murphy MP, Fernando M, Ronaghan M, Balce D, Yates R, Dicay M, Beck PL, MacNaughton WK, Söderholm JD, McKay DM. Targeting mitochondria-derived reactive oxygen species to reduce epithelial barrier dysfunction and colitis. Am J Pathol. 2014;184(9):2516–27.
68. Barbara G, Barbaro MR, Fuschi D, Palombo M, Falangone F, Cremon C, Marasco G, Standghellini V. Inflammatory and microbiota-related regulation of the intestinal epithelial barrier. Front Nutr. 2021;8:718356.
69. Motta JP, Flannigan KL, Agbor TA, Beaty JK, Blackler RW, Workentime ML, Da Silva GJ, Wang R, Buret AG, Wallace JL. Hydrogen sulfide protects from colitis and restores intestinal biofilm and mucus production. Inflamm Bowel Dis. 2015;21(5):1006–17.
70. Blachier F, Beaumont M, Kim E. Cysteine-derived hydrogen sulfide and gut health: a matter of endogenous or bacterial origin. Curr Opin Clin Nutr Metab Care. 2019;22(1):68–75.
71. Shimada Y, Kinoshita M, Harada K, Mizutani M, Masahata K, Kayama H, Takeda K. Commensal bacteria-dependent indole production enhances epithelial barrier function in the colon. PLos One. 2013;8(11):e80604.
72. Venkatesh M, Mukherjee S, Wang H, Li H, Sun K, Benechet AP, Qiu Z, Maher L, Redinbo MR, Phillips RS, Fleet JC, Kortagere S, Mukherjee P, Fasano A, Le Ven J, Nicholson JK, Dumas ME, Khanna KM, Mani S. Symbiotic bacterial metabolites regulate gastrointestinal barrier function via the xenobiotic sensor PXR and Toll-like receptor 4. Immunity. 2014;41(2): 296–310.
73. Zelante T, Iannitti RG, Cunha C, De Luca A, Giovannini G, Pieraccini G, Zecchi R, D'Angelo C, Massi-Benedetti C, Fallarino F, Carvalho A, Puccetti P, Romani L. Tryptophan catabolites from microbiota engage aryl hydrocarbon receptor and balance mucosal reactivity via interleukin-22. Immunity. 2013;39(2):372–85.
74. Jin UH, Lee SO, Sridahan G, Lee K, Davidson LA, Jayaraman A, Chapkin RS, Alaniz R, Safe S. Microbiome-derived tryptophan metabolites and their aryl hydrocarbon receptor-dependent agonist and antagonist activities. Mol Pharmacol. 2014;85(5):777–88.
75. Whitefield-Cargile C, Cohen ND, Chapkin RS, Weeks BR, Davidson LA, Goldsby JS, Hunt CL, Steinmeyer SH, Menon R, Suchodolski JS, Jayaraman A, Alaniz RC. The microbiota-derived metabolite indole decreases mucosal inflammation and injury in a murine model of NSAID enteropathy. Gut Microbes. 2016;7(3):246–61.
76. Bansal T, Alaniz RC, Wood TK, Jayaraman A. The bacterial signal indole increases epithelial-cell tight-junction resistance and attenuates indicators of inflammation. Proc Natl Acad Sci USA. 2010;107(1):228–33.
77. Lamas B, Richard ML, Leducq V, Pham HP, Michel ML, Da Costa G, Bridonneau C, Jegou S, Hoffmann TW, Natividad JM, Brot L, Taleb S, Couturier-Maillard A, Nion-Larmurier L, Merabtene F, Selsik P, Bourrier A, Cosnes J, Ryffel B, Beaugerie L, Launay JM, Langella P, Xavier RJ, Sokol H. CARD9 impacts colitis by altering gut microbiota metabolism of tryptophan into aryl hydrocarbon receptor ligands. Nat Med. 2016;22(6):598–605.
78. Xiao HW, Cui M, Li Y, Dong JL, Zhang SQ, Zhu CC, Jiang M, Zhu T, Wang B, Wang HC, Fan SJ. Gut microbiota-derived indole 3-propionic acid protects against radiation toxicity via retaining acyl-CoA-binding protein. Microbiome. 2020;8(1):69.
79. Jennis M, Cavanaugh CR, Leo GC, Mabus JR, Lenhard J, Hornby PJ. Microbiota-derived tryptophan indoles increase after gastric bypass surgery and reduce intestinal permeability in vitro and in vivo. Neurogastroenterol Motil. 2018;30(2) https://doi.org/10.1111/nmo.13178.

80. Alexeev EA, Lanis JM, Kao DJ, Campbell EL, Kelly CJ, Battista KD, Gerich ME, Jenkins BR, Walk ST, Kominsky DJ, Colgan SP. Microbiota-derived indole metabolites promote human and murine intestinal homeostasis through regulation of interleukin-10 receptor. Am J Pathol. 2018;188(5):1183–94.

81. Scott SA, Fu J, Chang PM. Microbial tryptophan metabolites regulate gut barrier function via the aryl hydrocarbon receptor. Proc Natl Acad Sci USA. 2020;117(32):19376–87.

82. Wlodarska M, Luo C, Kolde R, d'Hennezel E, Annand JW, Heim CE, Krastel P, Schmitt EK, Omar AS, Creasey EA, Garner AL, Mohammadi S, O'Connell DJ, Abubucker S, Arthur TD, Franzosa EA, Huttenhower C, Murphy LO, Haiser HJ, Vlamakis H, Porter JA, Xavier RJ. Indoleacrylic acid produced by commensal Peptostreptococcus species suppresses inflammation. Cell Host Microbe. 2017;22(1):25–37.

83. Hashimoto T, Perlot T, Rehman A, Trichereau J, Ishiguro H, Paolino M, Sigl V, Hanada T, Hanada R, Lipinski S, Wild B, Camargo SMR, Singer D, Richter A, Kuba K, Fukamisu A, Schreiber S, Clevers H, Verrey F, Rosenstiel P, Prenninger JM. ACE2 links amino acid metabolism malnutrition to microbial ecology andintestinal inflammation. Nature. 2012;487 (7408):477–81.

84. Kim CJ, Kovacs-Nolan JA, Yang C, Archbold T, Fan MZ, Mine Y. L-tryptophan exhibits therapeutic function in a porcine model of dextran sodium sulfate (DSS)-induced colitis. J Nutr Biochem. 2010;21(6):468–75.

85. Ellis RJ, Small DM, Ng KL, Vesey DA, Vitetta L, Francis RS, Gobe GC, Morais C. Indoxyl sulfate induces apoptosis and hypertrophy in human kidney proximal tubular cells. Toxicol Pathol. 2018;46(4):449–59.

86. Leong SC, Sirich TL. Indoxyl sulfate. Review of toxicity and therapeutic strategies. Toxins (Basel). 2016;8(12):358.

87. Ramezani A, Raj DS. The gut microbiome, kidney disease, and targeted interventions. J Am Soc Nephrol. 2014;25(4):657–70.

88. Tan X, Cao X, Zou J, Shen B, Zhang X, Liu Z, Lv W, Teng J, Ding X. Indoxyl sulfate, a valuable biomarker in chronic kidney disease and dialysis. Hemodial Int. 2017;21(2):161–7.

89. Levy M, Thaiss CA, Zeevi D, Dohnalova D, Zilberman-Shapira G, Mahdi JA, David E, Savidor A, Korem T, Herzig Y, Pevsner-Fischer M, Shapiro H, Christ A, Harmelin A, Halpern Z, Latz E, Flavell RA, Amit I, Segal E, Elinav E. Microbiota-modulated metabolites shape the intestinal microenvironment by regulating NLRP6 inflammasome signaling. Cell. 2015;163(6):1428–43.

90. Latz E, Xiao TS, Stutz A. Activation and regulation of the inflammasomes. Nat rev Immunol. 2013;13(6):397–411.

91. Macia L, Tan J, Vieira AT, Leach K, Stanley D, Luong S, Maruya M, McKenzie CI, Hijikata A, Wong C, Binge L, Thorburn AN, Chevalier N, Ang C, Marino E, Robert R, Offermanns S, Teixeira MM, Moore RJ, Flavell RA, Fagarasan S, Mackay CR. Metabolite-sensing receptors GPR43 and GPR109A facilitate dietary fibre-induced gut homeostasis through regulation of the inflammasome. Nat Commun. 2015;6:6734.

92. Malik A, Kanneganti TD. Inflammasome activation and assembly at a glance. J Cell Sci. 2017;130(23):3955–63.

93. Goodwin AC, Destefano Shields CE, Wu S, Huso DL, Wu XQ, Murray-Stewart TR, Hacker-Prietz A, Rabizadeh S, Woster M, Sears CL, Casero RA Jr. Polyamine catabolism contributes to enterotoxigenic Bacteroides fragilis-induced colon tumorigenesis. Proc Natl Acad Sci USA. 2011;108(37):15354–9.

94. Chang PV, Hao L, Offermanns S, Medzhitov R. The microbial metabolite butyrate regulates intestinal macrophages function via histone deacetylase inhibition. Proc Natl Acad Sci USA. 2014;111(6):2247–52.

95. Furusawa Y, Obata Y, Fukuda S, Endo TA, Nakato G, Takahashi D, Nakanishi Y, Uetake C, Kato K, Kato T, Takahashi M, Fukuda NN, Murakami S, Miyauchi E, Hino S, Atarashi K, Onawa S, Fujimura Y, Lockett T, Clarke JM, Topping DL, Tomita M, Hori S, Ohara O, Morita T, Koseki H, Kikuchi J, Honda K, Hase K, Ohno H. Commensal microbe-derived

butyrate induces the differentiation of colonic regulatory T cells. Nature. 2013;504(7480): 446–50.

96. Finnie IA, Dwarakanath AD, Taylor BA, Rhodes JM. Colonic mucin synthesis is increased by sodium butyrate. Gut. 1995;36(1):93–9.

97. Chen G, Ran X, Li B, Li Y, He D, Huang B, Fu S, Liu J, Wang W. Sodium butyrate inhibits inflammation and maintains epithelium barrier integrity in a TNBS-induced inflammatory bowel disease mice model. EBioMedicine. 2018;30:317–25.

98. Lee C, Kim BG, Kim JH, Chun J, Im JP, Kim JS. Sodium butyrate inhibits the NF-kappa B signaling pathway and histone deacetylation, and attenuates experimental colitis in an IL-10 independent manner. Int Immunopharmacol. 2017;51:47–56.

99. Zhou L, Zhang M, Wang Y, Dorfman RG, Liu H, Yu T, Chen X, Tang D, Xu L, Yin Y, Pan Y, Zhou Q, Zhou Y, Yu C. Fecalibacterium prausnitzii produces butyrate to maintain Th17/Treg balance and to ameliorate colorectal colitis by inhibiting histone deacetylase 1. Inflamm Bowel Dis. 2018;24(9):1926–40.

100. Steinhart AH, Hiruki T, Brzezinski A, Baker JP. Treatment of left-sided ulcerative colitis with butyrate enemas: a controlled study. Aliment Pharmacol Ther. 1996;10(5):729–36.

101. Breuer RI, Soergel KH, Lashner BA, Christ ML, Hanauer SB, Vanagunas A, Harig JM, Keshavarzian A, Robinson M, Sellin JH, Weinberg D, Vidican DE, Flemal KL, Rademaker AW. Short chain fatty acid rectal irrigation for left-sided ulcerative colitis: a randomized, placebo controlled trial. Gut. 1997;40(4):485–91.

102. Scheppach W. Treatment of distal ulcerative colitis with short-chain fatty acid enemas. A placebo-controlled trial. German-Austrian SCFA Study Group. Dig Dis Sci. 1996;41(11): 2254–9.

103. Lührs H, Gerke T, Müller JG, Melcher R, Schauber J, Boxberge F, Scheppach W, Menzel T. Butyrate inhibits NF-kappa B activation in lamina propria macrophages of patients with ulcerative colitis. Scand J Gastroenterol. 2002;37(4):458–66.

104. Scheppach W, Sommer H, Kirchner T, Paganelli GM, Bartram P, Christl S, Richter F, Dusel G, Kasper H. Effect of butyrate enemas on the colonic mucosa in distal ulcerative colitis. Gastroenterology. 1992;103(1):51–6.

105. Vernia P, Annese V, Bresci G, d'Albasio G, D'Incà R, Giaccari S, Ingrosso M, Mansi C, Riegler C, Valpiani D, Caprilli R, Gruppo Italiano per lo Studio del Colon and del Retto. Topical butyrate improves efficacy of 5-ASA in refractory distal ulcerative colitis: results of a multicentre trial. Eur J Clin Invest. 2003;33(3):244–8.

106. Vernia P, Marchegianno A, Caprilli R, Frieri G, Carrao G, Valpiani D, Di Paolo MC, Paoluzi P, Torsoli A. Short-chain fatty acid topical treatment in distal ulcerative colitis. Aliment Pharmacol Ther. 1995;9(3):309–13.

107. Jamka M, Kokot M, Kaczmarek N, Bermagambetova S, Nowak JK, Walkowiak J. The effect of sodium butyrate enemas compared with placebo on disease activity, endoscopic scores, and histological and inflammatory parameters in inflammatory bowel diseases: a systematic review of randomized controlled trials. Complement Med Res. 2021;28(4):344–56.

108. Thibault R, Blachier F, Darcy-Vrillon B, de Coppet P, Bourreille A, Segain JP. Butyrate utilization by the colonic mucosa in inflammatory bowel diseases: a transport deficiency. Inflamm Bowel dis. 2010;16(4):684–95.

109. Lux GD, Marton LJ, Baylin SB. Onithine decarboxylase is important in intestinal mucosal maturation and recovery from injury in rats. Science. 1980;210(4466):195–8.

110. McCormack SA, Viar MJ, Johnson LR. Migration of IEC-6 cells: a model for mucosal healing. Am J Physiol. 1992;263(3):G426–35.

111. McCormack SA, Wang JY, Viar MJ, Tague L, Davies PJ, Johnson LR. Polyamines influence transglutaminase activity and cell migration in two cell lines. Am J Physiol. 1994;267(3): C706–14.

112. Rao JN, Liu L, Zou T, Marasa BS, Boneva D, Wang SR, Malone DL, Turner DJ, Wang JY. Polyamines are required for phospholipase C-gamma1 expression promoting intestinal epithelial restitution after wounding. Am J Physiol. 2007;292(1):G335–43.

113. Bloemen JG, Schrienermacher MH, de Bruine AP, Buurman WA, Bouvy ND, Dejong CH. Butyrate enemas improve intestinal anastomotic strength in a rat model. Dis Colon Rectum. 2010;53(7):1069–75.

114. Mathew AJ, Wann VC, Abraham DT, Jacob PM, Selvan BS, Ramakrishna BS, Nair AN. The effect of butyrate on the healing of colonic anastomoses in rats. J Invest Surg. 2010;23(2): 101–4.

115. Rolandelli RH, Koruda MJ, Settle RG, Rombeau JL. Effects of intraluminal infusion of short-chain fatty acids on the healing of colonic anastomosis in the rat. Surgery. 1986;100(2): 198–204.

116. Hamer HM, Jonkers DMAE, Vanhoutvin SALW, Troost FJ, Rijkers G, de Bruïne A, Bast A, Venema K, Brummer RJM. Effect of butyrate enemas on inflammation and antioxidant status in the colonic mucosa of patients with ulcerative colitis in remission. Clin Nutr. 2010;29(6): 738–44.

117. Ford AC, Sperber AD, Corsetti M, Camilleri M. Irritable bowel syndrome. Lancet. 2020;396 (10263):1675–88.

118. Mayer EA, Savidge T, Shulman RJ. Brain-gut microbiome interactions and functional bowel disorders. Gastroenterology. 2014;146(6):1500–12.

119. Chey WD, Kurlander J, Eswaran S. Irritable bowel syndrome: a clinical review. JAMA. 2015;313(9):949–58.

120. Camilleri M, Lasch K, Zhou W. Irritable bowel syndrome: methods, mechanisms, and pathophysiology. The confluence of increased permeability, inflammation, and pain in irritable bowel syndrome. Am J Physiol. 2012;303(7):G775–85.

121. Matricon J, Meleine M, Gelot A, Piche T, Dapoigny M, Muller E, Ardid D. Review article: associations between immune activation, intestinal permeability and the irritable bowel syndrome. Aliment Pharmacol Ther. 2012;36(11–12):1009–31.

122. Simren M, Barbara G, Flint HJ, Spiegel BMR, Spiller RC, Vanner S, Verdu EF, Whorwell PJ, Zoetendal EG, Rome Foundation Committee. Intestinal microbiota in functional bowel disorders: a Rome foundation report. Gut. 2013;62(1):159–76.

123. Hughes PA, Zola H, Penttila IA, Blackshaw LA, Andrews JM, Krumbiegel D. Immune activation in irritable bowel syndrome: can neuroimmune interactions explain symptoms? Am J Gastroenterol. 2013;108(7):1066–74.

124. Ringel Y, Maharshak N. Intestinal microbiota and immune function in the pathogenesis of irritable bowel syndrome. Am J Physiol. 2013;305(8):G529–41.

125. Valdez-Morales EE, Overington J, Guerrero-Alba R, Ochoa-Cortes F, Ibeakanma CO, Spreadbury I, Bunnett NW, Beyak M, Vanner SJ. Sensitization of peripheral sensory nerves by mediators from colonic biopsies of diarrhea-predominant irritable bowel syndrome patients: a role for PAR2. Am J Gastroenterol. 2013;108(10):1634–43.

126. Raskov H, Burcharth J, Pommergaard HC, Rosenberg J. Irritable bowel syndrome, the microbiota and the gut-brain axis. Gut Microbes. 2016;7(5):365–83.

127. Chey WY, Jin HO, Lee MH, Sun SW, Lee KY. Colonic motility abnormality in patients with irritable bowel syndrome exhibiting abdominal pain and diarrhea. Am J Gastroenterol. 2001;96(5):1499–506.

128. Lebwohl B, Sanders DS, Green PHR. Coeliac disease. Lancet. 2018;391(10115):70–81.

129. Caminero A, Galipeau HJ, McCarville JL, Johnston CW, Bernier SP, Russell AK, Jury J, Herran AR, Casqueiro J, Tye-Din JA, Surette MG, Magarvey NA, Schuppan D, Verdu EF. Duodenal bacteria from patients with celiac disease and healthy subjects distinctly affect gluten breakdown and immunogenicity. Gastroenterology. 2016;151(4):670–83.

130. Lebwohl B, Rubio-Tapia A. Epidemiology, presentation, and diagnosis of celiac disease. Gastroenterology. 2021;160(1):63–75.

131. Caio G, Volta G, Sapone A, Leffler DA, De Giorgio R, Catassi C, Fasano A. Celiac disease: a comprehensive current review. BMC Med. 2019;17(1):142.

132. Wei G, Helmerhorst EJ, Darwish G, Blumenkranz G, Schuppan D. Gluten degrading enzymes for treatment of celiac disease. Nutrients. 2020;12(7):2095.

133. Chong PP, Chin VK, Looi CY, Wong WF, Madhavan P, Yong VC. The microbiome and irritable bowel syndrome. A review on the pathophysiology, current research and future therapy. Front Microbiol. 2019;10:1136.
134. Dionne J, Ford AC, Yuan Y, Chey WD, Lacy BE, Saito YA, Quigley EMM, Moayyedi P. A systematic review and meta-analysis evaluating the efficacy of a gluten-free diet and a low FODMAPs diet in treating symptoms of irritable bowel syndrome. Am J Gastroenterol. 2018;113(9):1290–300.
135. Altomare A, Di Rosa C, Imperia E, Emerenziani S, Cicala M, Guarino MPL. Diarrhea predominant-irritable bowel syndrome (IBS-D): effects of different nutritional patterns on intestinal dysbiosis and symptoms. Nutrients. 2021;13(5):1506.
136. Jalanka-Tuovinen J, Salonen A, Nikkilä J, Immonen O, Kekkonen R, Lahti L, Palva A, de Vos WM. Intestinal microbiota in healthy adults: temporal analysis reveals individual and common core and relation to intestinal symptoms. PLoS One. 2011;6(7):e23035.
137. Le Gall G, Noor SO, Ridgway K, Scovell L, Jamieson C, Johnson IT, Colquhoun IJ, Kemsley EK, Narbad A. Metabolomics of fecal extracts detects altered metabolic activity of gut microbiota in ulcerative colitis and irritable bowel syndrome. J Proteome Res. 2011;10(9): 4208–18.
138. Zaharia V, Varzescu M, Djavadi I, Newman E, Egnor RW, Alexander-Chacko J, Charney AN. Effects of short-chain fatty acids on colonic Na$^+$ absorption and enzyme activity. Comp Biochem Physiol. 2001;128(2):335–47.
139. Boudry G, Jamin A, Chatelais L, Gras-Le Guen C, Michel C, Le Huërou-Luron I. Dietary protein excess during neonatal life alters colonic microbiota and mucosal response to inflammatory mediators later in life in female pigs. J Nutr. 2013;143(8):1225–32.
140. Ahmed I, Greenwood R, De Lacy CB, Ratcliffe NM, Probert CS. An investigation of fecal volatile organic metabolites in irritable bowel syndrome. PLoS One. 2013;8(3):e58204.
141. Tana C, Umesaki Y, Imaoka A, Handa T, Kanazawa M, Fukudo S. Altered profiles of intestinal microbiota and organic acids may be at the origin of symptoms in irritable bowel syndrome. Neurogastroenterol Motil. 2010;22(5):512–9.
142. Baidoun F, Elshiwy K, Elkeraie Y, Merjaneh Z, Khoudari G, Sarmini MT, Gad M, Al-Husseini M, Saad A. Colorectal cancer epidemiology: recent trends and impact and outcomes. Curr Drug Targets. 2021;22(9):998–1009.
143. Center MM, Jemal A, Smith RA, Ward E. Worldwide variations in colorectal cancer. CA Cancer J Clin. 2009;59(6):366–78.
144. Parkin DM, Bray F, Ferlay J, Pisani P. Global cancer statistics, 2002. CA Cancer J Clin. 2005;55(2):74–108.
145. Stintzing S, Tejpar S, Gibbs P, Thiebach L, Lenz HJ. Understanding the role of primary tumour localization in colorectal cancer treatment and outcomes. Eur J Cancer. 2017;84:69–80.
146. Fearon ER, Volgenstein B. A genetic model for colorectal carcinogenesis. Cell. 1990;61(5): 759–67.
147. Arends JW. Molecular interactions in the Volgenstein model of colorectal carcinoma. J Pathol. 2000;190(4):412–6.
148. Markowitz SD, Bertagnolli MM. Molecular origins of cancer: molecular basis of colorectal cancer. N Engl J Med. 2009;361(25):2449–60.
149. Lamprecht SA, Lipkin M. Migrating colonic crypt epithelial cells: primary targets for transformation. Carcinogenesis. 2002;23(11):1777–80.
150. Merritt AJ, Potten CS, Watson AJ, Loh DY, Nakayama K, Nakayama K, Hickman JA. Differential expression of bcl-2 in intestinal epithelia. Correlation with attenuation of apoptosis in colonic crypts and the incidence of colonic neoplasia. J Cell Sci. 1995;108(6): 2261–71.
151. Romagnolo B, Berrebi D, Saadi-Keddoucci S, Porteu A, Pichard AL, Peuchmaur M, Vandewalle A, Kahn A, Perret C. Intestinal dysplasia and adenoma in transgenic mice after overexpression of an activated beta-catenin. Cancer Res. 1999;59(16):3875–9.

152. Wong WM, Mandir N, Goodlad RA, Wong BCY, Garcia SB, Lam SK, Wright NA. Histogenesis of human colorectal adenomas and hyperplastic polyps: the role of cell proliferation and crypt fission. Gut. 2002;50(2):212–7.
153. Hammoud SS, Cairns BR, Jones DA. Epigenetic regulation of colon cancer and intestinal stem cells. Curr Opin Cell Biol. 2013;25(2):177–83.
154. Puccini A, Berger MD, Naseem M, Tokunaga R, Battaglin F, Cao S, Hanna DL, McSkane M, Soni S, Zhang W, Lenz HJ. Colorectal cancer: epigenetic alterations and their clinical implications. Biochim Biophys Acta. 2017;1868(2):439–48.
155. Grady WM, Carethers JM. Genomic and epigenetic instability in colorectal cancer pathogenesis. Gastroenterology. 2008;135(4):1079–99.
156. Lengauer C, Kinzler KW, Volgenstein B. Genetic instabilities in human cancers. Nature. 1998;396(6712):643–9.
157. Kuipers EJ, Grady WM, Liberman D, Seufferlein T, Sung JJ, Boelens PG, van de Velde CJH, Watanabe T. Colorectal cancer. Nat Rev Dis Primers. 2015;1:15065.
158. Dekker E, Tanis PJ, Vleugels JLA, Kasi PM, Wallace MB. Colorectal cancer. Lancet. 2019;394(10207):1467–80.
159. Harada S, Morlote D. Molecular pathology of colorectal cancer. Adv Anat Pathol. 2020;27(1): 20–6.
160. Luo Y, Wong CJ, Kaz AM, Dzieciatkowski S, Carter KT, Morris SM, Wang J, Willis JE, Makar KW, Ulrich CM, Lutterbaugh JD, Shrubsole MJ, Zheng W, Markowitz SD, Grady WM. Differences in DNA methylation signatures reveal multiple pathways of progression from adenoma to colorectal cancer. Gastroenterology. 2014;147(2):418–29.
161. Van Engeland M, Derks S, Smits KM, Meijer GA, Herman JG. Colorectal cancer epigenetics: complex simplicity. J Clin Oncol. 2011;29(10):1382–91.
162. Bardelli A, Parsons DW, Silliman N, Ptak J, Szabo S, Saha S, Markowitz S, Willson JKV, Parmigiani G, Kinzler KW, Vogelstein B, Velculescu VE. Mutational analysis of the tyrosine kinase in colorectal cancers. Science. 2003;300(5621):949.
163. Chittenden TW, Howe EA, Culhane AC, Sultana R, Taylor JM, Holmes C, Quakenbush J. Functional classification analysis of somatically mutated genes in human breast and colorectal cancers. Genomics. 2008;91(6):508–11.
164. Jubb AM, Bell SM, Quirke P. Methylation and colorectal cancer. J Pathol. 2001;195(1): 111–34.
165. Parsons DW, Wang TL, Samuels Y, Bardelli A, Cummins JM, DeLong L, Silliman N, Ptak J, Szabo S, Willson JKV, Markowitz S, Kinzler KW, Vogelstein B, Lengauer C, Velculescu VE. Colorectal cancer: mutations in a signalling pathway. Nature. 2005;436(7052):792.
166. Starr TK, Allaei R, Silverstein KAT, Staggs RA, Sarver AL, Bergemann TL, Gupta M, O'Sullivan G, Matise I, Dupuy AJ, Collier LS, Powers C, Oberg AL, Asmann YW, Thibodeau SL, Tessarollo M, Copeland NG, Jenkins NA, Cormier RT, Largaespada DA. A transposon-based genetic screen in mice identifies genes altered in colorectal cancer. Science. 2009;323 (5922):1747–50.
167. Sinicrope FA. Lynch syndrome-associated colorectal cancer. N Engl J Med. 2018;379(8): 764–73.
168. Morin PJ, Sparks AB, Korinek V, Barker N, Clevers H, Vogelstein B, Kinzler KW. Activation of beta-catenin-Tcf signaling in colon cancer by mutations in beta-catenin or APC. Science. 1997;275(5307):1787–90.
169. Galiatsatos P, Foulkes WD. Familial adenomatous polyposis. Am J Gastroenterol. 2006;101 (2):385–98.
170. Vasen HFA, Tomlinson I, Castells A. Clinical management of hereditary colorectal cancer syndromes. Nat Rev Gastroenterol Hepatol. 2015;12(2):88–97.
171. Zhan T, Rindtorff N, Boutros M. Wnt signaling in cancer. Oncogene. 2017;36(11):1461–73.
172. Jess T, Rungoe C, Peyrin-Biroulet L. Risk of colorectal cancer in patients with ulcerative colitis: a meta-analysis of population-based cohort studies. Clin Gastroenterol Hepatol. 2012;10(6):639–45.

173. Terzic J, Grivennikov S, Karin E, Karin M. Inflammation and colon cancer. Gastroenterology. 2010;138(6):2101–14.
174. Castano-Milla C, Chaparro M, Gisbert JP. Systematic review with meta-analysis: the declining risk of colorectal cancer in ulcerative colitis. Aliment Pharmacol Ther. 2014;39(7):645–59.
175. Jess T, Simonsen J, Jorgensen KT, Pedersen BV, Nielsen NM, Frisch M. Decreasing risk of colorectal cancer in patients with inflammatory bowel disease over 30 years. Gastroenterology. 2012;143(2):375–81.
176. Li J, Ma X, Chakravarti D, Shalapour S, DePinho RA. Genetic and biological hallmarks of colorectal cancer. Genes Dev. 2021;35(11–12):787–820.
177. Lichtenstein P, Holm NV, Verkasalo PK, Iliadou A, Kaprio J, Koskenvuo M, Pukkala E, Skytthe A, Hemminki A. Environmental and heritable factors in the causation of cancer: analyses of cohorts of twins from Sweden, Denmark, and Finland. N Engl J Med. 2000;343(2): 78–85.
178. World Cancer Research Fund/American Institute for Cancer Research. Food, nutrition, physical activity, and the prevention of cancer: a global perspective. Washington DC: AICR; 2007.
179. Kim E, Coelho D, Blachier F. Review of the association between meat consumption and risk of colorectal cancer. Nutr Res. 2013;33(12):983–94.
180. Windey K, De Preter V, Verbeke K. Relevance of protein fermentation to gut health. Mol Nutr Food Res. 2012;56(1):184–96.
181. Song M, Garrett WS, Chan AT. Nutrients, foods, and colorectal cancer prevention. Gastroenterology. 2015;148(6):1244–60.
182. Botteri E, Iodice S, Bagnardi V, Raimondi S, Lowenfels AB, Maisonneuve P. Smoking and colorectal cancer: a meta-analysis. JAMA. 2008;300(23):2765–78.
183. Liang PS, Chen TY, Giovannucci E. Cigarette smoking and colorectal cancer incidence and mortality: systematic review and meta-analysis. Int J Cancer. 2009;124(10):2406–15.
184. Bardou M, Barkun AN, Martel M. Obesity and colorectal cancer. Gut. 2013;62(6):933–47.
185. Fedirko V, Tramacere I, Bagnardi V, Rota M, Scotti L, Islami F, Negri E, Straif K, Romieu I, La Vecchia C, Boffetta P, Jenab M. Alcohol drinking and colorectal cancer risk: an overall and dose-response meta-analysis of published studies. Ann Oncol. 2011;22(9):1958–72.
186. Akimoto N, Ugai T, Zhong R, Hamada T, Fujiyoshi K, Giannakis M, Wu K, Cao Y, Ng K, Ogino S. Rising incidence of early-onset colorectal cancer. A call to action. Nat Rev Clin Oncol. 2021;18(4):230–43.
187. Bultman SJ. Interplay between diet, gut microbiota, epigenetic events, and colorectal cancer. Mol Nutr Food Res. 2017;61(1):10.
188. O'Keefe SJD. Diet, microorganisms and their metabolites, and colon cancer. Nat Rev Gastroenterol Hepatol. 2016;13(12):691–706.
189. Vernia F, Longo S, Stefanelli G, Viscido A, Latella G. Dietary factors modulating colorectal carcinogenesis. Nutrients. 2021;13(1):143.
190. Chan DSM, Lau R, Aune D, Vieira R, Greenwood DC, Kampman E, Norat T. Red and processed meat and colorectal cancer incidence: meta-analysis of prospective studies. PLoS One. 2011;6(6):e20456.
191. Chao A, Thun MJ, Connell C, McCullough ML, Jacobs EJ, Dana Flanders W, Rodriguez C, Sinha R, Calle EE. Meat consumption and risk of colorectal cancer. JAMA. 2005;293(2): 172–83.
192. Larsson S, Wolk A. Meat consumption and risk of colorectal cancer: a meta-analysis of prospective studies. Int J Cancer. 2006;119(11):2657–64.
193. Norat T, Lukanova A, Ferrari P, Riboli E. Meat consumption and colorectal cancer risk: dose-response meta-analysis of epidemiological studies. Int J Cancer. 2002;98(2):241–56.
194. Sandhu MS, White IR, McPherson K. Systematic review of the prospective cohort studies on meat consumption and colorectal cancer risk: a meta-analytical approach. Cancer Epidemiol Biomarkers Prev. 2001;10(5):439–6.
195. Vieira AR, Abar L, Chan DSM, Vingeliene S, Polemiti E, Stevens C, Greenwood D, Norat T. Foods and beverages and colorectal cancer risk: a systematic review and meta-analysis of

cohort studies, an update of the evidence of the WCRF-AICR continuous update project. Ann Oncol. 2017;28(8):1788–802.

196. Smolinska K, Paluszkiewicz P. Risk of colorectal cancer in relation with frequency and total amount of red meat consumption. Systematic review and meta-analysis. Arch Med Sci. 2010;6 (4):605–10.

197. Bastide NM, Pierre FHF, Corpet DE. Heme iron from meat and risk of colorectal cancer: a meta-analysis and a review of the mechanisms involved. Cancer Prev Res (Phila). 2011;4(2): 177–84.

198. Alexander DD, Miller AJ, Cushing CA, Lowe KA. Processed meat and colorectal cancer: a quantitative review of prospective epidemiological studies. Eur J Cancer Prev. 2010;19(5): 328–41.

199. Ollberding NJ, Wilkens LR, Henderson BE, Kolonel LN, Le Marchand L. Meat consumption, heterocyclic amines and colorectal cancer risk: the multiethnic cohort study. Int J Cancer. 2012;131(7):E1125–33.

200. Parr CL, Hjartaker A, Lund E, Veierod MB. Meat intake, cooking methods and risk of proximal colon, distal colon and rectal cancer: the Norwegian Women and Cancer (NOWAC) cohort study. Int J Cancer. 2013;133(5):1153–63.

201. Zhang X, Giovannucci EL, Smith-Warner SA, Wu K, Fuchs CS, Pollak M, Willett WC, Ma J. A prospective study of intakes of zinc and heme iron and colorectal cancer risk in men and women. Cancer Causes Control. 2011;22(12):1627–37.

202. Kabat GC, Miller AB, Jain M, Rohan TE. A cohort study of dietary iron and heme iron intake and risk of colorectal cancer in women. Br J Cancer. 2007;97(1):118–22.

203. De Oliveira MJ, Boué G, Guillou S, Pierre F, Membré JM. Estimation of the burden of disease attributable to red meat consumption in France: influence on colorectal cancer and cardiovascular diseases. Food Chem Toxicol. 2019;130:174–86.

204. Lopez A, Cacoub P, Macdougall IC, Peyrin-Biroulet L. Iron deficiency anemia. Lancet. 2016;387(10021):907–16.

205. Saboor M, Zehra A, Hamadi HA, Mobarki AA. Revisiting iron metabolism, iron homeostasis and iron deficiency anemia. Clin Lab. 2021;67(3) https://doi.org/10.7754/Clin.Lab.2020. 200742.

206. Haenszel W, Kurihara M. Studies of Japanese migrants. I. Mortality from cancer and other diseases among Japanese in the United States. J Natl Cancer Inst. 1968;40(1):43–68.

207. Oba S, Shimizu N, Nagata C, Shimizu H, Kametani M, Takeyama N, Ohnuma T, Matsushita S. The relationship between the consumption of red meat, fat, and coffee and the risk of colon cancer: a prospective study in Japan. Cancer Lett. 2006;244(2):260–7.

208. Takachi R, Tsubono Y, Baba K, Inoue M, Sasazuki S, Iwasaki M, Tsugane S, Japan Public Health Center-based Prospective Study Group. Red meat intake may increase the risk of colon cancer in Japanese, a population with relatively low red meat consumption. Asia Pac J Clin Nutr. 2011;20(4):603–12.

209. Puccini A, Battaglin F, Lenz HJ. Management of advanced small bowel cancer. Curr Treat Options Oncol. 2018;19(12):69.

210. Errichiello E, Venesio T. Mitochondrial DNA variants in colorectal carcinogenesis: Drivers or passengers? J Cancer Res Clin Oncol. 2017;143(10):1905–14.

211. Nguyen LH, Goel A, Chung DC. Pathways of colorectal carcinogenesis. Gastroenterology. 2020;158(2):291–302.

212. Skonieczna K, Malyarchuk BA, Grzybowski T. The landscape of mitochondrial DNA variation in human colorectal cancer on the background of phylogenetic knowledge. Biochim Biophys Acta. 2012;1825(2):153–9.

213. Taylor RW, Barron MJ, Borthwick GM, Gospel A, Chinnery PF, Samuels DC, Taylor GA, Plusa SM, Needham SJ, Greaves LC, Kirkwood TB, Turnbull DM. Mitochondrial DNA mutations in human colonic crypt stem cells. J Clin Invest. 2003;112(9):1351–60.

214. Wood LD, Parsons DW, Jones S, Lin J, Sjöblom T, Leary RH, Shen D, Boca SM, Barber T, Ptak J, Silliman N, Szabo S, Dezso Z, Ustyanksky V, Nikolskaya T, Nikolsky Y, Karchin R,

Wilson PA, Kaminker JS, Zhang Z, Croshaw R, Willis J, Dawson D, Shipitsin M, Wilson JK, Sukumar S, Polyak K, Park BH, Pethiyagoda CL, Pant PV, Ballinger DG, Sparks AB, Hartigan J, Smith DR, Suh E, Papadopoulos N, Buckhaults P, Markowitz SD, Parmigiani G, Kinzler KW, Velculescu VE, Vogelstein B. The genomic landscapes of human breast and colorectal cancer. Science. 2007;318(5853):1108–13.

215. Gill CIR, Rowland IR. Diet and cancer: assessing the risk. Br J Nutr. 2002;88(S1):S73–87.
216. Barker N, Ridgway RA, van Es JH, van de Wetering M, Begthel H, van den Born M, Danenberg E, Clarke AR, Sansom OJ, Clevers H. Crypt stem cells as the cells-of-origin of intestinal cancer. Nature. 2009;457(7229):608–11.
217. Vermeulen L, Todaro M, de Sousa MF, Sprick MR, Kemper M, Perez Alea M, Richel DJ, Stassi G, Medema JP. Single-cell cloning of colon cancer stem cells reveals a multi-lineage differentiation capacity. Proc Natl Acad Sci USA. 2008;105(36):13427–32.
218. Zeki SS, Graham TA, Wright NA. Stem cells and their implications for colorectal cancer. Nat Rev Gastroenterol Hepatol. 2011;8(2):90–100.
219. Yousefi M, Linheng L, Lengner CJ. Hierarchy and plasticity in the intestinal stem cell compartment. Trends Cell Biol. 2017;27(10):753–64.
220. Wang X, Yang Y, Huycke MM. Microbiome-driven carcinogenesis in colorectal cancer: Models and mechanisms. Free Radic Biol Med. 2017;105:3–15.
221. Attene-Ramos MS, Nava GM, Muellner MG, Wagner ED, Plewa MJ, Gaskins HR. DNA damage and toxicogenomic analyses of hydrogen sulfide in human intestinal epithelial FHs 74 Int cells. Environ Mol Mutagen. 2010;51(4):304–14.
222. Aggarwal N, Donald ND, Malik S, Selvendran SS, McPhail MJ, Monahan KJ. The association of low-penetrance variants in DNA repair genes with colorectal cancer: a systematic review and meta-analysis. Clin Transl Gastroenterol. 2017;8(7):e109.
223. Laporte GA, Leguisamo NM, Kalil AN, Saffi J. Clinical importance of DNA repair in sporadic colorectal cancer. Crit Rev Oncol Hematol. 2018;126:168–85.
224. Tomasini PP, Guecheva TN, Leguisamo NM, Péricart S, Brunac AC, Hoffmann JS, Saffi J. Analysing the opportunities to target DNA double-strand breaks repair and replicative stress responses to improve therapeutic index of colorectal cancer. Cancers (Basel). 2021;13(13): 3130.
225. Szabo C, Coletta C, Chao C, Modis K, Szczesny B, Papapetropoulos A, Hellmich MR. Tumor-derived hydrogen sulfide, produced by cystathionine-β-synthase, stimulates bio-energetics, cell proliferation, and angiogenesis in colon cancer. Proc Natl Acad Sci. 2013;110 (30):12474–4479.
226. Phillips CM, Zatarain JR, Nicholls ME, Porter C, Widen SG, Thanki K, Johnson P, Jawad MU, Moyer MP, Randall JW, Hellmich JL, Maskey M, Qiu S, Wood TG, Druzhyna N, Szczesny B, Modis K, Szabo C, Chao C, Hellmich MR. Upregulation of cystathionine-β-synthase in colonic epithelia reprograms metabolism and promotes carcino-genesis. Cancer Res. 2017;77(21):5741–50.
227. Chao C, Zatarain JR, Ding Y, Coletta C, Mrazek AA, Druzhyna N, Johnson P, Chen H, Hellmich JL, Asimakopoulou A, Yanagi K, Olah G, Szoleczky P, Törö P, Bohanon FJ, Cheema M, Lewis R, Eckelbarger D, Ahmad A, Modis K, Untereiner A, Szczesny B, Papapetropoulos A, Zhou J, Hellmich MR, Szabo C. Cystathionine-beta-synthase inhibition for colon cancer: Enhancement of the efficacy of aminooxyacetic acid via the prodrug approach. Mol Med. 2016;22:361–79.
228. Hale VL, Jeraldo P, Mundy M, Yao J, Keeney J, Scott N, Cheek H, Davidson J, Greene M, Martinez C, Lehman J, Pettry C, Reed E, Lyke K, White BA, Diener C, Resendis-Antonio O, Gransee J, Dutta P, Petterson XM, Boardman L, Larson D, Nelson H, Chia N. Synthesis of multi-omic data and community metabolic models reveals insights into the role of hydrogen sulfide in colon cancer. Methods. 2018;149:59–68.
229. Bullman S, Pedamallu CS, Sicinska E, Clancy TE, Zhang X, Cai D, Neuberg D, Huang K, Guevara F, Nelson T, Chipashvili O, Hagan T, Walker M, Ramachandran A, Diosdado B, Serna G, Mulet N, Landolfi S, Ramon Y, Cajal S, Fasani R, Aguirre AJ, Ng K, Elez E,

Ogino S, Tabernero J, Fuchs CS, Hahn WC, Nuciforo P, Meyerson M. Analysis of Fusobacterium persistence and antibiotic response in colorectal cancer. Science. 2017;358 (6369):1443–8.

230. Kostic AD, Chun E, Robertson L, Glickman JN, Gallini CA, Michaud M, Clancy TE, Chung DC, Lochhead P, Hold GL, El-Omar EM, Brenner D, Fuchs CS, Meyerson M, Garrett WS. Fusobacterium nucleatum potentiates intestinal tumorigenesis and modulates the tumor-immune microenvironment. Cell Host Microbe. 2013;14(2):207–15.

231. Rubinstein MR, Baik JE, Lagana SM, Han RP, Raab WJ, Sahoo D, Dalerba P, Wang TC, Han HW. Fusobacterium nucleatum promotes colorectal cancer by inducing Wnt/β-catenin modulator Annexin A1. EMBO Rep. 2019;20(4):e47638.

232. Yachida S, Mizutani S, Shiroma H, Shiba S, Nakajima T, Sakamoto T, Watanabe H, Masuda K, Nishimoto Y, Kubo M, Hosoda F, Rokutan H, Matsumoto M, Takamaru H, Yamada M, Matsuda T, Iwasaki M, Yamaji T, Yachida T, Soga T, Kurokawa K, Toyoda A, Ogura Y, Hayashi T, Hatakeyama M, Nakagama H, Saito Y, Fukuda S, Shibata T, Yamada T. Metagnomic and metabolomic analyses reveal distinct stage-specific phenotypes of the gut microbiota in colorectal cancer. Nat Med. 2019;25(6):968–76.

233. Yang Y, Weng W, Peng J, Hong L, Yang L, Toiyama Y, Gao R, Liu M, Yin M, Pan C, Li H, Guo B, Zhu Q, Wei Q, Moyer MP, Wang P, Cai S, Goel A, Qin H, Ma Y. Fusobacterium nucleatum increases proliferation of colorectal cancer cells and tumor development in mice by activating Toll-like receptor-4 signaling to nuclear factor-κB, and up-regulating expression of microRNA-21. Gastroenterology. 2017;152(4):851–66.

234. Hale VL, Jeraldo P, Chen J, Mundy M, Yao J, Priya S, Keeney G, Lyke K, Ridlon J, White BA, French AJ, Thibodeau SN, Diener C, Resendis-Antonio O, Gransee J, Dutta T, Petterson XM, Sung J, Blekhman R, Boardman L, Larson D, Nelson H, Chia N. Distinct microbes, metabolites, and ecologies define the microbiome in deficient and proficient mismatch repair colorectal cancers. Genome Med. 2018;10(1):78.

235. Yasici C, Wolf PG, Kim H, Cross TWL, Vermillion K, Carroll T, Augustus GJ, Mutlu E, Tussing-Humphreys L, Braunschweig C, Xicola RM, Jung B, Llor X, Ellis NA, Gaskins HR. Race-dependent association of sulfidogenic bacteria with colorectal cancer. Gut. 2017;66 (11):1983–94.

236. Yamagishi K, Onuma K, Chiba Y, Yagi S, Aoki S, Sato T, Suguwara Y, Hosoya N, Saeki Y, Takahashi M, Fuji M, Ohsaka T, Okajima T, Akita K, Suzuki T, Senawongse P, Urushiyama A, Kawai K, Shoun H, Ishii Y, Ishikawa H, Sugiyama S, Nakajima M, Tsuboi M, Yamanaka T. Generation of gaseous sulfur-containing compounds in tumor tissue and suppression of gas diffusion as an antitumor treatment. Gut. 2012;61(4):554–61.

237. Andriamihaja M, Lan A, Beaumont M, Audebert M, Wong X, Yamada K, Yin Y, Tomé D, Carrasco-Pozo C, Gotteland M, Kong X, Blachier F. The deleterious metabolic and genotoxic effects of the bacterial metabolite p-cresol on colonic epithelial cells. Free Radic Biol Med. 2015;85:219–32.

238. Al Hinai EA, Kullamethee P, Rowland IR, Swann J, Walton GE, Commane GM. Modelling the role of microbial p-cresol in colorectal genotoxicity. Gut Microbes. 2019;10(3):398–411.

239. Gilbert MS, IjssennaggerN KAK, van Mil SWC. Protein fermentation in the gut: implications for intestinal dysfunction in humans, pigs, and poultry. Am J Physiol. 2018;315(2):G159–70.

240. Winter J, Nyskohus L, Young GP, Hu Y, Conlon MA, Bird AR, Topping DL, Le Leu RK. Inhibition by resistant starch of red meat-induced promutagenic adducts in mouse colon. Cancer Prev Res (Phila). 2011;4(11):1920–8.

241. Kikugawa K, Kato T. Formation of mutagenic diaziquinone by interaction of phenol with nitrite. Food Chem Toxicol. 1988;26(3):209–14.

242. Hinzman MJ, Novotny C, Ullah A, Shamsuddin AM. Fecal mutagen fecapentaene-12 damages mammalian colon epithelial DNA. Carcinogenesis. 1987;8(10):1475–9.

243. Bernstein C, Holubec H, Bhattacharyya AK, Nguyen H, Payne CM, Zaitlin B, Bernstein H. Carcinogenecity of deoxycholate, a secondary bile acid. Arch Toxicol. 2011;85(8):863–71.

244. Hamer HM, De Preter V, Windey K, Verbeke K. Functional analysis of colonic bacterial metabolism: relevant to health? Am J Physiol. 2012;302(1):G1–9.
245. Seitz HK, Simanovski UA, Garzon FT, Rideout JM, Peters TJ, Koch A, Berger MR, Einecke H, Maiwald M. Possible role of acetaldehyde in ethanol-related rectal cocarcinogenesis in the rat. Gastroenterology. 1990;98(2):406–13.
246. Armand L, Andriamihaja M, Gellenoncourt S, Bitane V, Lan A, Blachier F. In vitro impact of amino acid-derived bacterial metabolites on colonocyte mitochondrial activity, oxidative stress response and DNA integrity. Biochim Biophys Acta. 2019;1863(8):1292–301.
247. Blachier F, Davila AM, Benamouzig R, Tomé D. Chanelling of arginine in NO and polyamine pathways in colonocytes and consequences. Front Biosc (Landmark Ed). 2011;16(4): 1331–443.
248. Johnson CH, Dejea CM, Edler D, Hoang LT, Santidrian AF, Felding BH, Ivanisevic J, Cho K, Wick EC, Hechenbleikner EM, Uritboonthai W, Goetz L, Casero RA Jr, Pardoll DM, White JR, Patti GJ, Sears CL, Siuzdak G. Metabolism links bacterial biofilms and colon carcinogenesis. Cell Metab. 2015;21(6):891–7.
249. Costerton JW, Stewart PS, Greenberg EP. Bacterial biofilm: a common cause of persistent infections. Science. 1999;284(5418):1318–22.
250. Dejea CM, Wick EC, Heichenbleikner EM, White JR, Mark Welch JL, Rossetti BJ, Peterson SM, Snesrud EC, Borisy GG, Lazarev M, Stein E, Vadivelu J, Roslani AC, Malik AA, Wanyiri JW, Goh KL, Thevambiga A, Fu K, Wan F, Llosa N, Housseau F, Romans K, Wu XQ, McAllister FM, Wu S, Vogelstein B, Kinzler BW, Pardoll DM, Sears CL. Microbiota organization is a distinct feature of proximal colorectal cancers. Proc Natl Acad Sci USA. 2014;111 (51):18321–6.
251. Gerner EW, Meyskens FL Jr. Polyamines and cancer: old molecules, new understanding. Nat Rev Cancer. 2004;4(10):781–92.
252. Russell WR, Duncan SH, Scobbie L, Duncan G, Cantlay L, Graham Calder A, Anderson SE, Flint HJ. Major phenylpropanoid-derived metabolites in the human gut can arise from microbial fermentation of protein. Mol Nutr Food Res. 2013;57(3):523–35.
253. Yan C, Duanmu X, Zeng L, Liu B, Song Z. Mitochondrion DNA: distribution, mutations, and elimination. Cells. 2019;8(4):379.
254. Namslauer I, Brzezinski P. A mitochondrial DNA mutation linked to colon cancer results in proton leaks in cytochrome c oxidase. Proc Natl Acad Sci USA. 2009;106(9):3402–7.
255. Gulec S, Anderson GJ, Collins JF. Mechanistic and regulatory aspects of intestinal iron absorption. Am J Physiol. 2014;307(4):G397–409.
256. Young GP, Rose IS, St John DJ. Haem in the gut. I. Fate of haemoproteins and the absorption of heme. J Gastroenterol Hepatol. 1989;4(6):537–45.
257. Schaer DJ, Buehler PW, Alayash AI, Belcher JD, Vercelloti GM. Hemolysis and free hemoglobin revisited: exploring hemoglobin and hemin scavengers as a novel class of therapeutic proteins. Blood. 2013;121(8):1276–84.
258. Kodadoga Gamage SM, Cheng T, Lee KT, Dissabandara L, Lam AK, Gopalan V. Hemin, a major heme molecule, induced cellular and genetic alterations in normal colonic and colon cancer cells. Pathol Res Pract. 2021;224:153530.
259. Glei M, Klenow S, Sauer J, Wegewitz U, Richter K, Pool-Zobel BL. Hemoglobin and hemin induce DNA damage in human colon tumor cells HT29 clone 19A and in primary human colonocytes. Mutat Res. 2006;594(1–2):162–71.
260. de Vogel J, van Eck WB, Sesink AL, Jonker-Termont DS, Kleibeuker J, van der Meer R. Dietary heme injures surface epithelium resulting in hyperproliferation, inhibition of apoptosis and crypt hyperplasia in rat colon. Carcinogenesis. 2008;29(2):398–403.
261. Blachier F, Davila AM, Mimoun S, Benetti PH, Atanasiu C, Andriamihaja M, Benamouzig R, Bouillaud F, Tomé D. Luminal sulfide and large intestine mucosa: friend or foe? Amino Acids. 2010;39(2):335–47.

262. Velcich A, Yang W, Heyer J, Fragale A, Nicholas C, Viani S, Kucherlapati R, Lipkin M, Yang K, Augenlicht L. Colorectal cancer in mice genetically deficient in the mucin Muc2. Science. 2002;295(5560):1726–9.

263. Martin OCB, Olier M, Ellero-Simatos S, Naud N, Dupuy J, Huc L, Taché S, Graillot V, Levêque M, Bézirard V, Héliès-Toussaint C, Estrada FBY, Tondereau V, Lippi Y, Naylies C, Peyriga L, Canlet C, Davila AM, Blachier F, Ferrier L, Boutet-Robinet E, Guéraud F, Théodorou V, Pierre FHF. Haem iron reshapes colonic luminal environment: impact on mucosal homeostasis and microbiome through aldehyde formation. Microbiome. 2019;7(1): 72.

264. Constante M, Fagoso G, Calvé A, Samba-Mondonga M, Santos MM. Dietary heme induces gut dysbiosis, aggravates colitis, and potentiates the development of adenomas in mice. Front Microbiol. 2017;8:1809.

265. Saillant V, Lipuma D, Ostyn E, Joubert L, Boussac A, Guerin H, Brandelet G, Arnoux P, Lechardeur D. A novel Enterococcus faecalis heme transport regulator (FhtR) senses host heme to control its intracellular homeostasis. mBio. 2021;12(1):e03392–20.

266. Tong Y, Guo M. Bacterial heme-transport proteins and their heme-coordination modes. Arch Biochem Biophys. 2009;481(1):1–15.

267. Wilks A, Burkhard KA. Heme and virulence: how bacterial pathogens regulate, transport and utilize heme. Nat Prod Rep. 2007;24(3):511–22.

268. Archer MC. Mechanisms of action of N-nitroso compounds. Cancer Surv. 1989;8(2):241–50.

269. Carlström M, Moretti CH, Weitzberg E, Lundberg JO. Microbiota, diet and the generation of reactive nitrogen compounds. Free Radic Biol Med. 2020;161:321–5.

270. Barnes JL, Zubair M, John K, Poirier MC, Martin FL. Carcinogens and DNA damage. Biochem Soc Trans. 2018;46(5):1213–24.

271. Duncan SH, Iyer A, Russell WR. Impact of protein on the composition and metabolism of the human gut microbiota and health. Proc Nutr Soc. 2021;80(2):173–85.

272. Lundberg JO, Weitsberg E. Biology of nitrogen oxides in the gastrointestinal tract. Gut. 2013;62(4):616–29.

273. Thresher A, Foster R, Ponting DJ, Stalford SA, Tennant RE, Thomas R. Are all nitrosamines concerning? A review of mutagenicity and carcinogenicity data. Regul Toxicol Pharmacol. 2020;116:104749.

274. Loh YH, Jakszyn P, Luben RN, Mulligan AA, Mitrou PM, Khaw KT. N-nitroso compounds and cancer incidence: the European Prospective Investigation into Cancer and Nutrition (EPIC)-Norfolk study. Am J Clin Nutr. 2011;93(5):1053–61.

275. Hu Y, Le Leu R, Christophersen CT, Somashekar R, Conlon MA, Meng XQ, Winter JM, Woodman RJ, McKinnon R, Young GP. Manipulation of the gut microbiota using resistant starch is associated with protection against colitis-associated colorectal cancer in rats. Carcinogenesis. 2016;37(4):366–75.

276. Sealy L, Chalkey R. The effect of sodium butyrate on histone modification. Cell. 1978;14(1): 115–21.

277. Archer SY, Meng S, Shei A, Hodin RA. p21(WAF1) is required for butyrate-mediated growth inhibition of human colon cancer cells. Proc Natl Acad Sci USA. 1998;95(12):6791–6.

278. Nakano K, Mizuno T, Sowa Y, Orita T, Yoshino T, Okuyama Y, Fujita T, Ohtani-Fujita N, Matsukawa Y, Yamagishi H, Oka T, Nomura H, Sakai T. Butyrate activates the WAF1/Cip1 gene promoter through Sp1 sites in a p53-negative human colon cancer cell line. J Biol Chem. 1997;272(35):22199–206.

279. Siavoshian S, Blottière HM, Cherbut C, Galmiche JP. Butyrate stimulates cyclin D and p-21 and inhibits cyclin-dependent kinase 2 expression in HT-29 colonic epithelial cells. Biochem Biophys Res Commun. 1997;231(1):169–72.

280. Hinnebusch BF, Meng S, Wu JT, Archer SY, Hodin RA. The effects of short-chain fatty acids on human colon cancer cell phenotypes are associated with histone hyperacetylation. J Nutr. 2002;132(5):1012–7.

281. Leschelle X, Delpal S, Goubern M, Blottière HM, Blachier F. Butyrate metabolism upstream and downstream acetyl-CoA synthesis and growth control of human colon carcinoma cells. Eur J Biochem. 2000;267(21):6435–42.
282. Belcheva A, Irrazabal T, Robertson SJ, Streutker C, Maughan H, Rubino S, Moriyama EH, Copeland JK, Surendra A, Kumar S, Green B, Geddes K, Pezo RC, Navarre WW, Milosevic M, Wilson BC, Girardin SE, Wolever TMS, Edelmann W, Guttman DS, Philipot DJ, Martin A. Gut microbial metabolism drives transformation of MSH2-deficient colon epithelial cells. Cell. 2014;158(2):288–99.
283. Freeman HJ. Effects of different concentrations of sodium butyrate on 1,2-dimethylhydrazine-induced rat intestinal neoplasia. Gastroenterology. 1986;91(3):596–602.
284. Khatami F, Aghamir SMK, Tavangar SM. Oncometabolites: a new insight for oncology. Mol Genet Genomic Med. 2019;7(9):e873.
285. Bultman SJ, Jobin C. Microbial-derived butyrate: an oncometabolite or tumor-suppressive metabolite? Cell Host Microbe. 2014;16(2):143–5.
286. Ferguson LR, Harris PJ. The dietary fiber debate: more food for thought. Lancet. 2003;361 (9368):1487–8.
287. Camilleri M, Sellin JH, Barrett KE. Pathophysiology, evaluation, and management of chronic watery diarrhea. Gastroenterology. 2017;152(3):515–32.
288. Rose C, Parker A, Jefferson B, Cartmell E. The characterization of feces and urine: a review of the literature to inform advanced treatment technology. Crit Rev Environ Sci Technol. 2015;45 (17):1827–79.
289. Edmonds CJ. Absorption and secretion of fluid and electrolytes by the rectum. Scand J Gastroenterol. 1984;93:79–87.
290. Hogan DL, Ainsworth MA, Isenberg JI. Review article: gastroduodenal bicarbonate secretion. Aliment Pharmacol Ther. 1994;8(5):475–88.
291. Murek M, Kopic S, Geibel J. Evidence for intestinal chloride secretion. Exp Physiol. 2010;95 (4):471–8.
292. Kunzelmann K, Mall M. Electrolyte transport in the mammalian colon: mechanisms and implications for disease. Physiol Rev. 2002;82(1):245–89.
293. Barrett KE, Keely SJ. Chloride secretion by the intestinal epithelium: molecular basis and regulatory aspects. Annu Rev Physiol. 2000;62:535–72.
294. Geibel JP. Secretion and absorption by colonic crypts. Annu Rev Physiol. 2005;67:471–90.
295. Pacha J, Teisinger J, Popp M, Capek K. Na,K-ATPase and the development of Na+ transport in rat distal colon. J Membr Biol. 1991;120(3):201–10.
296. Yde J, Keely SJ, Moeller HB. Expression, regulation and function of Aquaporin-3 in colonic epithelial cells. Biochim Biophys Acta. 2021;1863(7):1836619.
297. Zhu C, Chen Z, Jiang Z. Expression, distribution and role of aquaporin water channels in human and animal stomach and intestines. Int J Mol Sci. 2016;17(9):1399.
298. Guttman JA, Samji FN, Li Y, Deng W, Lin A, Brett Finlay B. Aquaporins contribute to diarrhoea caused by attaching and effacing bacterial pathogens. Cell Microbiol. 2007;9(1): 131–41.
299. Laforenza U, Cova E, Gastaldi G, Tritto S, Grazioli M, LaRusso NF, Splinter PL, D'Adamo P, Tosco M, Ventura U. Aquaporin-8 is involved in water transport in isolated superficial colonocytes from rat proximal colon. J Nutr. 2005;135(10):2329–36.
300. Schiller LR, Pardi DS, Spiller R, Semrad CE, Surawicz CM, Giannella RA, Krejs GJ, Farthing MJG, Sellin JH. Gastro 2013 APDW/WCOG Shanghai working party report: chronic diarrhea: definition, classification, diagnosis. J Gastroenterol Hepatol. 2014;29(1):6–25.
301. Li Y, Xia S, Jiang X, Feng C, Gong S, Ma J, Fang Z, Yin J, Yin Y. Gut microbiota and diarrhea: an updated review. Front Cell Infect Microbiol. 2021;11:625610.
302. Anbazhagan AN, Priyamvada S, Alrefai WA, Dudeja PK. Pathophysiology of IBD associated diarrhea. Tissues Barriers. 2018;6(2):e1463897.
303. Eherer AJ, Fordtran JS. Fecal osmotic gap and pH in experimental diarrhea of various causes. Gastroenterology. 1992;103(2):545–51.

304. Hammer HF, Hammer J. Diarrhea caused by carbohydrate malabsorption. Gastroenterol Clin North Am. 2012;41(3):611–27.
305. Misselwitz B, Butter M, Verbeke K, Fox MR. Update on lactose malabsorption and intolerance: pathogenesis, diagnosis and clinical management. Gut. 2019;68(11):2080–91.
306. Vandenplas Y. Lactose intolerance. Asia Paci J Clin Nutr. 2015;24(S1):S9–S13.
307. Alexandre V, Even PC, Larue-Achagiotis C, Blouin JM, Blachier F, Benamouzig R, Tomé D, Davila AM. Lactose malabsorption and colonic fermentations alter host metabolism in rats. Br J Nutr. 2013;110(4):625–31.
308. Hammer HF, Fine KD, Santa Ana CA, Porter JL, Schiller LR, Fordtran JS. Carbohydrate malabsorption. Its measurement and its contribution to diarrhea. J Clin Invest. 1990;86(6): 1936–44.
309. Camilleri M, Linden DR. Measurement of gastrointestinal and colonic motor functions in humans and animals. Cell Mol Gastroenterol Hepatol. 2016;2(4):412–28.
310. Jia J, Frantz N, Khoo C, Gibson GR, Rastall RA, McCartney AL. Investigation of the faecal microbiota associated with canine chronic diarrhea. FEMS Microbiol Ecol. 2010;71(2): 304–12.
311. Suchodolski JS, Foster ML, Sohail MU, Leutenegger C, Queen EV, Steiner JM, Marks SL. The fecal microbiome in cats with diarrhea. PLoS One. 2015;19(10):e0127378.
312. Binder HJ. Role of colonic short-chain fatty acid transport in diarrhea. Annu Rev Physiol. 2010;72:297–313.
313. Zeissig S, Fromm A, Mankertz J, Weiske J, Zeitz M, Fromm M, Schultze JD. Butyrate induces intestinal sodium absorption via Sp3-mediated transcriptional up-regulation of epithelial sodium channels. Gastroenterology. 2007;132(1):236–48.
314. Sellin JH, De Soignie R. Short-chain fatty acids have polarized effects on sodium transport and intracellular pH in rabbit proximal colon. Gastroenterology. 1998;114(4):737–47.
315. Karaki S, Kuwahara A. Propionate-induced epithelial K(+) and Cl(−)/HCO3(−) secretion and free fatty acid receptor 2 (FFA2, GRP43) expression in the guinea pig distal colon. Pflugers Arch. 2011;461(1):141–52.
316. Munos MK, Walker CL, Black RE. The effect of oral rehydration solution and recommended home fluids on diarrhea mortality. Int J Epidemiol. 2010;39(S1):75–87.
317. Hahn S, Kim H, Garner P. Reduced osmolarity oral rehydration solution for treating dehydration due to diarrhea in children: systematic review. BMJ. 2001;323(7304):81–5.
318. Binder HJ, Brown I, Ramakrishna BS, Young GP. Oral rehydration therapy in the second decade of the twenty-first century. Curr Gastroenenterol Rep. 2014;16(3):376.
319. Raghupathy P, Ramakrishna BS, Oommen SP, Ahmed MS, Priyaa G, Dziura J, Young GP, Binder HJ. Amylase-resistant starch as adjunct to oral rehydration therapy in children with diarrhea. J Pediatr Gastroenterol Nutr. 2006;42(4):362–8.
320. Hoekstra JH, Szajewska H, Zikri MA, Micetic-Turk D, Weizman Z, Papadopoulou A, Guarino A, Dias JA, Oostvogels B. Oral rehydration solution containing a mixture of non-digestible carbohydrates in the treatment of acute diarrhea: a multicenter randomized placebo controlled study on behalf of the ESPGHAN working group on intestinal infections. J Pediatr Gastroenterol Nutr. 2004;39(3):239–45.
321. Abt MC, McKenney PT, Pamer EG. Clostridium difficile colitis: pathogenesis and host defence. Nat Rev Microbiol. 2016;14(10):609–20.
322. Sorg JA, Sonenshein AL. Bile salts and glycine as cogerminants for Clostridium difficile spores. J Bacteriol. 2008;190(7):2505–12.
323. Koenigsknecht MJ, Theriot CM, Bergin IL, Schumacher CA, Schloss PD, Young VB. Dynamics and establishment of Clostridium difficile infection in the murine gastrointestinal tract. Infect Immun. 2015;83(3):934–41.
324. Giel JL, Sorg JA, Sonenshein AL, Zhu J. Metabolism of bile salts in mice influences spore germination in Clostridium difficile. PLoS One. 2010;5(1):e8740.

325. Theriot CM, Bowman AA, Young VB. Antibiotic-induced alterations of the gut microbiota alter secondary bile acid production and allow for Clostridium difficile spore germination and outgrowth in the large intestine. mSphere. 2016;1(1):e00045–15.

326. Ferreyra JA, Wu KJ, Hryckowian AJ, Bouley DM, Weimer BC, Sonnenburg JL. Gut microbiota-produced succinate promotes C. difficile infection after antibiotic treatment or motility disturbance. Cell Host Microbe. 2014;16(6):770–7.

327. Ng KM, Ferreyra JA, Higginbottom SK, Lynch JB, Kashyap PC, Gopinath S, Naidu N, Choudhury B, Weimer BC, Monack DM, Sonnenburg JL. Microbiota-liberated host sugars facilitate post-antibiotic expansion of enteric pathogens. Nature. 2013;502(7469):96–9.

328. Davila AM, Blachier F, Gotteland M, Andriamihaja M, Benetti PH, Sanz Y, Tomé D. Intestinal luminal nitrogen metabolism: role of the gut microbiota and consequences for the host. Pharmacol Res. 2013;68(1):95–107.

329. Fernandez-Veledo S, Vendrell J. Gut microbiota-derived succinate: Friend or foe in human metabolic diseases? Rev Endocr Metab Disord. 2019;20(4):439–47.

330. Atkinson W, Lockardt S, Whorwell PJ, Keevil B, Houghton LA. Altered 5-hydroxytryptamine signaling in patients with constipation- and diarrhea-predominant irritable bowel syndrome. Gastroenterology. 2006;130(1):34–43.

331. Houghton LA, Atkinson W, Whitaker RP, Whorwell PJ, Rimmer MJ. Increased platelet depleted 5-hydroxytryptamine concentration following meal ingestion in symptomatic female subjects with diarrhea predominant irritable bowel syndrome. Gut. 2003;52(5):663–70.

332. Mawe GM, Hoffman JM. Serotonin signalling in the gut: functions, dysfunctions and therapeutic target. Nat Rev Gastroenterol Hepatol. 2013;10(8):473–86.

Modification of the Bacterial Metabolites by the Host after Absorption, and Consequences for the Peripheral Tissues' Metabolism, Physiology, and Physiopathology

5

Abstract

After synthesis by the intestinal bacteria, several metabolites are absorbed and metabolized by host tissues giving rise to bioactive co-metabolites. In liver, some bacterial metabolites like trimethylamine, indole, *p*-cresol, and phenylacetate are metabolized, and resulting co-metabolites like trimethylamine N-oxide, indoxyl sulfate, and *p*-cresyl sulfate impair hepatocyte energy metabolism and viability when produced in excess. In contrast, the bacterial metabolites indole and indole-3 acetate appear protective in situation of hepatic damage. Regarding the gut–kidney axis, the bacterial metabolite *p*-cresol, as well as the co-metabolites *p*-cresyl sulfate and indoxyl sulfate act as uremic toxins when synthesized in excess. These latter compounds have been shown to affect mitochondrial function in renal tubular cells and to provoke an inflammatory response. Trimethylamine N-oxide has been identified as a co-metabolite which at excessive concentration increases in experimental work platelet responsiveness to agonists, favors thrombus development within internal carotid artery, and provokes endothelial cell dysfunction. Phenylacetylglutamine is another co-metabolite that is involved in the gut–cardiovascular axis. This compound binds to adrenergic receptors, enhancing platelet activation and clot formation within carotid artery. Other compounds like *p*-cresol and indoxyl sulfate have been shown to provoke endothelial dysfunction at excessive concentrations. Regarding the gut–brain axis, from experimental works, surprisingly, there are reasons to believe that the metabolic activity of the gut microbiota can influence parameters characteristic of mood disorders. Although the mechanisms involved remain far from being well understood, emerging data indicate that host neurophysiology may be affected by both direct and indirect ways including actions on vagus nerve, enteric nervous system, intestinal immune and neuro-endocrine systems. Interestingly, in animals with no intestinal microbiota, norepinephrine, dopamine, and serotonin turnover are modified in brain suggesting complex relationships between intestinal microbiota and its host. Some compounds derived from the metabolic activity of the gut microbiota, like NH_3 may cross the blood–brain barrier and exert adverse effects on the central nervous system in case of liver failure. Tryptamine is another example of bacterial metabolite that can cross the blood–brain barrier function. Among indole-related compounds, isatin and oxindole can enter within the brain, and affect behavior and brain functions. Lastly, the co-metabolite indoxyl sulfate appears to exert, depending on the doses used, both beneficial and deleterious effects on the central nervous system in experimental works.

As indicated in previous parts, several bacterial metabolites can be transferred from the intestinal lumen to the bloodstream by being transported across the intestinal epithelial cells. This is typically the case for organic acids like short-chain fatty acids, succinate, and lactate which are transported through absorptive cells through dedicated transporters present in the apical and basolateral membranes. Some other bacterial metabolites can cross the epithelial layer by diffusion. Hydrogen sulfide (H_2S) is one example of luminal compound that easily diffuses across the epithelial cell membranes due to its chemical characteristics [1].

Some bacterial metabolites, after entry into colonocytes, are transferred from the lumen to the blood without being modified, while for some others, they can be partly metabolized by the colonocytes, giving rise to the so-called co-metabolites. Co-metabolites are thus compounds produced by the intestinal microbiota that are modified by the host. The bacterial metabolites and co-metabolites are then released in the portal vein that drains the blood from the intestine to the liver. The co-metabolites are not exclusively formed inside colonocytes but can be synthesized in the liver and in different peripheral tissues. Of major interest, as presented below, some bacterial metabolites and co-metabolites have been found to have biological activities toward host cells and tissues, either beneficial or deleterious, often depending on their concentrations. Of note, by comparing mice that are deprived of microbiota and those that are carrying a microbiota, it has been found that the presence of microbiota modifies to a large extent the bacterial metabolite composition in the different tissues examined [2]. Of equal importance, numerous bacterial metabolites are known to be involved in multiple metabolic pathways, strongly suggesting that the gut microbiota has a direct impact on host metabolism [2].

5.1 Bacterial Metabolites and the Gut–Liver Axis

Hepatocytes, which are by far the most abundant cells of the liver, are the first to be in contact with the numerous compounds released from the intestine, and these cells play a major role in their complex metabolic remodeling [3, 4]. Among this complex mixture of compounds, absorbed bacterial metabolites are known to be partly metabolized in the liver, giving rise to different molecules that intervene in the host metabolism and physiology. Other bacterial metabolites can also play a role on liver cells without being transformed as we will see in the following text.

Short-Chain Fatty Acids Produced by the Intestinal Microbiota that Have Not Been Used by the Colonic Epithelium Are Metabolized by the Liver Cells

Due to intense metabolic capacity of colonic epithelial cells to oxidize short-chain fatty acids, the concentration of these compounds in the portal vein is several orders of magnitude lower than in the luminal fluid. Indeed, the butyrate concentration in the portal blood is below 10 micromolar concentration, while being at the limit of detection in the systemic blood [5, 6]. Propionate concentration in the blood also does not exceed 10 micromolar. Only acetate, which represents the most abundant short-chain fatty acid in the intestinal luminal fluid, reaches higher concentrations in the peripheral blood, approximately 100 micromolar [7]. In humans, using capsules that release 13C-labeled acetate, propionate, and butyrate inside the colonic luminal fluid, it was found that the systemic availability in blood plasma of acetate was greater, averaging 36%, while those of propionate was only 9%, and those of butyrate was, as expected, negligible [8]. Indeed, this latter short-chain fatty acid is intensely metabolized in colonocytes [9]. Although acetate is a precursor for cholesterol synthesis in liver [10], when 13C-labeled acetate was introduced in the colon of healthy subjects, almost no incorporation of acetate in cholesterol was measured [8], indicating that acetate used for cholesterol synthesis does not originate from microbiota metabolic activity. In human volunteers, it has been shown that acetate released from the gut is largely taken up by the liver [11]. The acetate produced by the gut microbiota is used as a precursor for hepatic fatty acid synthesis [12]. In a rodent model, when 13C-labeled acetate and propionate were instilled in the cecum lumen, acetate, but not propionate, was used by the liver for fatty acid synthesis [13]. In a rodent model, dietary fructose was found to provide hepatic lipogenesis pathway through microbiota-derived acetate used for the synthesis of fatty acids [14]. Regarding propionate, it was found that a minor part of this bacterial metabolite is used for glucose production [8], while this short-chain fatty acid inhibits fatty acid synthesis in rat hepatocytes [15].

The Bacterial Metabolite Trimethylamine Is Metabolized by Liver Cells into Trimethylamine N-Oxide

Choline in the intestinal content can be converted by bacteria into trimethylamine [16, 17] (Fig. 5.1). The precursor choline is mainly provided by the diet [18] but can be also produced endogenously [19]. Trimethylamine has been shown to be produced by different bacterial species among different bacterial genus including *Clostridium*, *Desulfovibrio* spp., *Escherichia*, *Klebsiella*, and *Proteus* spp. [20, 21]. This bacterial metabolite is then transferred to the liver via the portal vein. In a murine model of ethanol-induced liver injury, reduction of trimethylamine production by the gut microbiota was found to ameliorate histological and biochemical indicators of liver disease [22], thus suggesting that trimethylamine in excess is a potential hepatotoxic (Fig. 5.1). Of note, 50% of patients suffering from primary or secondary liver diseases were found to have an increased urinary excretion of trimethylamine [23].

Liver cells can convert trimethylamine into trimethylamine N-oxide (also called TMAO) [24] (Fig. 5.1). Trimethylamine N-oxide is formed in the liver from the

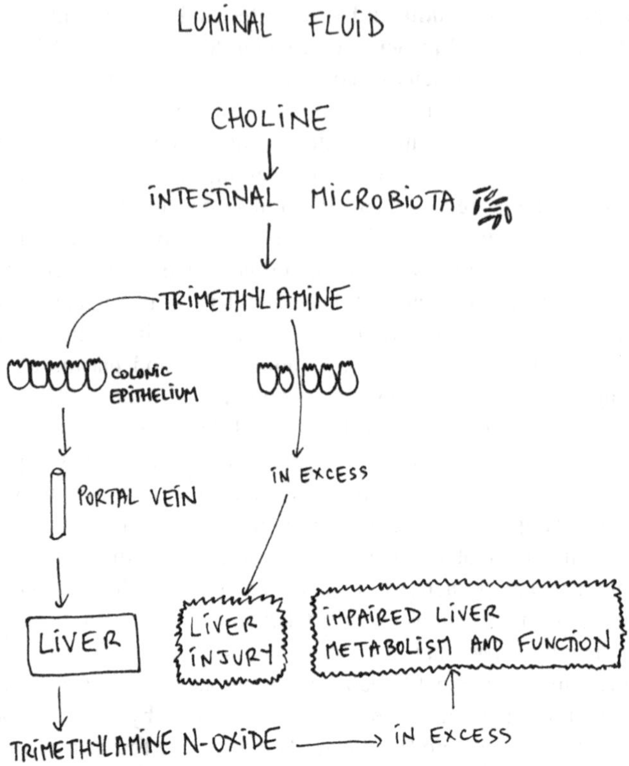

LUMINAL FLUID

CHOLINE
↓
INTESTINAL MICROBIOTA
↓
TRIMETHYLAMINE

COLONIC EPITHELIUM

IN EXCESS

PORTAL VEIN

LIVER   LIVER INJURY   IMPAIRED LIVER METABOLISM AND FUNCTION

TRIMETHYLAMINE N-OXIDE ——→ IN EXCESS

Fig. 5.1 Synthesis of trimethylamine by the intestinal microbiota, transformation of this bacterial metabolite into its co-metabolite trimethylamine N-oxide by the liver, and effects of this compound at excessive concentrations on the liver

choline-derived bacterial metabolites trimethylamine. Intestinal bacteria have been found to be able to use not only choline as a precursor for trimethylamine N-oxide production [16], but also from the phospholipid phosphatidylcholine [19], and from carnitine [25]. Carnitine is usually mainly provided by food but can be also synthesized endogenously from the amino acids lysine and methionine [26]. Trimethylamine N-oxide is suspected to be involved in the process of athero-genesis [27], as will be detailed in the paragraph dedicated to "The gut-cardiovascular axis." Regarding the liver, increased concentrations of trimethylamine N-oxide has been associated in patients with the severity of non-alcoholic fatty liver disease [28]. In mice fed with a high-fat diet, 18 weeks administration of trimethylamine N-oxide further impairs liver function and increased lipogenesis with a resultant accumulation of triglycerides in the liver [29] (Fig. 5.1). In subjects with non-alcoholic fatty liver disease, the plasma concentration of trimethylamine N-oxide is associated with all-cause mortality independently of classical risk factors [30]. In a cohort study, advancing age was strongly associated with the trimethylamine N-oxide plasma concentration [31], with an

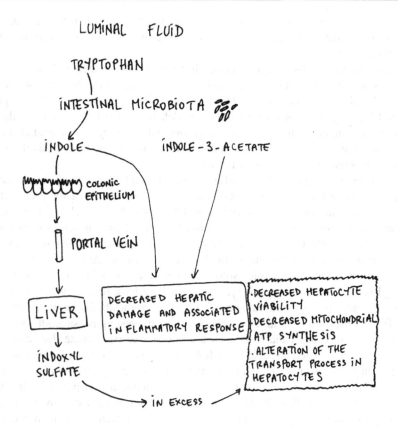

Fig. 5.2 Synthesis of indole and indole-related compounds by the intestinal microbiota, transformation of this bacterial metabolite into its co-metabolite indoxyl sulfate by the liver, and effects of this compound at excessive concentrations on the liver

implication of the gut microbiota metabolic activity being underlined in such association.

Tryptophan-Derived Bacterial Metabolites Exert Protective Effect on Liver Inflammation

Indole produced by the gut microbiota from tryptophan, after absorption in the portal vein, has been shown to exert some anti-inflammatory effects on liver cells. In a rodent model, indole was found to reduce the production of pro-inflammatory mediators by the liver [32]. In a model of obese mice, indole is able to reduce hepatic damage and the associated inflammatory response [33] (Fig. 5.2). The indole-related compound indole-3-acetate was found to alleviate in mice the high-fat diet induced hepatotoxicity, as based on histological examination and measurement of the liver inflammatory response [34] (Fig. 5.2). This latter bacterial metabolite reduces the expression of fatty acid synthase in hepatocytes [35], a characteristic that might prove to be clinically useful in case of excessive fatty acid production.

Indole in liver cells is partly metabolized by several enzymes belonging to the cytochrome (CYP) family, including notably CYP2E1 [36]. Several molecules are produced from indole, with indoxyl sulfate being the main co-metabolite produced [37] (Fig. 5.2). Of note, in rodent model, indoxyl sulfate is found only in conventional animals with intestinal microbiota, but not in germ-free animals [38], confirming the view that circulating indoxyl sulfate is originating exclusively from the intestinal microbiota metabolic activity.

As can be anticipated, subjects who consumed a high-protein diet show overall greater indoxyl sulfate urinary excretion than those who consumed a low-protein diet [39]. In addition, in a randomized, parallel, double-blind trial in overweight volunteers, protein supplementation provoked increased concentrations of indoxyl sulfate in urine [40]. CYP enzymes are presumably involved in the conversion of indole to indoxyl, an intermediate in the synthesis of indoxyl sulfate. CYP2E1 represents the major enzymatic isoform responsible for oxidation of indole to indoxyl [41]. The effects of indoxyl sulfate on liver have been little investigated, with one study showing a stimulatory effect of this co-metabolite on the activity of the efflux transporter P-gp in liver cells [42], thus suggesting possible alteration of the clearance of drugs handled by this transporter. Indoxyl sulfate decreased the uptake of the conjugated bile acid taurocholate by liver cells [43]. Taurocholic acid is secreted in the small intestine and returns to liver via the portal vein [44]. Thus, indoxyl sulfate in excess appears to interfere with the transport processes in liver (Fig. 5.2). Indoxyl sulfate in excess has been shown to reduce human primary hepatocytes viability, a process that is associated with a decrease of the mitochondrial membrane potential and of the ATP content [43] (Fig. 5.2), thus indicating alteration of the hepatocyte energy metabolism.

As will be detailed in the following paragraphs, indoxyl sulfate has been shown to exert activity on the kidney, on the cardiovascular system, and on the central nervous system, with reported signs of deleterious effects of excessive concentrations.

The Tyrosine-Derived Bacterial Metabolite *p*-cresol Is Further Metabolized in *p*-cresyl Sulfate in the Liver

The bacterial metabolite *p*-cresol can be absorbed through the intestinal epithelium, released in the portal vein, and then metabolized in the liver giving rise mainly to *p*-cresyl sulfate, and to a lesser extent to *p*-cresyl glucuronide [45, 46] (Fig. 5.3), and to other minor metabolites [47]. Little is known on the effect of the *p*-cresol co-metabolites on the liver, with one study showing that *p*-cresol sulfate and *p*-cresol glucuronide in excess affect human hepatocyte viability in association with an impaired mitochondrial energy metabolism in these cells [43] (Fig. 5.3).

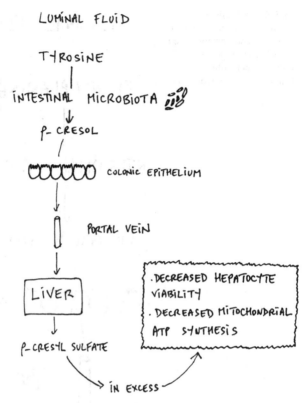

Fig. 5.3 Synthesis of *p*-cresol by the intestinal microbiota, transformation of this bacterial metabolite into its co-metabolite *p*-cresyl sulfate by the liver, and effects of this compound at excessive concentrations on hepatocytes

The Phenylalanine-Derived Bacterial Metabolite Phenylacetate Is Further Metabolized in Phenylacetylglutamine in the Liver

Phenylacetylglutamine (also called PAG) is a co-metabolite produced in the host liver from phenylacetate, a bacterial metabolite produced by the intestinal microbiota from the amino acid phenylalanine [48] (Fig. 5.4). The effects of phenylacetylglutamine on the liver have not been investigated, but as explained in the paragraph dedicated to the gut–cardiovascular system axis, this compound appears involved in this latter axis.

Key Points

- A minor part of acetate produced by the intestinal microbiota is used for the synthesis of fatty acids in the liver.
- Trimethylamine produced by the gut microbiota is converted in the liver into the co-metabolite trimethylamine N-oxide that is active on the cardiovascular system.
- Indole and indole-3 acetate produced by the intestinal microbiota are protective against liver inflammation.
- The liver converts indole to the co-metabolite indoxyl sulfate, *p*-cresol to the co-metabolite *p*-cresyl sulfate, and phenylacetate to phenylacetylglutamine.

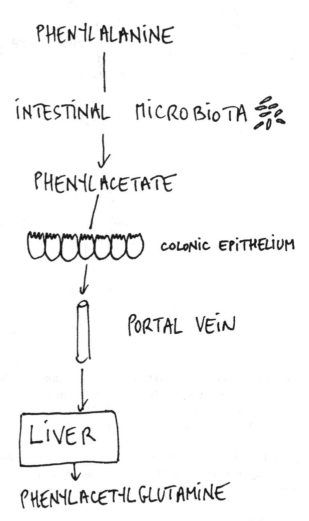

Fig. 5.4 Synthesis of phenylacetate by the intestinal microbiota and transformation of this bacterial metabolite into its co-metabolite phenylacetylglutamine by the liver

- Indoxyl sulfate and *p*-cresyl sulfate in excess impaired mitochondrial energy metabolism in hepatocytes and decreases their viability.

5.2   Bacterial Metabolites and the Gut–Endocrine Pancreas Axis

The search for bacterial metabolites that can interfere with the process of insulin secretion is an emerging field of research [49]. Insulin secretion is primarily the result of an influx of nutrients from the intestinal luminal fluid to the blood circulation after a meal, with glucose being well known to be central for such a secretion [50]. Several other nutrients, including amino acids, further increase the stimulus-secretion coupling provoked by an increase in blood glucose concentration. Particularly, arginine, lysine, and histidine are known to potentiate insulin secretion provoked by glucose [51–53]. Several studies have shown that among metabolites produced by the intestinal microbiota, some of them can also modulate insulin secretion.

Acetate Produced by the Intestinal Microbiota Further Increases Insulin Secretion

In rodents fed with a high-fat diet, it has been shown that increased production of acetate by an altered gut microbiota leads to the activation of the parasympathetic nervous system, which, in turn, promotes increased glucose-stimulated insulin secretion [54].

The Co-metabolite Hippurate Can Enhance Insulin Secretion and Glucose Tolerance

Hippurate is one of the most abundant microbial-host co-metabolites that is produced by the conjugation of glycine and microbial benzoate in the liver and kidney through phase 2 detoxification enzymes [55] (Fig. 5.5). In human volunteers, urinary hippurate was positively associated with microbial gene richness and with functional capacity of the intestinal microbiota to synthesize benzoate [56]. In rodents fed with a high-fat diet, it has been shown that chronic infusion of hippurate results in enhanced insulin secretion and improved glucose tolerance [56], thus suggesting that the microbiota-derived hippurate modulates insulin secretion (Fig. 5.5).

Agmatine Potentiates Insulin Secretion Provoked by Glucose in Pancreatic Beta-cells

As detailed in Chap. 3, agmatine is a metabolite produced by the intestinal microbiota from arginine. Bacteria capacity for agmatine production and release in the surrounding medium differs considerably according to the bacterial species examined [57], suggesting that the composition of the intestinal microbiota has a great impact on the agmatine availability in the gut for subsequent absorption.

Agmatine has been shown to increase insulin secretion by pancreatic islet cells when tested in the presence of an intermediate concentration of glucose [58]. This stimulation was dose-dependent and observed at an initial concentration of 100 micromolar agmatine. In this study, agmatine was found to accumulate in islet cells and to stimulate calcium uptake.

Fig. 5.5 Synthesis of benzoate by the intestinal microbiota, transformation of this bacterial metabolite into its co-metabolite hippurate by the liver and kidney, and effects of this compound on pancreatic beta cells

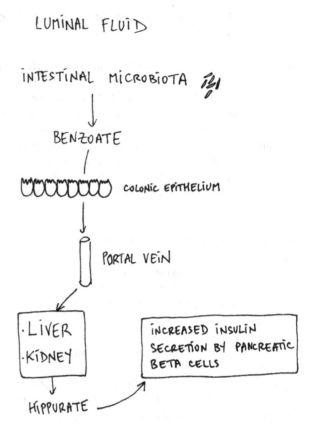

Several Bacterial Metabolites Are Associated with an Increased Risk of Type 2 Diabetes

In a cohort of subjects with different characteristics in terms of clinical and biochemical parameters, several metabolites produced by the intestinal microbiota, including 3-(4-hydroxyphenyl) lactate (derived from phenylalanine) and 2-hydroxyhippurate (a glycine conjugate), were associated with an increased risk of type 2 diabetes as well as decreased insulin secretion and/or decreased insulin sensitivity [59]. Although this study was purely observational, it raises the possibility of a potential role of these bacterial metabolites in the etiology of type 2 diabetes.

Key Points
- The bacterial metabolites acetate and agmatine potentiate insulin secretion provoked by glucose.
- The co-metabolite hippurate modulates insulin secretion.

5.3   Bacterial Metabolites and the Gut–Kidney Axis

During the progression of kidney diseases, uremic toxins may accumulate in body fluids leading to the so-called uremic syndrome [60–62]. In patients with chronic kidney disease, a progression of kidney dysfunction is associated with the accumulation in plasma of protein-bound and water-soluble uremic solutes. Chronic kidney disease is defined as abnormalities of kidney structure and/or function that last over 3 months [63]. Among the uremic solutes, more than one hundred uremic toxins have been identified [64, 65]. Among these uremic toxins, several of them appear to originate from the metabolic activity of the intestinal microbiota.

p-Cresol, *p*-cresyl Sulfate, and Chronic Kidney Disease: The Dangerous Relationships

Among these uremic toxins, the bacterial metabolite *p*-cresol derived from L-tyrosine and its co-metabolite *p*-cresyl sulfate have been largely suspected to aggravate chronic kidney disease [66, 67]. *p*-cresol and *p*-cresyl sulfate refer to the category of protein-bound and low molecular weight compounds that are only partly removed by conventional dialysis techniques [68], mainly because of their strong capacity to bind to serum proteins [69–72]. These compounds thus accumulate in blood. In contrast, the other *p*-cresol co-metabolite *p*-cresyl glucuronide appears much less protein-bound in serum than *p*-cresyl sulfate [73].

Accordingly, *p*-cresol and *p*-cresyl sulfate are found in numerous studies at much higher concentrations in blood from chronic kidney disease and hemodialysis patients than in blood originating from healthy volunteers [70, 73–88]. Of note, in rodent model, *p*-cresyl sulfate is found only in the blood of animals with an intestinal microbiota, but not in the blood of germ-free animals [38], thus confirming that *p*-cresol and its derivatives are exclusively originating from the metabolic activity of the intestinal microbiota.

p-cresyl sulfate dialytic clearance rate by tubular secretion is much lower in hemodialysis patients than in the healthy counterpart [88, 89], suggesting altered secretion of *p*-cresyl sulfate in chronic kidney disease. The handling of *p*-cresyl sulfate implicates the organic anion transporter OAT3 since mice invalidated for this transporter show increased *p*-cresyl sulfate plasma concentration [90].

The proposition that *p*-cresol and *p*-cresyl sulfate act as uremic toxins during chronic kidney disease is derived from several clinical and experimental observations. A prospective and observational study in patients with chronic kidney disease indicates that the baseline level of *p*-cresyl sulfate is a predictor of chronic kidney disease progression [91], and *p*-cresol-free concentration in serum of hemodialysis patients is associated with mortality [92]. From in vitro data with renal tubular cells, and in vivo data from partially nephrectomized mice treated with *p*-cresyl sulfate for 4 weeks, this latter co-metabolite appears to be involved in kidney fibrosis [93]. Knowing that renal handling of *p*-cresyl sulfate mainly depends on tubular secretion [94], it is worth considering that this co-metabolite causes renal tubular cell damage by inducing oxidative stress [95]. This co-metabolite used in vitro at micromolar concentrations modifies gene expression in cultured proximal

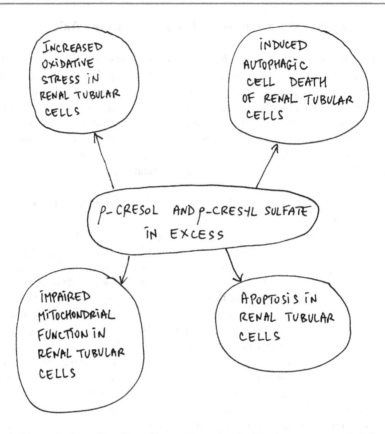

Fig. 5.6 Schematic view of the effects of *p*-cresol or *p*-cresyl sulfate in excess on renal tubular cells

renal tubular cells in a way that is compatible with an inflammatory response [96]. In addition, from studies with renal tubular cells, *p*-cresyl sulfate when tested at micromolar concentrations impairs mitochondrial function and biogenesis [97]. The parent compound *p*-cresol exerts at micromolar concentrations an apoptotic effect on renal tubular cells [98], while higher concentrations induce autophagic cell death [99]. Figure 5.6 recapitulates schematically the deleterious in vitro effects of *p*-cresol and its co-metabolite on kidney cells when present in excess.

Interestingly, in patients with end-stage renal disease on maintenance hemodialysis, the consumption for 8 weeks of a diet containing resistant starch can reduce *p*-cresol concentration in serum [100]. Furthermore, in dialysis patients, supplementation of the diet with non-digestible carbohydrates lowers the plasma concentration of *p*-cresyl sulfate [101], thus suggesting that such dietary intervention is efficient in reducing the production of colon-derived *p*-cresol [102]. The fact that low-protein diets (0.6–0.8 g/kg/day) are often recommended for retarding the progression of chronic kidney disease and delaying initiation of maintenance dialysis therapy [103] is in accordance with the view that higher protein intake increases notably the availability of L-tyrosine for the synthesis of *p*-cresol by the intestinal microbiota. Further works are obviously required to define, among the numerous uremic toxins, the specific role played by p-cresol (and p-cresyl sulfate) in the alterations of kidney functions recorded during chronic kidney disease progression.

Indoxyl Sulfate Is another Uremic Toxin Involved in Chronic Kidney Disease Progression

As explained in Chap. 3, indole is produced by the intestinal microbiota from tryptophan. Indoxyl sulfate after production in the liver is released in the peripheral blood and then excreted in urine. Total indoxyl sulfate concentration in blood can be increased from micromolar concentrations in healthy individuals, up to 1.1 millimolar in severe chronic kidney disease [104]. Indoxyl sulfate represents another uremic toxin that may accumulate in body fluids leading to the so-called uremic syndrome [105]. Indeed, indoxyl sulfate at excessive concentration aggravates chronic kidney disease in patients [106]. In healthy individuals, indoxyl sulfate is almost entirely bound to proteins in blood (approximately 93% of this co-metabolite is in bound form) [107]. The main binding protein in blood is albumin [108]. Circulating indoxyl sulfate in free form is then efficiently excreted in the urine by proximal tubular cells through basolateral organic anion transporters [109].

However, in patients with chronic kidney disease, only 85% of this co-metabolite is protein-bound [110], and thus a higher part of indoxyl sulfate is in free form. As kidney function declines, indoxyl sulfate total concentration increases in the blood and this elevation contributes to further progression of chronic kidney disease [111]. In a cohort study, it was found that blood indoxyl sulfate concentrations are higher in patients with chronic kidney failure progression than in stabilized patients [112]. Then, indoxyl sulfate concentration in blood has been proposed as an indicator of chronic kidney disease progression in dialyzed patients [113].

From experimental in vitro and in vivo studies, indoxyl sulfate has been shown to have deleterious effects on kidney cells when present in excess. Indoxyl sulfate increases the expression of inflammation-associated genes in cultured proximal renal tubular cells [96]. This effect coincides with a capacity of this bacterial co-metabolite to increase the net production of reactive oxygen species in the proximal tubular cells [114, 115], and this effect appears at the origin of an oxidative stress in these cells [116] (Fig. 5.7). In addition, indoxyl sulfate reduces the glutathione concentration in renal tubular cells [117]. Considering the central role of reduced glutathione in the process of reactive oxygen species disposal [118], it is plausible that a reduced intracellular concentration of glutathione will render renal tubular cells more vulnerable to oxidative stress. The fact that indoxyl sulfate also reduces the superoxide scavenging activity in the kidney of normal and uremic rodents [119] is another element that will make the kidney cells sensitive to oxidative stress (Fig. 5.7). Indeed, superoxide is one of the reactive oxygen species that is deleterious to cells when its intracellular concentration exceeds a threshold value [120]. These effects are of major importance when considering that reactive oxygen and nitrogen species are elevated in renal tubular cells in the process of chronic kidney disease progression [121].

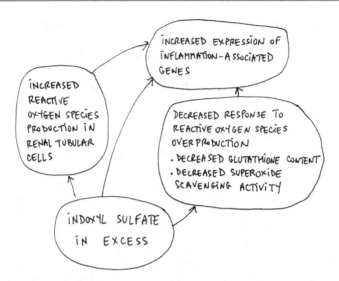

Fig. 5.7 Schematic view of the effects of indoxyl sulfate in excess on renal tubular cells

Indoxyl sulfate in excess has been shown to be involved in renal fibrosis. Briefly, renal fibrosis results from excessive accumulation of extracellular matrix after renal insult [122]. Administration of indoxyl sulfate in experimental models of chronic kidney disease leads to glomerular sclerosis and interstitial fibrosis [123]. These effects can be explained by the capacity of this co-metabolite to promote the transformation of kidney fibroblasts into matrix-producing phenotype, thus increasing collagen deposition, a process that is linked to interstitial fibrosis [124]. This proposition is reinforced by the fact that indoxyl sulfate increases the expression of genes that are known to be implicated in kidney fibrosis [125].

Trimethylamine N-oxide and Kidney Diseases: An Aggravating Co-metabolite
As explained in previous paragraphs, trimethylamine N-oxide is formed in the liver from the choline-derived bacterial metabolite trimethylamine. In healthy subjects, the circulating concentration of trimethylamine N-oxide is low being in the 3–7 micromolar range [126]. Several concordant studies have found that trimethylamine N-oxide is elevated in the blood and urine of subjects with impaired renal function when compared with healthy subjects [127–130]. Such an association urged studies aiming at searching for a possible causative link between increased trimethylamine N-oxide synthesis and renal dysfunction. By supplementing rodents with either choline or trimethylamine N-oxide, it was found that such dietary supplementation for 6 weeks led to tubulointerstitial fibrosis with collagen deposition in the kidneys of animals [129], thus indicating deleterious effects of this co-metabolite when present in excess on renal histopathological structure (Fig. 5.8). This conclusion was reinforced by another study showing in a mouse model of kidney disease that inhibition of trimethylamine N-oxide production by the intestinal microbiota attenuated the development of renal injury and fibrosis [131]. In a mice model of

Fig. 5.8 Effects of the co-metabolite trimethylamine N-oxide in excess on kidney injury and fibrosis

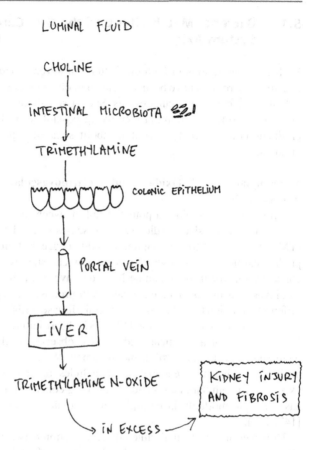

LUMINAL FLUID

CHOLINE

INTESTINAL MICROBIOTA

TRIMETHYLAMINE

COLONIC EPITHELIUM

PORTAL VEIN

LIVER

TRIMETHYLAMINE N-OXIDE

KIDNEY INJURY AND FIBROSIS

IN EXCESS

calcium oxalate crystal deposition in renal tubular cells, trimethylamine N-oxide was found to aggravate kidney injury provoked by such crystal deposition [132]. These converging studies have led to consider trimethylamine N-oxide not only as a biomarker of renal disease but also as a plausible player in chronic kidney disease aggravation [133].

Key Points
- The bacterial metabolite *p*-cresol and its co-metabolite *p*-cresyl sulfate at excessive concentrations act as uremic toxins during chronic kidney disease.
- The co-metabolite indoxyl sulfate is deleterious at excessive concentrations for kidney tubular cells and is involved in renal fibrosis.
- The co-metabolite trimethylamine N-oxide aggravates chronic kidney disease at excessive concentrations.

5.4   Bacterial Metabolites and the Gut–Cardiovascular System Axis

Higher concentrations in biological fluids of several metabolites produced by the intestinal microbiota have been associated in observational studies with an increased risk of adverse cardiovascular events. Such associations have motivated in vitro and in vivo experiments aiming at identifying possible causal links between excessive production of suspected bacterial metabolites and the aggravation of cardiovascular diseases.

Trimethylamine   N-Oxide   and   Cardiovascular   Diseases: A   Role in Atherosclerosis

The first studies revealing a potential link between the intestinal microbiota metabolic activity and risk of cardiovascular diseases focused on trimethylamine N-oxide (TMAO). Elevated trimethylamine N-oxide concentration in plasma was shown to predict cardiovascular disease risk in clinical studies [48]. Indeed, circulating concentrations of trimethylamine N-oxide have been shown to be associated with cardiovascular disease incidence, and even proposed as a predictor of outcomes in different situations, including peripheral disease [134], coronary artery disease [135], acute coronary syndrome [135–137], heart failure [19, 138–141], and stroke [142]. While not all clinical studies have observed a clear relationship between increased concentrations of trimethylamine N-oxide in plasma, examination of available studies in meta-analyses concludes on a robust relationship between excessive trimethylamine N-oxide circulating concentration and cardiovascular disease risk and mortality in multiple cohort studies performed in different countries [143, 144].

However, it must be underlined that association between trimethylamine N-oxide elevated plasma concentration and the incidence of cardiovascular disease does not allow to postulate on any causal link between this biochemical parameter and the final clinical outcomes. To make advances in relationship with this important question, both experimental in vivo and in vitro studies are necessary. Overall, in most studies, data obtained from animal models and cells show that trimethylamine N-oxide enhances atherosclerosis and atherosclerosis-related biochemical events [16, 25, 145–148]. However, one study found that elevated concentrations of trimethylamine N-oxide did not enhance atherosclerosis in a model of rodent that developed spontaneously atherogenesis in basal condition [149], likely because the atherogenesis processes operate rapidly in this model.

Several mechanisms that would link trimethylamine N-oxide excessive circulating concentrations and the atherosclerotic process include the reported effect of this co-metabolite on cholesterol metabolism. Briefly, it is well known that elevated plasma concentrations of LDL cholesterol, cholesterol, and triglycerides, and low concentration of HDL cholesterol are leading contributors to an increased risk for cardiovascular diseases [150–152]. In mice, trimethylamine N-oxide reduces cholesterol clearance in the host [25] (Fig. 5.9). Consistent with this finding, decreased conversion of trimethylamine into trimethylamine N-oxide in the liver,

Fig. 5.9 Schematic view of the effects of trimethylamine N-oxide at excessive concentrations on endothelial cells and platelets, and effects of this compound on cholesterol metabolism

and thus reduction of circulating concentration of this latter co-metabolite, restores cholesterol balance [153]. In addition, reduction of trimethylamine N-oxide synthesis in mice prevented diet-driven hepatic cholesterol accumulation [154].

Increased platelet activity has been associated with cardiometabolic dysfunction and modification of parameters known to be involved in thrombotic events [155, 156]. Platelet activation, aggregation, and subsequent generation of an occlusive intra-arterial thrombus are central in atherothrombotic disease [157]. In that context, it has been shown that mice supplemented with trimethylamine N-oxide or with the precursor choline display an increased responsiveness of platelet to agonists that activate platelets [158] (Fig. 5.9). In this latter study, in vitro exposure of platelets recovered from healthy volunteers to trimethylamine N-oxide enhances the responsiveness of platelets to different activators of platelets. Of equal importance, treatment of mice with this latter co-metabolite provokes thrombus development within the internal carotid artery [158]. In accordance with these results, by diminishing hepatic trimethylamine N-oxide synthesis, decreased platelet responsiveness and thrombus formation was observed in a mice model of carotid artery injury [159].

Vascular endothelial cell dysfunction is another component of vascular inflammation, a process that is critically involved in atherosclerosis and thrombotic events [160, 161]. In in vitro experiments, trimethylamine N-oxide activates human coronary artery endothelial cells [162] (Fig. 5.9). This latter result was confirmed in vivo in a rodent model where trimethylamine N-oxide was acutely injected. In that experimental situation, this microbiota-derived co-metabolite increases indicators of vascular inflammation with activation of aortic endothelial cells and increased expression of genes coding for pro-inflammatory factors [163]. From in vitro experiments with human umbilical vein endothelial cells, it has been shown that trimethylamine N-oxide induces an oxidative stress in these cells [164]. In this latter study, it was observed from in vitro and in vivo experiments that this bacteria-derived metabolite induces vascular senescence and dysfunction with impaired endothelial cell proliferation and migration. In addition to increasing the production of reactive oxygen species, trimethylamine N-oxide induces activation of inflammasome in endothelial cells [165] (Fig. 5.9). As shortly explained in the previous paragraph, inflammasomes are a set of intracellular complexes that drives innate immune and inflammatory responses by releasing pro-inflammatory mediators including pro-inflammatory cytokines under inappropriate activation [166, 167]. Of note, activation of inflammasome by trimethylamine N-oxide was found to be associated with endothelial dysfunction [168, 169].

Phenylacetylglutamine and Cardiovascular Diseases: A Role in Platelet Activation
As previously presented, phenylacetylglutamine is a co-metabolite produced in the host liver from phenylacetate, a bacterial metabolite produced by the intestinal microbiota from the amino acid phenylalanine. The association between increased concentration of phenylacetylglutamine in plasma with incident risk of major adverse cardiac events, like myocardial infarction and stroke, was validated in a large cohort, and this association was shown to be independent of classical cardiovascular disease risk factors [170]. In isolated platelets, it was shown that phenylacetylglutamine enhances platelet activation and responsiveness to different agonists [170] (Fig. 5.10). In an animal model of carotid injury, phenylacetylglutamine accelerates platelet clot formation within carotid artery, and thrombus formation. This latter bacteria-derived metabolite appears to promote its deleterious effects by binding to adrenergic receptors [170], thus identifying phenylacetylglutamine as an adrenergic agonist (Fig. 5.10). These results are of major importance when considering that adrenergic receptors are crucially involved in platelet activation [171] and heart diseases [172].

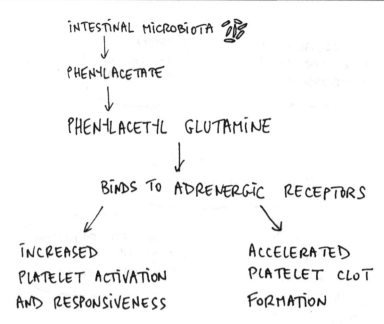

INTESTINAL MICROBIOTA

↓

PHENYLACETATE

↓

PHENYLACETYL GLUTAMINE

↓

BINDS TO ADRENERGIC RECEPTORS

INCREASED
PLATELET ACTIVATION
AND RESPONSIVENESS

ACCELERATED
PLATELET CLOT
FORMATION

Fig. 5.10 Schematic view of the effects of phenylacetylglutamine on platelets

p-cresol: A Role in the Increased Risk of Cardiovascular Dysfunction in Patients with Chronic Kidney Disease

Cardiovascular diseases represent one of the major causes of mortality in uremic [173], and chronic kidney disease patients [174]. This led to investigate how the accumulation of uremic toxins may play a role in chronic kidney disease-associated cardiovascular events. Among the uremic toxins, attention has been notably paid to the tyrosine-derived bacterial metabolite p-cresol and associated co-metabolites. Suspicion of deleterious effects of these microbiota-derived substances on the cardiovascular system originates mostly from observational studies in patients with renal diseases, and from in vitro experiments with endothelial cells.

Regarding observational studies, in the work by Meijers and collaborators [175], free p-cresol concentration in serum represents a risk predictor of cardiovascular disease in chronic kidney disease patients, while in hemodialysis patients, serum-free and total p-cresol concentrations are related to cardiovascular events [82]. High urinary excretion of p-cresyl sulfate is directly associated with cardiovascular events in patients with chronic kidney disease [176]. In addition, free serum p-cresol concentration is associated with cardiovascular disease when the primary endpoint is the time of the first cardiovascular event [177]. Lastly, serum-free p-cresyl sulfate concentration predicts cardiovascular mortality in the elderly [178]. However, in one study with hemodialysis patients, p-cresol plasma concentration was not associated with cardiovascular mortality [179]; and in another study, p-cresyl sulfate was not associated with cardiovascular outcomes [180]. However, in this latter study, sub-group analysis among patients with lower serum albumin indicates that a

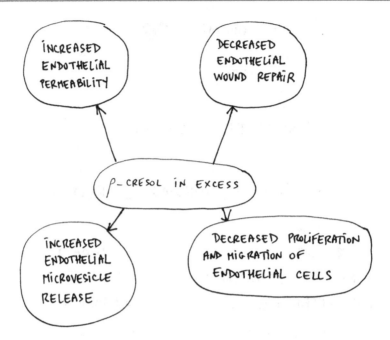

Fig. 5.11 Schematic view of the effects of p-cresol in excess on vascular endothelial cells

twofold higher *p*-cresyl sulfate concentration is associated with a higher risk of sudden cardiac death.

Since endothelial dysfunction plays an important role in the development of cardiovascular diseases [181], since patients with chronic kidney disease may show endothelial dysfunction [182]; and since in the case of endothelial injury, an active process implicating proliferation and migration of endothelial cells plays an important role in the healing of endothelial injury [183], several experimental studies have tested the effects of *p*-cresol on endothelial cell physiology. When *p*-cresol at increasing concentrations was tested in the presence of albumin on endothelial cells, this compound inhibits dose-dependently the endothelial cell proliferation, and the endothelial wound repair in an in vitro test [173, 184] (Fig. 5.11). Li et al. showed that inhibition of endothelial cell proliferation by *p*-cresol was partly due to an accumulation of cells in the G0/G1 phase of the cell cycle [185]. In the presence of albumin, *p*-cresol causes a strong increase in endothelial permeability in the human umbilical vein endothelial cells (HUVEC) model [186] (Fig. 5.11). Using *p*-cresol tested in the presence of albumin, Zhu and colleagues found that *p*-cresol at high concentrations impairs endothelial progenitor cell proliferation and found in addition that this bacterial metabolite affects cell migration [187]. *p*-Cresol inhibition of endothelial progenitor cell proliferation is associated with an activation of the signaling pathways p-38 and with mitogen-activated protein kinases Erk1 and 2 phosphorylation [188]. At the highest concentration tested, *p*-cresol reduces the capacity of endothelial progenitor cells to form a tube-like structure.

p-cresol, either in the free form or bound to albumin, increases the endothelial microvesicle release from endothelial cells in the extracellular milieu [189] (Fig. 5.11). Endothelial microvesicles, in the form of membrane vesicles ranging in size from 0.1 to 1.0 µm, are released from activated or apoptotic endothelial and blood cells and reflect the severity of endothelial cell damages [190]. These microvesicles, which are produced by cytoplasmic membrane blebbing and shedding, are considered as active intracellular messengers that are involved in the regulation of vascular physiology [191]. Increased endothelial microvesicle release is believed to alter the endothelial repair process by reducing the capacity of endothelial cells to migrate, and by increasing the senescence of mature endothelial cells [192]. Enhanced circulating endothelial microvesicles have been associated with endothelial dysfunction in patients with end-stage renal failure [193]. Interestingly, in hemodialysis patients, free serum p-cresol concentration is associated with the quantity of circulating endothelial microvesicles [177]. Recently, it has been shown that p-cresol in excess enhanced the release of microvesicles from endothelial cells [192]. Lastly, p-cresol in the endothelial cells activates the integrin-linked kinase [194]. This kinase represents a key component in the integrin signaling complex that is involved in vascular vessel integrity and angiogenesis [195–197]. This bulk of experimental evidence obtained in vitro clearly indicates that p-cresol in excess may affect endothelial cell biology and repair.

p-cresol has been also tested on cardiomyocytes, the cells that drive heart contraction [198]. Using neonatal cardiomyocytes, Peng and collaborators have shown that p-cresol reduces in a reversible way the cellular spontaneous contraction rate and provokes disassembly of gap junction, thus supporting the association between an excess of p-cresol and cardiomyocyte dysfunction [199]. Indeed, in cardiac muscles, gap junctions contribute to the electrical cell-to-cell coupling and impulse propagation between cardiomyocytes. In addition, at high concentrations, p-cresol induces disruption of cardiomyocyte adherens junctions, thus likely affecting the intercellular junctions between cardiomyocytes [200].

Further studies are needed to test if the lowering of p-cresol and/or p-cresyl sulfate circulating concentrations may allow to lower the risk of cardiovascular outcomes in chronic kidney disease patients [68].

Indoxyl Sulfate Is Deleterious for Endothelial Cells in Patients with Chronic Kidney Disease

Indoxyl sulfate can be included in the family of uremic endotheliotoxins, meaning that this compound induces endothelial dysfunction, one central element implicated in cardiovascular morbidity and mortality [201]. From in vivo and in vitro preclinical studies, indoxyl sulfate has been shown to promote both pro-thrombotic processes, through different mechanisms including aryl hydrocarbon receptor activation [202–204], and pro-oxidant processes in endothelial cells [205–208]. Clinical data obtained in patients with chronic kidney disease indicate that indoxyl sulfate is likely to represent one of the links between impaired renal function and adverse cardiovascular events, notably regarding hemostatic disorders [209] and thrombotic events [210].

Key Points

- The co-metabolite trimethylamine N-oxide favors endothelial dysfunction, vascular inflammation, atherogenesis, and thrombus formation at excessive concentrations.
- The co-metabolite phenylacetylglutamine favorizes platelet clot formation within carotid artery and thrombus formation at excessive concentrations.
- The bacterial metabolite *p*-cresol in excess favors endothelial cell and cardiomyocyte dysfunction.
- The co-metabolite indoxyl sulfate represents likely one of the links between impaired renal function and adverse cardiovascular events.

5.5   Bacterial Metabolites and the Gut–Bone Axis

The evidence that link metabolites produced by the gut microbiota and bone physiology are relatively scarce. Most available data are related to the positive effects that short-chain fatty acids appear to exert on bone, notably in the situation of impaired bone quality.

Short-Chain Fatty Acids: An Emerging Beneficial Role for Increasing Bone Mass

Bone loss induced by ovariectomy or inflammation in rodent model can be partially rescued by supplying a mixture of butyrate, propionate, and acetate in the drinking water [211]. Although in this study the way of administration of short-chain fatty acids was not the way these compounds are supplied in real-life situations (these bacterial metabolites are produced by the intestinal bacteria and absorbed by the intestinal epithelium), the results obtained point out the potential effect of these compounds. Then, it is worth to note that in physiological situations, butyrate and propionate are almost completely metabolized during the transfer from the luminal content to the blood [6]. Among short-chain fatty acids, only acetate is detected at higher concentrations (approximately 100 micromolar) in peripheral blood. The effects of short-chain fatty acids taken by the oral way in the mice model were associated with the inhibition of osteoclast differentiation and bone resorption, while bone formation was not affected [211]. Briefly, overall bone mass and quality determine bone strength that can be evaluated by different parameters that include bone mineral density and bone microarchitecture [212]. These parameters depend on relative osteoblast and osteoclast activity to maintain adequate bone remodeling. Osteoblasts are the bone-forming cells [213] while osteoclasts are bone-resorbing cells [214]. For instance, low bone mass and altered microarchitecture can be associated with estrogen deficiency, as observed after menopause in animal models [215] and human subjects [216].

When mice were treated with antibiotics to drastically reduce the abundance of their intestinal microbiota, the animals displayed reduced circulating concentration of insulin-like growth factor-1 (also called IGF-1) and reduced bone formation [217]. Insulin-like growth factor-1 is a compound with both endocrine and paracrine actions that promotes bone growth and density [218–220]. Supplementation of antibiotic-treated mice with short-chain fatty acids provided in the drinking water allowed to restore insulin-like growth factor-1 in blood [217].

In a prospective cohort study gathering more than 200,000 participants, greater intake of total dietary fibers, and subtypes from various food sources, was associated with higher heel bone mineral density [221], reinforcing the potential role of intestinal microbiota-derived acetate as one beneficial element for bone health maintenance.

Key Points
- Short-chain fatty acids provided orally can counteract bone loss.
- Reduced abundance of intestinal microbiota leads to reduced bone formation.

5.6 Bacterial Metabolites and the Gut–Brain Axis

Surprisingly, accumulating evidence suggest that the intestinal microbiota exerts some effects on brain development [222], neurogenesis [223], and interacts with the peripheral and central nervous system [224]. Most of the works on that topic have been performed on animal models, showing that the intestinal microbiota can apparently modulate parameters characteristics of mood disorders and emotional behavior [225–229]. Although the metabolic capacity of bacteria to synthesize neurotransmitters and other metabolites with effective or presumed action on the central and peripheral nervous system of the host is much intriguing, the precise physiological roles of this bacterial metabolic activity and the underlying mechanisms remain in its early infancy, opening a very excitatory new field of research [230] with considerable potential applications.

Experimental Data Indicate that Some Bacterial Metabolites Which Are Known to Act as Neuroactive Compounds in the Host Are Used in the First Place for Communication between Different Bacterial Species
As illustrated in Chap. 3, in regards with the role of bacteria-derived gamma-amino butyric acid and norepinephrine on bacterial physiology, a first consideration that is worth to be considered is related to the fact that microbes can communicate with each other via some of the metabolites known to act as neurotransmitters in the host [231]. These bacterial metabolites can indeed modulate several characteristics of bacterial activity. For instance, enterohemorrhagic *E.Coli* can sense the luminal norepinephrine to express some of its virulence traits [232], and this bacterial metabolite can stimulate the proliferation of several strains of enteric pathogens [233]. Luminal norepinephrine is also known to increase the virulence properties of *Campylobacter jejuni* [234]. However, the effects of norepinephrine in the context of nonpathogenic microorganism dialogue remains poorly documented [235]. Gamma-aminobutyric acid (also called GABA) is another illustrating example of bacterial metabolite with known neurotransmitter properties in the host which is used in the bacterial world to support the growth of specific bacterial species. Indeed, gamma-

aminobutyric acid produced by *Bacteroides fragilis* has been shown to be used as a nutrient in support of the growth of a Gram-positive bacterium of the *Ruminococcaceae* family [236]. Thus, importantly, metabolites produced by the intestinal bacteria with neurotransmitter function in the host nervous system are used in the first place as molecules for communication within the intestinal microbial world. In other words, this is a fascinating example of the utilization of the same molecules for completely different functions in the bacterial and mammalian world. These bacterial metabolites are also likely involved, as some experimental data presented below suggest, in the communication between the intestinal microbiota and its host.

Several Bacterial Metabolites with Neuroactive Properties Can Modulate Host Neurophysiology by either Binding to Receptors Present in the Intestinal Wall or by Being Absorbed in the Circulating Blood

An increasing amount of preclinical evidence indicate that several bacterial metabolites can modulate host nervous system functioning, thus pointing out an intriguing interkingdom signaling between intestinal microbiota and the host [237]. This metabolic crosstalk is dependent on a mixture of microbiota-derived compounds that can influence host neurophysiology either through direct interactions with receptors present in different cell phenotypes present within the intestinal tract, or following absorption and diffusion through the gut epithelium, and entry into the portal circulation [235]. The bacterial metabolites may influence host neurophysiology via different routes including the vagus nerve activity, the enteric nervous system, the intestinal immune system, and the neuro-endocrine system [238, 239]. Information about the characteristics of the luminal environment, such as the osmolarity and bacterial product composition, are transmitted to the central nervous system by the vagus nerve [231]. At the interface between the intestinal microbiota and the host lies a network of neurons known as the enteric nervous system, which has been found to be modulated either directly or indirectly by several bacterial metabolites [238] as will be illustrated in the following paragraphs.

How Do the Bacterial Metabolites Find their Way between the Intestinal Lumen and the Central Nervous System?

Regarding bacterial metabolites with known neurotransmitter function in the host, to postulate on possible direct effects of these bacterial compounds on the central nervous system, several prerequisites must be established. Firstly, the bacterial metabolites must be transferred across the intestinal epithelium from the intestinal lumen to the portal bloodstream without being fully metabolized during this transcellular journey (Fig. 5.12). The transfer through the intestinal epithelium may depend on the permeability of the intestinal epithelium, which can be itself modulated by bacterial metabolites [240].

Then, the compounds must not be fully degraded during liver cell metabolism. Then, if so, the bacterial metabolites may be released in the peripheral blood. From peripheral blood, the compounds must then enter the brain to exert some putative roles there. As will be detailed in the next paragraphs, only few information is available regarding these prerequisites. Thus, following the entry into the circulating

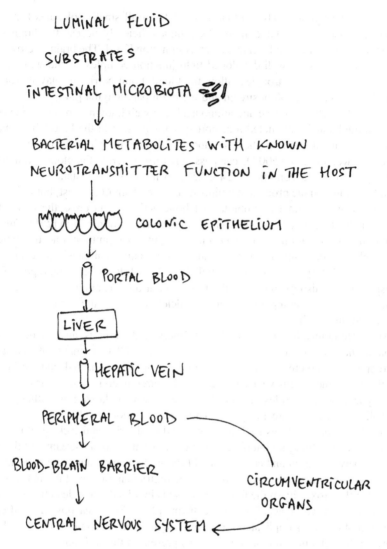

Fig. 5.12 Schematic view of the transfer of bacterial metabolites into the central nervous system after colonic absorption, and transfer into peripheral blood

blood, one given bacterial metabolite will have to be transferred across the blood–brain barrier (also called BBB) to enter the central nervous system and to act there (Fig. 5.12). The blood–brain barrier is a vascular structure that separates the central nervous system from the peripheral blood circulation. By tightly controlling the passage of molecules and ions, by delivering nutrients and oxygen according to neuronal needs, and by protecting the brain from deleterious compounds, the blood–brain barrier maintains an environment that is compatible with a proper functioning of the different brain structural zones [241]. This barrier is indeed one of the tightest barriers in the human body, so that about 98% of all small molecules tested, and nearly all macromolecular compounds cannot efficiently access the brain [242], except in the case of blood–brain barrier dysfunction [243]. The brain microvascular endothelial cells, with well-developed tight junction complexes, and with the support of pericytes and astrocytes, allow the blood–brain barrier to play its role as a boundary between the bloodstream and the brain parenchyma [244, 245].

Nutrients such as glucose and amino acids are well known to be able to enter the brain through carrier-mediated transport systems present in the blood–brain barrier [246–248]. In addition to these compounds, some lipophilic molecules with a molecular weight below 500 Da can passively diffuse across the blood–brain barrier under physiological conditions [249, 250]. Ammonia, for instance, a compound formed not only in the process of amino acid metabolism in the host, but also largely resulting from intestinal microbiota metabolic activity, can cross the blood–brain barrier in the diffusible form NH_3, but not in the NH_4^+ cationic form [251]. This is an important consideration to be taken into account since ammonia accumulation in blood plasma in case of severe hepatic insufficiency can lead to severe brain dysfunction [252], as will be detailed in the following paragraphs. Efflux transporters are also expressed at the blood–brain barrier, allowing clearance from brain to blood of compounds that are deleterious for the brain at excessive concentrations [247].

The communication between the periphery and brain is also made possible through the circumventricular organs (also called CVOs), made of specialized neuroepithelial structures found in the midline of the brain, and gathered around the third and the fourth ventricles [253]. The circumventricular organs play both sensory and secretory roles facilitated by increased vascularization. Although these specialized structures are morphologically and functionally diverse, they share a range of common features such as fenestrated capillaries and lack of blood–brain barrier, thus providing an interface not only between the bloodstream and the brain but also between brain and cerebrospinal fluid in the ventricular system of the brain [254–256]. For instance, among the circumventricular organs, the area postrema located at the proximity of the fourth ventricle is involved in the detection of several toxins that may circulate in the bloodstream [257]. The area postrema, after the detection of toxins, is implicated for instance in the emesis (vomiting) process [258], allowing the potential elimination of toxins present in the stomach.

Overall, as will be detailed in the next chapters, the passage of microbiota-derived neurotransmitter-like compounds from blood to brain appears plausible since, as detailed in the next chapters, in experimental models several bacterial metabolites when injected in the intestinal lumen are recovered within the brain. The passage of these bacterial metabolites, including those with neuroactive properties, either through the blood–brain barrier or through the circumventricular organs remains however poorly documented [259].

Germ-free animal models (without microbes) have been largely used to decipher the part played by the intestinal microbiota in various physiological and pathophysiological situations as already presented in some previous paragraphs of this book. Although enabling to build new working hypothesis, these experimental models have several inherent important limitations [259], since numerous physiological functions and metabolic pathways are altered in animals without microbes [260], rendering the comparison and interpretation of data recorded in germ-free versus normal counterpart sometimes difficult. It is however interesting to note that germ-free animals are characterized notably by important modifications regarding brain and gut physiology, in a presumed gut–brain axis, when compared with normal animals with an intestinal microbiota [222, 261, 262]. Also, the animal behavior shows some modifications when comparing animals with a microbiota, and those without it [222, 263–266]. Observed changes in brain functioning and behavior in germ-free animals could be mediated by the lack of microbiota in a direct or indirect fashion through brain-related and non-brain-related alterations.

Gamma-Aminobutyric Acid Is Absorbed by the Intestinal Epithelium but Entry through the Blood–Brain Barrier Is Observed Only in Situation of Slight Disruption
As briefly introduced in Chap. 3, gamma-aminobutyric acid is the major inhibitory neurotransmitter in the central nervous system [267]. Gamma-aminobutyric acid is found in different specialized structures of the brain [268]. This neurotransmitter is notably known to be implicated in the reduction of anxiety disorders [269, 270]. Gamma-aminobutyric acid is also involved in the modulation of intestinal motility by the enteric nervous system [271]. Gamma-aminobutyric acid is produced by intestinal bacteria and released in the intestinal content [272]. Although this compound is absorbed by the mammal intestine [273], by comparing normal and germ-free rats, and by measuring its concentration in arterial blood and venous effluents from small and large bowel, it was found that gamma-aminobutyric acid produced by the microbiota represents only a minor fraction of the gamma-aminobutyric acid produced by the intestinal ecosystem, the vast majority of this compound originating presumably from the metabolism of glutamine in the intestinal epithelium [274]. In addition, gamma-aminobutyric acid whatever its origin does not cross the blood–brain barrier in rodent models [275], so that blood–brain barrier must be slightly disrupted to allow delivery of circulating gamma-aminobutyric acid in experimental model [276].

The Absence of Microbiota in the Host Increases Norepinephrine Turnover in Brain
Norepinephrine (also called noradrenaline) is notably produced in a small nucleus of cells located in the lateral wall of the brainstem which produce most of the norepinephrine released in the brain [277]. Norepinephrine is considered as a participant in the shaping and the wiring in the central nervous system during development [278], and as a player for cognitive functions in healthy state. Impaired norepinephrine production is also involved in cognitive dysfunctions in several neurodegenerative diseases [279]. The effects of bacterial norepinephrine produced and released from intestinal bacteria are not known. Norepinephrine does not appear to cross the blood–brain barrier in normal situations, since systemic administration of norepinephrine has no effect on cerebral circulation [280]. However, in primate model, when the blood–brain barrier is opened by osmotic disruption and norepinephrine

infused in blood, this compound increases cerebral blood flow, oxygen consumption, and glucose uptake [281].

Interestingly, mice with no microbiota (germ-free animals) show an increased turnover rate of norepinephrine in the brain when compared with normal animals [222], suggesting that the intestinal microbiota through still unknown mechanisms can modulate norepinephrine metabolism in the brain of mammals.

The Absence of Microbiota in the Host Alters Dopamine Turnover in Brain
Dopamine is a neurotransmitter synthesized in central and peripheral nervous system [282]. Dopamine is a critical modulator of learning and motivation [283]. Bacterial tyrosinases which catalyze the conversion of tyrosine to L-dihydroxyphenylalanine (also known as DOPA), the direct precursor of dopamine, are widely found in many bacterial genera [284]. The possible effects of dopamine produced by the intestinal bacteria on the host nervous system remain to be explored. Dopamine from the circulating blood usually does not enter the brain through the blood–brain barrier [285].

Germ-free mice display an altered turnover of dopamine in the striatum when compared with animals with a normal gut microbiota [222, 263]. The striatum acts as an integrative hub for information processing in the brain as revealed by analysis of multiple connections between central nervous system zones and neuroimaging data [286].

Excessive Histamine Production by Bacteria in Food, and Maybe by Intestinal Bacteria, May Lead to Intoxication but Histamine Does Not Cross the Blood-Brain Barrier
Histamine and its receptors were first described as parts of the immune and gastro-intestinal systems, but their presence in the central nervous system and their implication in behavior and energy homeostasis have gained increased attention [287]. The role of histamine in allergic diseases has been largely studied [288].

Apart from synthesis in specialized blood cells, histamine is synthesized by histaminergic neurons and enteroendocrine cells [289]. Then, histamine is considered as a neuromodulator [290]. Histamine can modulate the activity of the enteral nervous system [291]. Then, histamine is considered as a neuromodulator (Nomura H, Shimizume R, Ikegaya Y. Histamine: a key neuromodulator of memory consolidation and retrieval. Curr Top Behav Neurosci. 2022;59:329-53). Excessive production of histamine from the bacteria in food not appropriately stored may lead to histamine intoxication [292]. Of note, *Lactobacillus paracasei* has been shown to degrade histamine, pointing out its potential capacity for decreasing the histamine content in contaminated food [293]. Histamine in normal situation does not cross the blood–brain barrier [294]. Further research is needed to study the possible effects of excessive histamine production by the intestinal microbiota on the host peripheral nervous system [291, 295].

Serotonin Produced by the Gut Microbiota Contributes to the Circulating Concentration of this Compound and to Serotonin Turnover in the Brain
Serotonin, a neurotransmitter involved in numerous processes involved in behavior, learning, and appetite, is produced in specialized cells of the human brain, but also in

intestinal enteroendocrine cells [296]. Serotonin is also centrally implicated in mood and cognition [297]. The effects of serotonin synthesized by the intestinal bacteria on the host nervous system, if any, is not known. Of major significance, comparison between germ-free and conventional animals (with an intestinal microbiota) reveals a large decrease of serotonin in the blood circulation of germ-free animals when compared with their normal counterpart [38], thus suggesting that the production of serotonin by the intestinal microbiota and its subsequent intestinal absorption plays a significant role in fixing the circulating concentration of this neurotransmitter. However, the serotonin originating from the enteroendocrine cells likely accounts for a major part of the circulating serotonin [298]. It is generally considered that circulating serotonin does not cross the blood–brain barrier to any significant extent [299]. However, in anesthetized rats, it has been shown that circulating serotonin was able to transiently enter the brain parenchyma by inducing a short-term break-down of the blood–brain barrier, influencing then the spontaneous cerebral cortical activity [300].

Serotonin produced by the enteroendocrine cells is likely to play a central role on the intestinal peripheral nervous system. Indeed, nerve terminals of vagal afferents are near enteroendocrine cells, and these terminals express serotonin receptors [301]. As afferent nerves are not exposed to the luminal side of the intestinal mucosa, sensory neurons are believed to be indirectly activated by stimuli in the intestinal content via paracrine signaling, which can be mediated by compounds like serotonin, this latter compound being released from neuroendocrine cells in the mucosa [302]. However, the possible role of enteroendocrine cells to vagus nerve signaling in the modulation of brain function has received little attention, and thus further experimental works are needed [231].

Of note, the fact that in germ-free animals the turnover of serotonin in brain is increased suggests microbiota-host metabolic interactions regarding this neurotransmitter [222].

Circulating Tryptamine Can Enter the Brain

Tryptamine can be produced from tryptophan by the intestinal microbiota [303, 304]. Tryptamine, which can also be synthesized in different structures of the brain, has been proposed to act as a neurotransmitter [305]. In animal model, intravenous injection of labeled tryptamine was followed by rapid intake into the brain [306], indicating passage of this compound through the blood–brain barrier. Little information is available regarding the role, if any, played by the tryptamine originating from the microbial activity on the brain physiology.

Decreasing Ammonia Net Production by the Intestinal Bacteria: Is it Efficient for Limiting Brain Damages in Hepatic Encephalopathy?

As explained in Chap. 3 of this book, the colon is a major site of production of ammonia. In the colon, the bacteria produce ammonia from the urea that diffuses from the blood capillaries to the intestinal lumen. This conversion of urea to ammonia is made through the catalytic activity of bacterial ureases. These enzymes are not expressed in the host's cells, and then only the bacteria lodged by the host are

Fig. 5.13 Production of ammonia (considered as the sum of NH₄⁺ and NH₃) by the intestinal microbiota and by the intestinal epithelial cells, disposal of this neurotoxic compound by the liver, and excretion of urea

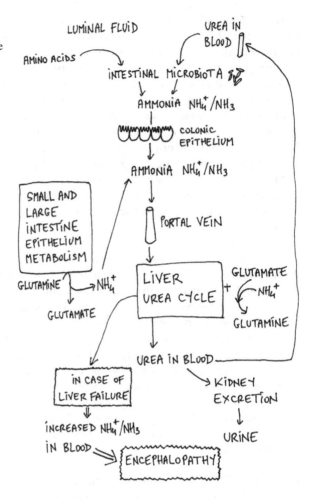

able to perform this conversion. In addition to be produced from urea, this latter metabolite is also produced by the colonic bacteria, by deamination of amino acids originating from undigested proteins [307, 308]. Ammonia in the colonic content is largely absorbed through colonocytes [309] (Fig. 5.13). Enterocytes of the small intestine and colonocytes are also important contributors for ammonia production, since these cells use glutamine as major energy substrate, and then produce ammonia when glutamine is converted into ammonia and glutamate [310] (see Chap. 1). Ammonia from intestinal origin is released in the portal vein, reaching the liver where this metabolite is transformed in urea in the liver urea cycle, and eliminated in urine (Fig. 5.13). Urea cycle represents the principal elimination route for ammonia, allowing to maintain the circulating concentration of ammonia below toxic threshold [311]. High ammonia concentration in the brain leads to inflammation, elevation of cerebral glutamine, and astroglial swelling, in relationship with increased intracranial pressure, cerebral edema, and disturbance of brain functions [312]. Glutamine synthetase found mainly in the perivenous hepatocytes of the liver (and in muscles) represents a secondary metabolic resource for ammonia disposal [313] (Fig. 5.13). Hepatic encephalopathy represents a frequent complication of liver diseases with a complex spectrum of neurological cognitive disturbances and altered level of consciousness [314, 315]. High circulating concentration of ammonia, due to failure of liver to eliminate this metabolic waste (from the host point of view) is considered as the leading cause of encephalopathy [316] (Fig. 5.13). Then, any strategy that would allow to decrease ammonia production by the colonic intestinal microbiota can represent a valuable way to decrease blood ammonia concentration, and thus hopefully the clinical signs of hepatic encephalopathy. This can be theoretically done by decreasing microbial ammonia production and/or increasing ammonia utilization by the bacteria for their nitrogen metabolism. This can also be done by limiting glutamine conversion to ammonia and glutamate in intestinal epithelial cells, and/or increasing ammonia utilization for glutamine synthesis in tissues equipped with this metabolic capacity.

Although the non-absorbable disaccharide lactulose was initially found in two small trials to show some efficiency in the treatment of portal-systemic encephalopathy [317, 318], a systematic review of the available data concluded that lactulose shows insufficient evidence of efficiency to recommend its use for the treatment of hepatic encephalopathy [319]. Protein restriction has also been proposed as a possible treatment for hepatic encephalopathy. However, such a restriction appears also insufficient for significantly ameliorating the course of the disease [320]. A mixture of ornithine and aspartate, which provides intermediates for glutamate production, and thus supply a substrate for glutamine synthesis from ammonia and glutamate, decreased serum ammonia concentration and improved hepatic encephalopathy in animal models [321], and in clinical trials [322, 323]). It has been proposed, as a strategy for hepatic encephalopathy treatment, to focus both on the ways to reduce the circulating ammonia (notably by reducing its production by the intestinal microbiota), and on the ways to limit its endogenous production by the host tissues, notably at the level of intestine [324].

The Bacterial Metabolite *p*-Cresol Is Suspected to Be One Element that Would Be Involved in the Etiology of Autism Spectrum Disorders

Autism spectrum disorders (also known as ASD) are diagnosed within the first 3 years of life and are defined notably as persistent deficits in social communication and interaction, as well as restrictive and repetitive behavior [325]. The pathogenesis of autism spectrum disorders is far from being understood, and such disorders without a known specific cause (named idiopathic form) represent most cases [326]. Autism spectrum disorders gather heterogeneous forms of the disease with multiple etiologies, different subtypes, and various developmental trajectories [327]. Post-mortem examination of brain from individuals with autism spectrum disorders may reveal region-specific abnormalities in neuronal morphology and cytoarchitectural organization, with consistent findings reported for abnormalities in the prefrontal cortex, fusiform gyrus, frontoinsular and cingulate cortex, hippocampus, amygdala, cerebellum, and brainstem [326].

Autism candidate genes have been identified regarding their prevalence, inheritance, and mutations involved in genotype-phenotype correlations [328]. From several studies including studies with twins [329], several environmental factors have been identified as putative risk factors for autism spectrum disorders [330, 331]. In that complicated context, what are then the available evidence that allow to propose that elevated bacterial metabolites in urine and feces may represent valuable early biomarkers of autism spectrum disorders, and even maybe candidates that would play a partial role in the etiology of autism?

To answer that question, it is of interest to consider that a subset of children with autism spectrum disorders exhibits high concentrations in biological fluids of diverse amino acid-derived bacterial metabolites in several studies [332–340]. These observations have led to the challenging hypothesis that excessive protein fermentation by the large intestine microbiota would play a role in the etiology of autism disorders in some children diagnosed with autism spectrum disorders [341]. Among the different amino acid-derived bacterial metabolites, special attention has been paid to p-cresol.

As indicated in Chap. 3, p-cresol is produced by the intestinal microbiota from the amino acid tyrosine but cannot be synthesized by the host. P-Cresol in excess is suspected to exert deleterious effects on the central nervous system, such effects being possibly related to the etiology of autism spectrum disorders. Firstly, in the study by Altieri et al. [334], higher p-cresol concentrations in the morning spot urine are measured in autistic children when compared to age- and sex-matched control children. Incidentally, the recovery of p-cresol in the host urine is a clear indication of the absorption of this compound from the intestine to the blood circulation. These differences in urine concentrations are detectable up to the age of seven, while the urinary p-cresol levels normalized at the age of 8 years and thereafter. In the study by Gabriele et al. [337], the total urinary p-cresol and its co-metabolites p-cresyl sulfate and p-cresyl glucuronide are higher in young children diagnosed with autism spectrum disorder, here again up to the age of 8 years, when compared with control children of same age and sex, but not in older children. Interestingly, urinary levels of p-cresol and p-cresyl sulfate are associated with stereotypic and compulsive/ repetitive behavior, but not with the overall signs of autism severity. Measurement of p-cresol in feces of autistic and control children has also been performed in two other studies, and in accordance with the data obtained from urine analysis, the results indicate that p-cresol is higher in feces recovered in autistic children when compared to control children [335, 338]. These results have motivated experimental works to determine if p-cresol may play a partial role in the etiology of autism.

Several evidence from preclinical studies support this suspicion. Of major interest, mice exposed to p-cresol for 4 weeks in their drinking water are characterized by social behavior deficits, stereotypies, and perseverative behaviors, but by no changes in anxiety, locomotion, or cognition [342]. In this latter study, abnormal social behavior induced by p-cresol was associated with decreased activity of central dopamine neurons involved in the social reward circuit. Pascucci et al. show that a single intravenous injection of p-cresol at a dose of 1 mg/kg in the BTBR mice model (a model genetically homogenous that displays three behavioral traits used to diagnose autism [343, 344]) induces a behavior characterized by anxiety and increased locomotor activity [345]. At the higher dose of 10 mg/kg, p-cresol exacerbates the core symptoms of autism in mice, and notably thwarts preference for social interaction between mice. In addition, in this study, the authors performed a brain region-specific neurochemical analysis and showed that these modifications of behavior are paralleled by a dose-dependent increase of the dopamine turnover in some regions of the brain, namely the amygdala, nucleus accumbens, and dorsal caudate putamen.

However, at this point of discussion, it is worth indicating that autism spectrum disorders should not be viewed only as a "brain disease" since abnormalities in the immune system and digestive tract are also described in such disorders [346]. Regarding the digestive tract, elevated levels of p-cresol in urine is associated with the anomaly of intestinal transit in autistic children [336]. Such anomaly of transit is relatively frequent in these children, although with very different incidences, since it represents between 9 and 70% of children diagnosed autistic according to the different studies considered [347, 348]. In addition, in one study, 43% of children with autism spectrum disorder show altered intestinal permeability [349], leading to the hypothesis that an increased transfer of luminal components (including p-cresol) from the intestinal content to the bloodstream may occur in these children.

From in vitro and in vivo experiments, several mechanisms may account for the deleterious effect of excessive circulating p-cresol on neural functions. These effects include the effects of p-cresol on the brain Na/K ATPase activity [350], and on the blunted conversion of dopamine to norepinephrine due to the inhibition of dopamine-beta hydroxylase activity [351]. The inhibition of this latter enzymatic activity is due to the action of p-cresol on one residue at the active site of the enzyme. Intraperitoneal injection of p-cresol (30 mg/kg) in rats modifies the expression of N-methyl-D-aspartate glutamate receptors (also known as NMDARs) subunits in the nucleus accumbens and hippocampus, raising the view that p-cresol may impair N-methyl-D-aspartate glutamate receptors-dependent activity in these structures [352]. These receptors are known to play a central role in the process of learning, memory, and synaptic development [353, 354]. Lastly, p-cresol inhibits oligodendrocyte differentiation in vitro [355], these cells being the myelin-forming cells of the central nervous system that develops from glial progenitor cells [356].

Overall, the relatively high frequency of intestinal transit and permeability impairment in autistic children, together with an increased production of p-cresol by the intestinal microbiota could plausibly lead to deleterious effects of these bacterial metabolites on specialized areas of the central nervous system in a subset of predisposed children. To confirm or invalidate this proposition, clinical studies are required to test in children with an early diagnosis of autism, if the lowering of p-cresol and/or p-cresyl sulfate production by the intestinal microbiota by dietary and/or pharmacological means could by itself improve the intestinal/neurological outcomes.

Indole and Indole-Related Compounds Are Bacterial Metabolite and Co-Metabolites that Are Active on the Central Nervous System

There is emerging evidence indicating the influence of indole on brain metabolism and physiology and on host behavior [357]. Chronic overproduction of indole in rats monocolonized with indole-producing *E. coli* has been shown to enhance anxiety-like behaviors and depression in these animals [358]. This latter study also found that giving intra-caecal indole to conventional rats activated a cerebral nucleus called the dorsal vagal complex. By comparing mice mono-associated with a non-indole-producing *E. coli* strain, or with an indole-producing *E. coli* strain, it was found that chronic high indole production by the intestinal microbiota increased the vulnerability to the adverse effects of chronic stress on overall emotional behavior [359]. It has been shown that a rise in plasma indole derived from intestinal microbiota metabolic activity is associated with hepatic encephalopathy, a neuropsychiatric trouble caused by hepatic dysfunction [360]. Furthermore, the findings of the observational prospective NutriNet-Santé Study revealed a positive link between urine indole and indole-related compound concentrations and recurrent depression symptoms. This correlation raises the hypothesis that the synthesis of these compounds by the gut microbiota in excessive amounts might possibly play a role in the emergence of mood disorders in humans [361].

Some beneficial effects of indole have also been reported. Indeed, in the model of mice treated with antibiotics (to diminish the concentration of bacteria in the intestine), intraperitoneal injection of indole reduced central nervous system inflammation [362]. Of note, in this latter experimental study, indole was not the sole compound that reduces central nervous system inflammation since indole-3-propionate and indole-3-aldehyde also display some protective effects. Regarding indole-3-propionate, this bacterial metabolite was found able to protect primary neurons against the oxidative damages provoked by reactive oxygen species when present in excess [363]. The beneficial versus deleterious effects of indole on the central nervous system and behavior, as observed in experimental works, are likely related to the different exposure levels and experimental designs according to the different studies.

Some additional information on the effect of indole-related compounds on the central nervous system are available. Isatin, oxindole, and indoxyl sulfate are the most studied indole-related compounds for their neuroactive effects.

The Co-Metabolite Isatin which Is Produced from the Bacterial Metabolite Indole Enters the Brain and Affects Behavior and Brain Functions

Regarding isatin (1H-indole-2,3-dione), this compound is a co-metabolite that is produced by the host from indole [36]. Isatin can be detected in different regions of the brain [364]. Systemic administration of isatin to rodents leads to the accumulation of this compound in the brain [358]. In addition, administration of indole in the large intestine results in the accumulation of isatin in the brain [358], thus importantly suggesting that this co-metabolite of indole enters the brain by still unknown processes. The hippocampus and cerebellum have the largest concentrations of isatin in the rat brain, whereas the prefrontal cortex and brainstem have the lowest quantities [365]. The impact of isatin on rodent behavior and brain function, either beneficial or deleterious, appears to depend on the doses used [366, 367]. At 10 mg/ kg, isatin when given intraperitoneally reduces the anxiolytic effect of intracerebroventricular atrial natriuretic peptide (also known as ANP) injection [368]. Higher amounts of isatin, over 40 mg/kg, when given here again by the intraperitoneal way result in a reduction in locomotor activity in the open field test and mobility in the forced swimming test in rats [369], indicating that isatin may exert a sedative effect.

An intraperitoneal injection of isatin (100 mg/kg) improves parkinsonian symptoms in a parkinsonian rat model induced by 6-hydroxydopamine lesion [370]. Isatin intraperitoneal administration (100 mg/kg) increases dopamine concentration in the striatum of rats and motor activity in the model of Parkinson's disease induced by Japanese encephalitis virus [371]. Isatin inhibits in a dose-dependent manner the monoamine oxidase activities in extracts obtained from rat brain [372]. Monoamine oxidase activities are implicated in the degradation of neurotransmitters including serotonin, dopamine, and norepinephrine [373, 374].

The Bacterial Metabolite Oxindole Enters the Brain and Exerts Deleterious Effects on the Central Nervous System at Excessive Doses
Regarding oxindole, this indole-related compound has been found in the brain, and, in the rat model, oral administration of indole resulted in the accumulation of oxindole in the brain [375]. In addition, systemic administration of oxindole increased the concentration of this compound in the rat brain [358], thus indicating that this compound may, as isatin, enter the brain by a still unknown process. Importantly, oral administration in rodents of neomycin, a broad-spectrum antibiotic, decreased the brain oxindole content [375], thus suggesting that oxindole is originating, at least partly, from the intestinal microbiota metabolic activity. Interestingly, oxindole is found in human fecal samples and in mouse caecal contents [376], thus confirming that intestinal microbes are a source of this metabolite. Among the bacterial metabolites present in human stool, oxindole was found to be one of the dominant aryl hydrocarbon receptor activators [376]. The mechanism of action responsible for oxindole's effects on the central nervous system remains unknown. In vitro experiments on rat hippocampus slices demonstrate that oxindole may interact with voltage-gated sodium channels, increasing the threshold for producing action potentials and therefore drastically reducing neuron excitability [377]. These results led several authors to propose that oxindole may share some characteristics with known neurodepressant compounds [375, 377].

The Co-Metabolite Indoxyl Sulfate which Is Produced from the Bacterial Metabolite Indole Exerts both Beneficial and Deleterious Effects on Brain Depending on the Doses Used
In respect to indoxyl sulfate, this co-metabolite derived from the indole produced by the intestinal microbiota is detected in the mammalian brain [378]. Indoxyl sulfate is also detected in the cerebrospinal fluid of mice, and animals with no intestinal microbiota display low level of this compound in this fluid when compared with

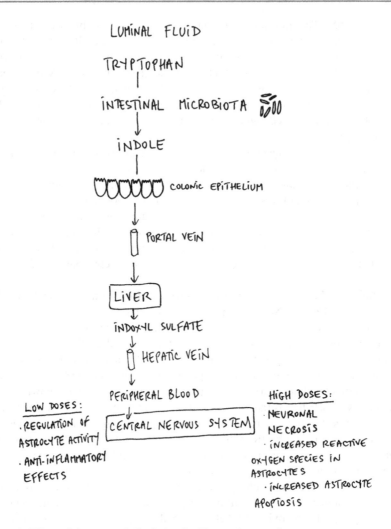

Fig. 5.14 Effects of the co-metabolite indoxyl sulfate on the central nervous system

normal mice hosting microbes [379] (Fig. 5.14). By comparing volunteers suffering from depression with healthy participants, 22 urine metabolites were identified for which abundance differed between the two groups of subjects. The Hamilton depression scale questionnaire score was used to measure the severity of depression. According to this investigation, the urine indoxyl sulfate concentration in individuals suffering from severe depression was highly significantly lower than in healthy counterparts [380]. In volunteers, serum indoxyl sulfate concentrations have been associated with psychic anxiety and with the related functional magnetic resonance imaging-based neurological signature [381]. Obviously, although of major interest, these association analyses do not allow to conclude on any causal link between indoxyl sulfate concentration in biological fluids and mood disorders but ask for other preclinical studies.

In a mouse model of experimental autoimmune encephalomyelitis, indoxyl sulfate delivered daily (10 mg/kg) via the intraperitoneal way modulated astrocyte activity and exerted anti-inflammatory action on the central nervous system via the aryl hydrocarbon receptor [362] (Fig. 5.14). Astrocytes are the most abundant cell population in the central nervous system and participate in several functions including the blood–brain barrier and the modulation of neuronal transmission [241, 382].

Conversely, when rats were exposed by the oral way to indoxyl sulfate used at higher doses (100 and 200 mg/kg), impairment of spatial memory and reduced locomotor and exploratory activities were observed [383]. In this latter study, after indoxyl sulfate administration, this compound was mainly recovered in the brainstem. Another study found that a single intraperitoneal injection of indoxyl sulfate at an even greater dose (800 mg/kg) caused histological changes in the brain compatible with neuronal necrosis [384] (Fig. 5.14). In this latter study, indoxyl sulfate, when used in the micromolar range, induced radical oxygen species production in primary astrocytes, and cell death in hippocampal neurons [384]. At a 10 micromolar concentration, indoxyl sulfate-induced apoptosis through oxidative stress in human astrocytes [385]. Thus, the effect of indoxyl sulfate on the brain, either beneficial or deleterious, appears to depend on the dose used. Unfortunately, no information is available on the physiological concentrations of indoxyl sulfate and of the other indole-related compounds that are reaching the central nervous system.

Key Points
- Bacterial metabolites like norepinephrine and gamma-aminobutyric acid, with known function of neurotransmitter in the host, are involved in communication between bacterial species.
- Animals with no intestinal microbiota are characterized by modifications of brain physiology.
- Animals with no intestinal microbiota display altered turnover of the neurotransmitters norepinephrine, dopamine, and serotonin in brain suggesting communication between the intestinal microbiota and the host central nervous system.
- Some bacterial metabolites and co-metabolites like tryptamine, indoxyl sulfate, isatin, and oxindole can enter the brain and affect behavior and brain functions.
- The bacterial metabolite p-cresol is suspected to be implicated in the etiology of autism spectrum disorders.
- Decreased ammonia net production by the intestinal microbiota, together with decreased ammonia endogenous production by the host, represent a potential strategy for the treatment of hepatic encephalopathy.

References

1. Cuevasanta E, Möller MN, Alvarez B. Biological chemistry of hydrogen sulfide and persulfides. Arch Biochem Biophys. 2017;617:9–25.
2. Zarei I, Koistinen VM, Kokla M, Klavus A, Babu AF, Lehtonen M, Auriola S, Hanhineva K. Tissue-wide metabolomics reveals wide impact of gut microbiota on mice metabolite composition. Sci Rep. 2022;12(1):15018.
3. Hou Y, Hu S, Li X, He W, Wu G. Amino acid metabolism in the liver: nutritional and physiological significance. Adv Exp Med Biol. 2020;1265:21–37.
4. Jones JG. Hepatic glucose and lipid metabolism. Diabetologia. 2016;59(6):1098–103.
5. Blaak EE, Canfora EE, Theis S, Frost G, Groen AK, Mithieux G, Nauta A, Scott K, Stahl B, van Harsselaar J, van Tol VEE, Verbeke K. Short chain fatty acids in human gut and metabolic health. Benef Microbes. 2020;11(5):411–55.
6. Hamer HM, Jonkers D, Venema K, Vanhoutvin S, Troost FJ, Brummer RJ. Review article: the role of butyrate on colonic function. Aliment Pharmacol Ther. 2008;27(2):104–19.
7. Wolever TM, Chiasson JL. Acarbose raises serum butyrate in human subjects with impaired glucose tolerance. Br J Nutr. 2000;84(1):57–61.
8. Boets E, Gomand S, Deroover L, Preston T, Vermeulen K, De Preter V, Hamer HM, Van den Mooter G, De Vuyst L, Courtin CM, Annaert P, Delcour JA, Verbeke KA. Systemic availability and metabolism of colonic-derived short-chain fatty acids in healthy subjects: a stable isotope study. J Physiol. 2017;595(2):541–55.
9. Cherbuy C, Darcy-Vrillon B, Morel MT, Pégorier JP, Duée PH. Effect of germ-free state on the capacities of isolated rat colonocytes to metabolize n-butyrate, glucose, and glutamine. Gastroenterology. 1995;109(6):1890–9.
10. Russell DW. Cholesterol biosynthesis and metabolism. Cardiovasc Drugs Ther. 1992;6(2):103–10.
11. Bloemen JG, Venema K, van de Poll MC, Olde Damink SW, Buurman WA, Dejong CH. Short chain fatty acids exchange across the gut and liver in humans measured at surgery. Clin Nutr. 2009;28(6):657–61.
12. Kindt A, Liebisch G, Clavel T, Haller D, Hörmannsperger G, Yoon H, Kolmeder D, Sigruener A, Krautbauer S, Seeliger C, Ganza A, Schweiser S, Morisset R, Strowig T, Daniel H, Helm D, Küster B, Krumsiek J, Ecker J. The gut microbiota promotes hepatic fatty acid desaturation and elongation in mice. Nat Commun. 2018;9(1):3760.
13. den Besten G, Lange K, Havinga R, van Dijk TH, Gerding A, van Eunen K, Müller M, Groen AK, Hooiveld GJ, Bakker BM, Reijngoud DJ. Gut-derived short-chain fatty acids are vividly assimilated into host carbohydrates and lipids. Am J Phys. 2013;305(12):G900–10.
14. Zhao S, Jang C, Liu J, Uehara K, Gilbert M, Izzo L, Zeng X, Trefely S, Fernandez S, Carrer A, Miller KD, Schug ZT, Snyder NW, Gade TP, Titchenell TP, Rabinowitz JD, Wellen KE. Dietary fructose feeds hepatic lipogenesis via microbiota-derived acetate. Nature. 2020;579(7800):586–91.
15. Demigné C, Morand C, Levrat MA, Besson C, Moundras C, Rémézy C. Effect of propionate on fatty acid and cholesterol synthesis and on acetate metabolism in isolated hepatocytes. Br J Nutr. 1995;74(2):209–19.
16. Wang Lv, Christophersen CT, Sorich MJ, Gerber JP, Angley MT, Conlon MA. Elevated fecal shortchain fatty acid and ammonia concentrations in children with autism spectrum disorder. Dig Dis Sci. 2012;57(8):2096-2102
17. Zeisel SH, Wishnok JS, Blusztajn JK. Formation of methylamines from ingested choline and lecithin. J Pharmacol Exp Ther. 1983;225(2):320–4.
18. Wiedeman AM, Barr SI, Green TJ, Xu Z, Innis SM, Kitts DD. Dietary choline intake: current state of knowledge across the life cycle. Nutrients. 2018;10(10):1513.
19. Tang WH, Wang Z, Levison BS, Koeth RA, Britt EB, Fu X, Wu Y, Hazen SL. Intestinal microbial metabolism of phosphatidylcholine and cardiovascular risk. N Engl J Med. 2013;368(17):1575–84.

20. Arias N, Arboleya S, Allison J, Kaliszewska A, Higarza SG, Gueimonde M, Arias JL. The relationship between choline bioavailability from diet, intestinal microbiota composition, and its modulation in human diseases. Nutrients. 2020;12(8):2340.
21. Romano KA, Vivas EI, Amador-Noguez D, Rey FE. Intestinal microbiota composition modulates choline bioavailability from diet and accumulation of the proatherogenic metabolite trimethylamine-N-oxide. MBio. 2015;6(2):e02481.
22. Helsley RN, Miyata T, Kadam A, Varadharajan V, Sangwan N, Huang EC, Banerjee R, Brown AL, Fung KK, Massey WJ, Neumann C, Orabi D, Osborn LJ, Schugar RC, McMullen MR, Bellar A, Poulsen KL, Kim A, Pathak V, Mrdjen M, Anderson JT, Willard B, McClain CJ, Mitchell M, McCullough AJ, Radaeva S, Barton B, Szabo G, Dasarathy S, Garcia-Garcia JC, Rotroff DM, Allende DS, Wang Z, Hazen SL, Nagy LE, Brown JM. Gut microbial trimethylamine is elevated in alcohol-associated hepatitis and contributes to ethanol-induced liver injury in mice. elife. 2022;11:e76554.
23. Mitchell S, Ayesh R, Barrett T, Smith R. Trimethylamine and foetor hepaticus. Scand J Gastroenterol. 1999;34(5):524–8.
24. Velasquez MT, Ramezani A, Manal A, Raj DS. Trimethylamine N-oxide: the good, the bad and the unknown. Toxins (Basel). 2016;8(11):326.
25. Koeth RA, Wang Z, Levison BS, Buffa JA, Org E, Sheehy BT, Britt EB, Fu X, Wu Y, Li L, Smith JD, DiDonato JA, Chen J, Li H, Wu GD, Lewis JD, Warrier M, Brown JM, Krauss RM, Tang WH, Bushman FD, Lusis AJ, Hazen SL. Intestinal microbiota metabolism of L-carnitine, a nutrient in red meat, promotes atherosclerosis. Nat Med. 2013;19(5):576–85.
26. Vaz FM, Wanders RJ. Carnitine biosynthesis in mammals. Biochem J. 2002;136(3):417–29.
27. Zeisel SH, Warrier M. Trimethylamine-N-oxide, the microbiome, and heart and kidney disease. Annu Rev Nutr. 2017;37:157–81.
28. Chen YM, Liu Y, Zhou RF, Chen XL, Wang C, Tan XY, Wang LJ, Zheng RD, Zhang HW, Ling WH, Zhu HL. Association of gut flora-dependent metabolite trimethylamine-N-oxide, betaine and choline with non-alcoholic fatty liver disease in adults. Sci Rep. 2016;6:19076.
29. Tan X, Liu Y, Long J, Chen S, Liao G, Wu S, Li C, Wang L, Ling W, Zhu H. Trimethylamine-N-oxide aggravates liver steatosis through modulation of bile acid metabolism and inhibition of farnesoid X receptor signaling in nonalcoholic fatty liver disease. Mol Nutr Food Res. 2019a;63(17):e1900257.
30. Flores-Guerrero JL, Post A, van Dijk PR, Connelly MA, Garcia E, Navis G, Bakker SJL, Dullaart RPF. Circulating trimethylamine-N-oxide is associated with all-cause mortality in subjects with nonalcoholic fatty liver disease. Liver Int. 2021;41(10):2371–82.
31. Rath S, Rox K, Bardenhorst SK, Schminke U, Dörr M, Mayerle J, Frost F, Lerch MM, Karch A, Brönstrup M, Pieper DH, Vital M. Higher trimethylamine-N-oxide plasma levels with increasing age are mediated by diet and trimethylamine-forming bacteria. mSystems. 2021;6(5):e0094521.
32. Beaumont M, Neyrinck AM, Olivares M, Rodriguez J, de Rocca SA, Roumain M, Bindels LB, Cani PD, Evenpoel P, Muccioli GG, Demoulin JB, Delzenne NM. The gut microbiota metabolite indole alleviates liver inflammation in mice. FASEB J. 2018;32(12):fj201800544.
33. Knudsen C, Neyrinck AM, Leyrolle Q, Baldin P, Leclercq S, Rodriguez J, Beaumont M, Cani PD, Bindels LB, Lanthier N, Delzenne NM. Hepatoprotective effects of indole, a gut microbial metabolite, in leptin-deficient obese mice. J Nutr. 2021;151(6):1507–16.
34. Ji Y, Gao Y, Chen H, Yin Y, Zhang W. Indole-3-acetic acid alleviates nonalcoholic fatty liver disease in mice via attenuation of hepatic lipogenesis, and oxidative and inflammatory stress. Nutrients. 2019;11(9):2062.
35. Krishnan S, Ding Y, Saedi N, Choi M, Sridharan GV, Sherr DH, Yarmush ML, Alaniz RC, Jayaraman A, Lee K. Gut microbiota-derived tryptophan metabolites modulate inflammatory response in hepatocytes and macrophages. Cell Rep. 2018;23(4):1099–111.
36. Gillam EM, Notley LM, Cai H, De Voss JJ, Guengerich FP. Oxidation of indole by cytochrome P450 enzymes. Biochemistry. 2000;39(45):13817–24.

37. King LJ, Parke DV, Williams RT. The metabolism of (2-14C) indole in the rat. Biochem J. 1966;98(1):266–77.
38. Wikoff WR, Anfora AT, Liu J, Schultz PG, Lesley SA, Peters EC, Siuzdak G. Metabolomics analysis reveals large effects of gut microflora on mammalian blood metabolites. Proc Natl Acad Sci U S A. 2009;106(10):3698–703.
39. Poesen R, Mutsaers HAM, Windey K, van den Broek PH, Verweij V, Augustijns P, Kuypers D, Jansen J, Evenepoel P, Verbeke K, Meijers B, Masereeuw R. The influence of dietary protein intake on mammalian tryptophan and phenolic metabolites. PLoS One. 2015;10(10):e0140820.
40. Beaumont M, Portune KJ, Steuer N, Lan A, Cerrudo V, Audebert M, Dumont F, Mancano G, Khodorova N, Andriamihaja M, Airinei G, Tomé D, Benamouzig R, Davila AM, Claus SP, Sanz Y, Blachier F. Quantity and source of dietary protein influence metabolite production by gut microbiota and rectal mucosa gene expression: a randomized, parallel, double-blind trial in overweight humans. Am J Clin Nutr. 2017;106(4):1005–19.
41. Banoglu E, Jha GG, King RS. Hepatic microsomal metabolism of indole to indoxyl, a precursor of indoxyl sulfate. Eur J Drug Metab Pharmacokinet. 2001;26(4):235–40.
42. Santana Machado T, Poitevin S, Paul P, McKay N, Jourde-Chiche N, Legris T, Mouly-Bandini A, Dignat-George F, Brunet P, Masereeuw R, Burtey S, Cerini C. Indoxyl sulfate upregulates liver P-glycoprotein expression and activity through aryl hydrocarbon receptor signaling. J Am Soc Nephrol. 2018;29(3):906–18.
43. Weigand KM, Schirris TJJ, Houweling M, van der Heuvel JJMW, Koenderink JB, Dankers ACA, Russel FGM, Greupink R. Uremic solutes modulate hepatic bile acid handling and induce mitochondrial toxicity. Toxicol In Vitro. 2019;56:52–61.
44. Agellon LB, Torchia EC. Intracellular transport of bile acids. Biochim Biophys Acta. 2000;1486(1):198–209.
45. Gryp T, Vanholder R, Vaneechoutte M, Glorieux G. p-cresyl sulfate. Toxins (Basel). 2017;9 (2):52.
46. Rong Y, Kiang TKL. Characterization of human UDP-glucuronosyltransferase enzymes in the conjugation of p-cresol. Toxicol Sci. 2020;176(2):285–96.
47. Yan Z, Zhong HM, Maher N, Torres R, Leo GC, Caldwell GW, Huebert N. Bioactivation of 4-methylphenol (p-cresol) via cytochrome P450-mediated aromatic oxidation in human liver microsomes. Drug Metab Dispos. 2005;33(12):1867–76.
48. Witkowski M, Weeks TL, Hazen SL. Gut microbiota and cardiovascular disease. Circ Res. 2020;127(4):553–70.
49. Schertzer JD, Lam TKT. Peripheral and central regulation of insulin by the intestine and microbiome. Am J Phys. 2021;320(2):E234–9.
50. Malaisse WJ, Sener A, Koser M, Herchuelz A. Stimulus-secretion coupling of glucose-induced insulin release. Metabolism of alpha- and beta-D-glucose in isolated islets. J Biol Chem. 1976;251(19):5936–43.
51. Blachier F, Mourtada A, Sener A, Malaisse WJ. Stimulus-secretion coupling of arginine-induced insulin release. Uptake of metabolized and nonmetabolized cationic amino acids by pancreatic islets. Endocrinology. 1989;124(1):134–41.
52. Sener A, Blachier F, Rasschaert J, Malaisse WJ. Simulus-secretion coupling of arginine-induced insulin release: comparison with histidine-induced insulin release. Endocrinology. 1990;127(1):107–13.
53. Sener A, Blachier F, Rasschaert J, Mourtada A, Malaisse-Lagae F, Malaisse WJ. Stimulus-secretion coupling of arginine-induced insulin release: comparison with lysine-induced insulin secretion. Endocrinology. 1989a;124(5):2558–67.
54. Perry RJ, Peng L, Barry NA, Cline GW, Zhang D, Cardone RL, Petersen KF, Kibbey RG, Goodman AL, Shulman GI. Acetate mediates a microbiome-brain-β-cell axis to promote metabolic syndrome. Nature. 2016;534(7606):213–7.
55. Pallister T, Jackson MA, Martin TC, Zierer J, Jennings A, Mohney RP, MacGregor A, Steves CJ, Cassidy A, Spector TD, Menni C. Hippurate as a metabolic marker of gut microbiome

diversity: modulation by diet and relationship to metabolic syndrome. Sci Rep. 2017;7(1): 13670.

56. Brial F, Chilloux J, Nielsen T, Vieira-Silva S, Falony G, Andrikopoulos P, Olanipekun M, Hoyles L, Djouadi F, Neves AL, Rodriguez-Martinez A, Mouawad GI, Pons N, Forslund S, Le-Chatelier E, Le Lay A, Nicholson J, Hansen T, Hyötyläinen T, Clement K, Oresic M, Bork P, Ehrlich SD, Raes J, Pedersen OB, Gauguier D, Dumas ME. Human and preclinical studies of the host-gut microbiome co-metabolite hippurate as a marker and mediator of metabolic health. Gut. 2021;70(11):2105–14.

57. Haenisch B, von Kügelgen I, Bönisch H, Göthert M, Sauerbruch T, Schepke M, Marklein G, Höfling K, Schröder D, Molderings GJ. Regulatory mechanisms underlying agmatine homeostasis in humans. Am J Phys. 2008;295(5):G1104–10.

58. Sener A, Lebrun P, Blachier F, Malaisse WJ. Stimulus-secretion of arginine-induced insulin release. Insulinotropic action of agmatine. Biochem Pharmacol. 1989b;38(2):327–30.

59. Vangipurapu J, Fernades Silva L, Kuulasmaa T, Smith U, Laasko M. Microbiota-related metabolites and the risk of type 2 diabetes. Diabetes Care. 2020;43(6):1319–25.

60. Glassock RJ. Uremic toxins: what are they? An integrated overview of pathobiology and classification. J Renal Nutr. 2008;18(1):2–6.

61. Vanholder R, Baurmeister U, Brunet P, Cohen G, Glorieux G, Jankowski J, European Uremic Toxin Work Group. A bench to bedside view of uremic toxins. J Am Soc Nephrol. 2008a;19 (5):863–70.

62. Vanholder R, De Smet R, Glorieux G, Argilés A, Baurmeister U, Brunet P, Clark W, Cohen G, De Deyn PP, Deppisch R, Descamps-Latscha B, Henle T, Jörres A, Lemke HD, Massy ZA, Paaslick-Deetjen J, Rodriguez M, Stegmayr D, Stenvinkel P, Tetta C, Wanner C, Zidek W, European Uremic Toxin Work Group (EU Tox). Review on uremic toxins: classification, concentration, and interindividual variability. Kidney Int. 2003;63(5):1934–43.

63. Stevens PE, Levin A, Kidney Disease: Improving Global Outcomes Chronic Kidney Disease Guideline Development Work Group Members. Evaluation and management of chronic kidney disease: synopsis of the kidney disease: improving global outcomes 2012 clinical practice guideline. Ann Intern Med. 2013;158(11):825–30.

64. Duranton F, Cohen G, De Smet R, Rodriguez M, Jankowski J, Vanholder R, Argiles A, European Uremic Toxin Work Group. Normal and pathologic concentrations of uremic toxins. J Am Soc Nephrol. 2012;23(7):1258–70.

65. Lim YJ, Sidor NA, Tonial NC, Che A, Urquhart BL. Uremic toxins in the progression of chronic kidney disease and cardiovascular disease: mechanisms and therapeutic targets. Toxins (Basel). 2021;13(2):142.

66. Blachier F, Andriamihaja M. Effects of L-tyrosine-derived bacterial metabolite p-cresol on colonic and peripheral cells. Amino Acids. 2022;54(3):325–38.

67. Nigam SK, Bush KT. Uraemic syndrome of chronic kidney disease: altered remote sensing and signalling. Nat Rev Nephrol. 2019;15(5):301–16.

68. Mair RD, Sirich TL, Meyer TW. Uremic toxin clearance and cardiovascular toxicities. Toxins (Basel). 2018;10(6):226.

69. Atherton JG, Hains DS, Bissler G, Pendley BD, Lindner E. Generation, clearance, toxicity, and monitoring possibilities of unaccounted uremic toxins for improved dialysis prescriptions. Am J Phys. 2018;315(4):F890–902.

70. Lesaffer G, De Smet R, Lameire N, Dhondt A, Duym P, Vanholder R. Intradialytic removal of protein-bound uraemic toxins: role of solute characteristics and of dialyser membrane. Nephrol Dial Transplant. 2000;15(1):50–7.

71. Liu WC, Tomino Y, Lu KC. Impacts of indoxyl sulfate and p-cresol sulfate on chronic kidney disease and mitigating effect of AST-120. Toxins (Basel). 2018;10(9):367.

72. Martinez AW, Recht NS, Hostetter TM, Meyer TW. Removal of p-cresol sulfate by hemodyalisis. J Am Soc Nephrol. 2005;16(11):3430–6.

73. Meert N, Schepers E, Glorieux G, Van Landschoot M, Goeman JL, Waterloos MA, Dhondt A, Van der Eycken J, Vanholder R. Novel method for simultaneous determination of

p-cresylsulphate and p-cresylglucuronide: clinical data and pathophysiological implications. Nephrol Dial Transplant. 2012;27(6):2388–96.

74. Calaf R, Cerini C, Génovésio C, Verhaeghe P, Jourde-Chiche N, Bergé-Lefranc D, Gondouin B, Dou L, Morange S, Argilés A, Rathelot P, Dignat-George F, Brunet P, Charpiot P. Determination of uremic solutes in biological fluids of chronic kidney disease patients by HPLC assay. J Chromatogr B Analyt Technol Biomed Life Sci. 2011;879(23):2281–6.

75. Chen TC, Wang CY, Hsu CY, Wu CH, Kuo CC, Wang KC, Yang CC, Wu MT, Chuang FR, Lee CT. Free p-cresol sulfate is associated with survival and function of vascular access in chronic hemodialysis patients. Kidney Blood Press Res. 2012;36(6):583–8.

76. De Smet R, David F, Sandra P, Van Kaer J, Lesaffer G, Dhondt A, Lameire N, Vanholder R. A sensitive HPLC method for uqntification of free and total p-cresol in patients with chronic renal failure. Clin Chim Acta. 1998;278(1):1–21.

77. Faguli RM, De Smet R, Buoncristiani U, Lameire N, Vanholder R. Behavior of non-protein-bound and protein-bound uremic solutes during daily hemodialysis. Am J Kidney Dis. 2002;40(2):339–47.

78. Gryp T, De Paepe K, Vanholder R, Kerckhof FM, Van Biesen W, Van de Wiele T, Verbeke F, Speeckaert M, Joossens M, Couttenye MM, Veneechoutte M, Glorieux G. Gut microbiota generation of protein-bound uremic toxins and related metabolites is not altered at different stages of chronic kidney disease. Kidney Int. 2020;97(6):1230–42.

79. Hsu HJ, Yen CH, Wu IW, Hsu KH, Chen CK, Sun CY, Chou CC, Chen CY, Tsai CJ, Wu MS, Lee CC. The association of uremic toxins and inflammation in hemodialysis patients. PLoS One. 2014;9(7):e102691.

80. Ikematsu N, Kashiwagi M, Hara K, Waters B, Matsusue A, Takayama M, Kubo SI. Organ distribution of endogenous p-cresol in hemodialysis patients. J Med Investig. 2019;66(1.2): 81–5.

81. Leong SC, Sao JN, Taussig A, Plummer NS, Meyer TW, Sirich TL. Residual function effectively controls plasma concentrations of secreted solutes in patients on twice weekly hemodialysis. J Am Soc Nephrol. 2018;29(7):1992–9.

82. Lin CJ, Wu CJ, Pan CF, Chen YC, Sun FJ, Chen HH. Serum protein-bound uraemic toxins and clinical outcomes in haemodialysis patients. Nephrol Dial Transplant. 2010;25(11):3693–700.

83. Nakabayashi I, Nakamura M, Kawakami K, Ohta T, Kato I, Uchida K, Yoshida M. Effects of symbiotic treatment on serum level of p-cresol in haemodialysis patients: a preliminary study. Nephrol Dial Transplant. 2011;26(3):1094–8.

84. Poesen R, Evenepoel P, de Loor H, Kuypers D, Augustijns P, Meijers B. Metabolism, protein binding, and renal clearance of microbiota-derived p-cresol in patients with CKD. Clin J Am Soc Nephrol. 2016;11(7):1136–44.

85. Prokopienko AJ, West RE 3rd, Stubbs JR, Nolin TD. Development and validation of a UHPLC-MS/MS method for measurement of a gut-derived uremic toxin panel in human serum: an application in patients with kidney disease. J Pharm Biomed Anal. 2019;174:618–24.

86. Salmean YA, Segal MS, Palii SP, Dahl WJ. Fiber supplementation lowers plasma p-cresol in chronic kidney disease patients. J Ren Nutr. 2015;25(3):316–20.

87. Sirich TL, Fong K, Larive B, Beck GJ, Chertow GM, Levin NW, Kliger AS, Plummer NS, Meyer TW. Limited reduction in uremic solute concentrations with increased dialysis frequency and time in the frequent Hemodialysis network daily trial. Kidney Int. 2017;91(5): 1186–92.

88. Sirich TL, Funk BA, Plummers NS, Hostetter TH, Meyer TW. Prominent accumulation in hemodialysis patients of solutes normally cleared by tubular secretion. J Am Soc Nephrol. 2014;25(3):615–22.

89. Sirich TL, Aronov PA, Plummer NS, Hostetter TH, Meyer TW. Numerous protein-bound solutes are cleared by the kidney with high efficiency. Kidney Int. 2013;84(3):585–90.

90. Wu W, Bush KT, Nigam SK. Key role for the organic anion transporters, OAT1 and OAT3, in the in vivo handling of uremic toxins and solutes. Sci Rep. 2017;7:4939.

91. Wu IW, Hsu KH, Lee CC, Sun CY, Hsu HJ, Tsai CJ, Tzen CY, Wang YC, Lin CY, Wu MS. p-cresyl sulphate and indoxyl sulphate predict progression of chronic kidney disease. Nephrol Dial Transplant. 2011;26(3):938–47.

92. Bammens B, Evenepoel P, Keuleers H, Verbeke K, Vanrenterghem Y. Free serum concentrations of the protein-bound retention solute p-cresol predict mortality in hemodialysis patients. Kidney Int. 2006;69:1081–7.

93. Sun CY, Chang SC, Wu MS. Uremic toxins induce kidney fibrosis by activating intrarenal renin-angiotensin-aldosterone system associated epithelial-to-mesenchymal transition. PLoS One. 2012;7:e34026.

94. Poesen R, Viaene L, Verbeke K, Claes K, Bammens B, Sprangers B, Naesens M, Vanrenterghem Y, Kuypers D, Evenepoel P, Meijers B. Renal clearance and intestinal generation of p-cresyl sulfate and indoxyl sulfate in CKD. Clin J Am Soc Nephrol. 2013;8 (9):1508–14.

95. Watanabe H, Miyamoto Y, Honda D, Tanaka H, Wu Q, Endo M, Noguchi T, Kadowaki D, Ishima Y, Kotani S, Nakajima M, Kataoka K, Kim-Mitsuyama S, Tanaka M, Fukagawa M, Otagiri M, Maruyama T. p-cresyl sulfate causes renal tubular cell damage by inducing oxidative stress by activation by NADPH oxidase. Kidney Int. 2013;83(4):582–92.

96. Sun CY, Hsu HH, Wu MS. p-cresol sulfate and indoxyl sulfate induce similar cellular inflammatory gene expression in cultured proximal renal tubular cells. Nephrol Dial Transplant. 2013;28(1):70–8.

97. Sun CY, Cheng ML, Pan HC, Lee JH, Lee CC. Protein-bound uremic toxins impaired mitochondrial dynamics and functions. Oncotarget. 2017;8:77722–33.

98. Brocca A, Virzi GM, de Cal M, Cnataluppi V, Ronco C. Cytotoxic effects of p-cresol in renal epithelial tubular cells. Blood Purif. 2013;36(3–4):219–25.

99. Lin HH, Huang CC, Lin TY, Lin CY. P-cresol mediates autophagic cell death in renal proximal tubular cells. Toxicol Lett. 2015;234:20–9.

100. Khosroshahi HT, Abedi B, Ghoiazadeh M, Samadi A, Jouyban A. Effects of fermentable high fiber diet supplementation on gut-derived and conventional nitrogenous product in patients on maintenance hemodialysis: a randomized controlled trial. Nutr Metab. 2019;16:18.

101. Meijers BK, De Preter V, Verbeke K, Vanrenterghem Y, Evenpoel P. p-cresyl sulfate serum concentrations in haemodialysis patients are reduced by the prebiotic oligofructose-enriched inulin. Nephrol Dial Transplant. 2010b;25(1):219–24.

102. Meyer TW, Hostetter TH. Uremic solutes from colon microbes. Kidney Int. 2012;81(10): 949–54.

103. Ko GJ, Obi Y, Tortorici AR, Kalantar-Zadeh K. Dietary protein intake and chronic kidney disease. Curr Opin Clin Nutr Metab. 2017;20(1):77–85.

104. Vanholder R, Schepers E, Pletinck A, Nagler EV, Glorieux G. The uremic toxicity of indoxyl sulfate and p-cresyl sulfate: a systematic review. J Am Soc Nephrol. 2014;25(9):1897–907.

105. Vanholder R, Meert N, Schepers E, Glorieux G. Uremic toxins: do we know enough to explain uremia? Blood Purif. 2008b;26(1):77–81.

106. Cheng TH, Ma MC, Liao MT, Zheng CM, Lu KC, Liao CH, Hou YC, Liu WC, Lu CL. Indoxyl sulfate, a tubular toxin, contributes to the development of chronic kidney disease. Toxins (Basel). 2020;12(11):684.

107. Devine E, Krieter DH, Ruth M, Jankovski J, Lemke HD. Binding affinity and capacity for the uremic toxin indoxyl sulfate. Toxins (Basel). 2014;6(2):416–29.

108. Viaene L, Annaert P, de Loor H, Poesen R, Evenepoel P, Meijers B. Albumin is the main plasma binding protein for indoxyl sulfate and p-cresyl sulfate. Biopharm Drug Dispos. 2013;34(3):165–75.

109. Enomoto A, Takeda M, Tojo A, Sekine T, Cha SH, Khamdang S, Takayama F, Aoyama I, Nakamura S, Endou H, Niwa T. Role of organic anion transporters in the tubular transport of indoxyl sulfate and the induction of its nephrotoxicity. J Am Soc Nephrol. 2002;13(7): 1711–20.

110. Hobby GP, Karaduta O, Dusio GF, Singh M, Zybailov BL, Arthur JM. Chronic kidney disease and the gut microbiome. Am J Phys. 2019;316(6):F1211–7.
111. Fujii H, Goto S, Fukagawa M. Role of uremic toxins for kidney, cardiovascular and bone dysfunction. Toxins (Basel). 2018;10(5):202.
112. Wang W, Hao G, Pan Y, Ma S, Yang T, Shi P, Zhu Q, Xie Y, Ma S, Zhang Q, Ruan H, Ding F. Serum indoxyl sulfate is associated with mortality in hospital-acquired acute kidney injury: a prospective cohort study. BMC Nephrol. 2019;20(1):57.
113. Tan X, Cao X, Zhou J, Shen B, Zhang X, Liu Z, Lv W, Teng J, Ding X. Indoxyl sulfate, a valuable biomarker in chronic kidney disease and dialysis. Hemodial Int. 2017;21(2):161–7.
114. Bolati D, Shimizu H, Yisireyili M, Nishijima F, Niwa T. Indoxyl sulfate, a uremic toxin, downregulates renal expression of Nrf2 through activation of NF-κB. BMC Nephrol. 2013;14: 56.
115. Shimizu H, Yisireyili M, Higashiyama Y, Nishijima F, Niwa T. Indoxyl sulfate upregulates renal expression of ICAM-1 via production of ROS and activation of NF-kappaB and p53 in proximal tubular cells. Life Sci. 2013;92(2):143–8.
116. Motojima M, Hosokawa A, Yamato H, Muraki T, Yoshioka T. Uremic toxins of organic anions up-regulate PAI-1 expression by induction of NF-kappaB and free radical in proximal tubular cells. Kidney Int. 2003;63(5):1671–80.
117. Edamatsu T, Fujieda A, Itoh Y. Phenyl sulfate, indoxyl sulfate and p-cresyl sufate decrease glutathione level to render cells vulnerable to oxidative stress in renal tubular cells. PLoS One. 2018;13(2):e0193342.
118. D'Autréaux B, Toledano MB. ROS as signalling molecules: mechanisms that generate specificity in ROS homeostasis. Nat Rev Mol Cell Biol. 2007;8(10):813–24.
119. Owada S, Goto S, Bannai K, Hayashi H, Nishijima F, Niwa T. Indoxyl sulfate reduces superoxide scavenging activity in the kidneys of normal and uremic rats. Am J Nephrol. 2008;28(3):446–54.
120. Peskin AV. Cu,Zn-superoxide dismutase gene dosage and cell resistance to oxidative stress: a review. Biosci Rep. 1997;17(1):85–9.
121. Ratliff BB, Abdulmahdi W, Pawar R, Wolin MS. Oxidant mechanisms in renal injury and disease. Antioxid Redox Signal. 2016;25(3):119–46.
122. Humphreys BD. Mechanisms of renal fibrosis. Annu Rev Physiol. 2018;80:309–26.
123. Miyazaki T, Ise M, Hirata M, Endo K, Ito Y, Seo H, Niwa T. Indoxyl sulfate stimulates renal synthesis of transforming growth factor-beta 1 and progression of renal failure. Kidney Int. 1997;63:S211–4.
124. Milanesi S, Garibaldi S, Saio M, Ghigliotti G, Picciotto D, Ameri P, Garibotto G, Barisione C, Verzola D. Indoxyl sulfate induces renal fibroblast activation through a targetable heat shock protein 90-dependent pathway. Oxidative Med Cell Longev. 1997;2019:2050183.
125. Shimizu H, Yisireyili M, Nishijima F, Niwa T. Stat3 contributes to indoxyl sulfate-induced inflammatory and fibrotic gene expression and cellular senescence. Am J Nephrol. 2012;36(2): 184–9.
126. Hamaya R, Ivey KL, Lee DH, Wang M, Li J, Franke A, Sun Q, Rimm EB. Association of diet with circulating trimethylamine-N-oxide concentration. Am J Clin Nutr. 2020;112(6): 1448–55.
127. Bain MA, Faull R, Fornasini G, Milne RW, Evans AM. Accumulation of trimethylamine and trimethylamine-N-oxide in end-stage renal disease patients undergoing haemodialysis. Nephrol Dial Transplant. 2006;21(5):1300–4.
128. Bell JD, Lee JA, Lee HA, Sadler PJ, Wilkie DR, Woodham RH. Nuclear magnetic resonance studies of blood plasma and urine from subjects with chronic renal kidney failure: identification of trimethylamine-N-oxide. Biochim Biophys Acta. 1991;1096(2):101–7.
129. Tang WH, Wang Z, Kennedy DJ, Wu Y, Buffa JA, Agatisa-Boyle B, Li XS, Levison BS, Hazen SL. Gut microbiota-dependent trimethylamine N-oxide (TMAO) pathway contributes to both development of renal insufficiency and mortality risk in chronic kidney disease. Circ Res. 2015a;116(3):448–55.

130. Zeng Y, Guo M, Fang X, Teng F, Tan X, Li X, Wang M, Long Y, Xu Y. Gut microbiota-derived trimethylamine N-oxide and kidney function: a systematic review and meta-analysis. Adv Nutr. 2021;12(4):1286–304.

131. Zhang W, Miikeda A, Zuckerman J, Jia X, Charugundla S, Zhou Z, Kaczor-Urbanowicz KE, Magyar C, Guo F, Wang Z, Pelligrini M, Hazen SL, Nicholas SB, Lusis AJ, Shih DM. Inhibition of microbiota-dependent TMAO production attenuates chronic kidney disease in mice. Sci Rep. 2021;11(1):518.

132. Dong F, Jiang S, Tang C, Wang X, Ren X, Wei Q, Tian J, Hu W, Guo J, Fu X, Liu L, Patzak A, Persson PB, Gao F, Lai EY, Zhao L. Trimethylamine N-oxide promotes hyperoxaluria-induced calcium oxalate deposition and kidney injury by activating autophagy. Free Radic Biol Med. 2022;179:288–300.

133. Fogelman AM. TMAO is both a biomarker and a renal toxin. Circ Res. 2015;116(3):396–7.

134. Senthong V, Wang Z, Fan Y, Wu Y, Hazen SL, Tang WH. Trimethylamine N-oxide and mortality risk in patients with peripheral artery disease. J Am Heart Assoc. 2016b;5(10):e004237.

135. Senthong V, Li XS, Hudec T, Coughlin J, Wu Y, Levison B, Wang Z, Hazen SL, Tang WH. Plasma trimethylamine N-oxide, a gut microbe-generated phosphatidylcholine metabolite, is associated with atherosclerotic burden. J Am Coll Cardiol. 2016a;67(22):2620–8.

136. Li XS, Obeid S, Klingenberg R, Gencer B, Mach F, Räber L, Windecker S, Rodondi N, Nanchen D, Muller O, Miranda MX, Matter CM, Wu Y, Li L, Wang Z, Alamri HS, Gogonea V, Chung YM, Tang WH, Hazen SL, Lüscher TF. Gut microbiota-dependent trimethylamine N-oxide in acute coronary syndromes: a prognostic marker for incident cardiovascular events beyond traditional risk factors. Eur Heart J. 2017b;38(11):814–24.

137. Tan Y, Sheng Z, Zhou P, Liu C, Zhao H, Song L, Li J, Zhou J, Chen Y, Wang L, Qian H, Sun Z, Qiao S, Xu B, Gao R, Yan H. Plasma trimethylamine N-oxide as a novel biomarker for plaque rupture in patients with ST-segment-elevation myocardial infarction. Circ Cardiovasc Interv. 2019b;12(1):e007281.

138. Lever M, George PM, Slow S, Bellamy D, Young JM, Ho M, McEntyre CJ, Elmslie JL, Atkinson W, Molyneux SL, Troughton RW, Frampton CM, Richards AM, Chambers ST. Betaine and trimethylamine-N-oxide as predictors of cardiovascular outcomes show different patterns in diabetes mellitus: an observational study. PLoS One. 2014;9(12):e114969.

139. Suzuki T, Heaney LM, Bhandari SS, Jones DJL, Ng LL. Trimethylamine N-oxide and prognosis in acute heart failure. Heart. 2016;102(11):841–8.

140. Tang WH, Wang Z, Shrestha K, Borowski AG, Wu Y, Troughton RW, Klein AL, Hazen SL. Intestinal microbiota-dependent phosphatidylcholine metabolites, diastolic dysfunction, and adverse clinical outcomes in chronic systolic heart failure. J Card Fail. 2015b;21(2):91–6.

141. Troseid M, Ueland T, Hov JR, Svardal A, Gregersen I, Dahl CP, Aaklus S, Gude E, Bjorndal B, Halvorsen B, Karlsen TH, Aukrust P, Gullestad L, Berge RK, Yndestad A. Microbiota-dependent metabolite trimethylamine-N-oxide is associated with disease severity and survival of patients with chronic heart failure. J Intern Med. 2015;277(6):717–26.

142. Haghikia A, Li XS, Liman TG, Bledau N, Schmidt D, Zimmermann F, Kränkel N, Widera C, Sonnenschein K, Haghikia A, Weissenborn K, Fraccarollo D, Heimesaat MM, Bauersachs J, Wang Z, Zhu W, Bavendiek U, Hazen SL, Endres M, Landmesser U. Gut microbiota-dependent trimethylamine N-oxide predicts risk of cardiovascular events in patients with stroke and is related to proinflammatory monocytes. Arterioscler Thromb Vasc Biol. 2018;38(9):2225–35.

143. Heianza Y, Ma W, Manson JE, Rexrode KM, Qi L. Gut microbiota metabolites and risk of major adverse cardiovascular disease events and death: a systematic review and meta-analysis of prospective studies. J Am Heart Assoc. 2017;6(7):e004947.

144. Qi J, You T, Li J, Pan T, Xiang L, Han Y, Zhu L. Circulating trimethylamine N-oxide and the risk of cardiovascular diseases: a systematic review and meta-analysis of 11 prospective cohort studies. J Cell Mol Med. 2018;22(1):185–94.

145. Ding L, Chang M, Guo Y, Zhang L, Xue C, Yanagita T, Zhang T, Wang Y. Trimethylamine-N-oxide (TMAO)-induced atherosclerosis is associated with bile acid metabolism. Lipids Health Dis. 2018;17(1):286.
146. Wang Z, Roberts AB, Buffa JA, Levison BS, Zhu W, Org E, Gu X, Huang Y, Zamanian-Daryoush M, Culley MK, DiDonato AJ, Fu X, Hazen JE, Krajcik D, DiDonato JA, Lusis AJ, Hazen SL. Non-lethal inhibition of gut microbial trimethylamine production for the treatment of atherosclerosis. Cell. 2015;163(7):1585–95.
147. Wu P, Chen JN, Chen JJ, Tao J, Wu SY, Xu GS, Wang Z, Wei DH, Yin WD. Trimethylamine N-oxide promotes apoE−/− mice atherosclerosis by inducing vascular endothelial cell pyroptosis via SDHB/ROS pathway. J Cell Physiol. 2020;235(10):6582–91.
148. Xu J, Zhou D, Poulsen O, Imamura T, Hsiao YH, Smith TH, Malhotra A, Dorrestein P, Knight R, Haddad GG. Intermittent hypoxia and hypercapnia accelerate atherosclerosis, partially via trimethylamine-oxide. Am J Respir Cell Mol Biol. 2017;57(5):581–8.
149. Aldana-Hernandez P, Leonard KA, Zhao YY, Curtis JM, Field CJ, Jacobs RL. Dietary choline or trimethylamine N-oxide supplementation does not influence atherosclerosis development in Ldlr−/− and Apoe−/− male mice. J Nutr. 2020;150(2):249–55.
150. Blachier F, Andriamihaja M, Blais A. Sulfur-containing amino acids and lipid metabolism. J Nutr. 2020;150(S1):2524S–31S.
151. Chang Y, Robidoux J. Dyslipidemia management update. Curr Opin Pharmacol. 2017;33:47–55.
152. Schaefer EJ, Geller AS, Endress G. The biochemical and genetic diagnosis of lipid disorders. Curr Opin Lipidol. 2019;30(2):56–62.
153. Warrier M, Shih DM, Burrows AC, Ferguson D, Gromovsky AD, Brown AL, Marshall S, McDaniel A, Schugar RC, Wang Z, Sacks J, Rong X, de Aguiar Vallim T, Chou J, Ivanova PT, Myers DS, Brown HA, Lee RG, Crooke RM, Graham MJ, Liu X, Parini P, Tontonoz P, Lusis AJ, Hazen SL, Temel RE, Brown JM. The TMAO-generating enzyme flavin monooxygenase 3 is a central regulator of cholesterol balance. Cell Rep. 2015;10(3):326–38.
154. Pathak P, Helsley RN, Brown AL, Buffa JA, Choucair I, Nemet I, Gogonea CB, Gogonea V, Wang Z, Garcia-Garcia JC, Cai L, Temel R, Sangwan N, Hazen SL, Brown JM. Small molecule inhibition of gut microbial choline trimethylamine lyase activity alters host cholesterol and bile acid metabolism. Am J Phys. 2020;318(6):H1474–86.
155. Frossard M, Fuchs I, Leitner JM, Hsieh K, Vlcek M, Losert H, Domanovits H, Schreiber W, Laggner AN, Jilma B. Platelet function predicts myocardial damage in patients with acute myocardial infarction. Circulation. 2004;110(11):1392–7.
156. Tantry US, Bonello L, Aradi D, Price MJ, Jeong YH, Angiolillo DJ, Stone GW, Curzen N, Geisler T, Ten Berg J, Kirtane A, Siller-Matula J, Mahla E, Becker RC, Bhatt DL, Waksman R, Rao SV, Alexopoulos D, Marcucci R, Reny JL, Trenk D, Sibbing D, Gurbel PA, Working Group on On-Treatment Platelet Reactivity. Consensus and debate on the definition of on-treatment platelet reactivity to adenosine diphosphate associated with ischemia and bleeding. J Am Coll Cardiol. 2013;62(24):2261–73.
157. Jennings LK. Mechanisms of platelet activation: need for new strategies to protect against platelet-mediated atherosclerosis. Thromb Haemost. 2009;102(2):248–57.
158. Zhu W, Gregory JC, Org E, Buffa JA, Gupta N, Wang Z, Li L, Fu X, Wu Y, Mehrabian M, Balfour Sartor R, McIntyre TM, Silverstein RL, Tang WHW, DiDonato JA, Brown JM, Lusis AJ, Hazen SL. Gut microbial metabolite TMAO enhances platelet hyperreactivity and thrombosis risk. Cell. 2016;165(1):111–24.
159. Zhu W, Buffa JA, Wang Z, Warrier M, Schugar R, Shih DM, Gupta N, Gregory JC, Org E, Fu X, Li L, DiDonato JA, Lusis AJ, Brown JM, Hazen SL. Flavin monooxygenase 3, the host hepatic enzyme in the metaorganismal trimethylamine N-oxide-generating pathway, modulates platelet responsiveness and thrombosis risk. J Thromb Haemost. 2018;16(9):1857–72.
160. Gimbrone MA Jr, Garcia-Cardena G. Endothelial cell dysfunction and the pathobiology of atherosclerosis. Circ Res. 2016;118(4):620–36.

161. Witkowski M, Landmesser U, Rauch U. Tissue factor as a link between inflammation and coagulation. Trends Cardiovasc Med. 2016;26(4):297–303.

162. Cheng X, Qiu X, Liu Y, Yuan C, Yang X. Trimethylamine-N-oxide promotes tissue factor expression and activity in vascular endothelial cells: a new link between trimethylamine N-oxide and atherosclerotic thrombosis. Thromb Res. 2019;177:110–6.

163. Seldin MM, Meng Y, Qi H, Zhu W, Wang Z, Hazen SL, Lusis AJ, Shih DM. Trimethylamine N-oxide promotes vascular inflammation through signaling of mitogen-activated protein kinase and nuclear factor-κB. J Am Heart Assoc. 2016;5(2):e002767.

164. Ke Y, Li D, Zhao M, Liu C, Liu J, Zeng A, Shi X, Cheng S, Pan B, Zheng L, Hong H. Gut flora-dependent metabolite trimethylamine-N-oxide accelerates endothelial cell senescence and vascular aging through oxidative stress. Free Radic Biol Med. 2018;116:88–100.

165. Chen ML, Zhu XH, Ran L, Lang HD, Yi L, Mi MT. Trimethylamine-N-oxide induces vascular inflammation by activating the NLRP3 inflammasome through the SIRT3-SOD2-mtROS signaling pathway. J Am Heart Assoc. 2017b;6(9):e006347.

166. Lamkanfi M, Dixit VM. Inflammasomes and their roles in health and disease. Annu Rev Cell Dev Biol. 2012;28:137–61.

167. Rathinam VA, Fitzgerald KA. Inflammasome complexes: emerging mechanisms and effector functions. Cell. 2016;165(4):792–800.

168. Boini KM, Hussain T, Li PL, Koka S. Trimethylamine-N-oxide instigates NLRP3 inflammasome activation and endothelial dysfunction. Cell Physiol Biochem. 2017;44(1):152–62.

169. Sun X, Jiao X, Ma Y, Liu Y, Zhang L, He Y, Chen Y. Trimethylamine N-oxide induces inflammation and endothelial dysfunction in human umbilical vein endothelial cells via activating ROS-TXNIP-NLRP3 inflammasome. Biochem Biophys Res Commun. 2016;481(1–2):63–70.

170. Nemet I, Saha PP, Gupta N, Zhu W, Romano KA, Skye SM, Cajka T, Mohan ML, Li L, Wu Y, Funabashi M, Ramer-Tait AE, Naga Prasad SV, Fiehn O, Rey FE, Tang WHW, Fischbach MA, DiDonato JA, Hazen SL. A cardiovascular disease-linked gut microbial metabolite acts via adrenergic receptors. Cell. 2020;180(5):862–77.

171. Offermanns S. Activation of platelet function through G protein-coupled receptors. Circ Res. 2006;99(12):1293–304.

172. Wang J, Gareri C, Rockman HA. G-protein-coupled receptors in heart disease. Circ Res. 2018;123(6):716–35.

173. Dou L, Bertrand E, Cerini C, Faure V, Sampol J, Vanholder R, Berland Y, Brunet P. The uremic solutes p-cresol and indoxyl sulfate inhibit endothelial proliferation and wound repair. Kidney Int. 2004;65(2):442–51.

174. Wheeler DC. Cardiovascular disease in patients with chronic renal failure. Lancet. 1996;348(9043):1673–4.

175. Meijers BK, Claes K, Bammens B, de Loor H, Viaene L, Verbeke K, Kuypers D, Vanrenterghem Y, Evenepoel P. p-cresol and cardiovascular risk in mild- to-moderate kidney disease. Clin J Am Soc Nephrol. 2010a;5(7):1182–9.

176. Poesen R, Viaene L, Verbeke K, Augustijns P, Bammens B, Claes K, Kuypers D, Evenepoel P, Meijers B. Cardiovascular diseases relates to intestinal uptake of p-cresol in patients with chronic kidney disease. BMC Nephrol. 2014;15:87.

177. Meijers BKI, Bammens B, De Moore B, Verbeke K, Vanrenterghem Y, Evenepoel P. Free p-cresol is associated with cardiovascular disease in hemodialysis patients. Kidney Int. 2008;73(10):1174–80.

178. Wu IW, Hsu KH, Hsu HJ, Lee CC, Sun CY, Tsai CJ, Wu MS. Serum free p-cresyl sulfate levels predict cardiovascular and all-cause mortality in the elderly hemodialysis patients: a prospective cohort study. Nephrol Dial Transplant. 2012;27(3):1169–75.

179. Melamed ML, Plantinga L, Shafi T, Parekh R, Meyer TW, Hostetter TH, Coresh J, Powe NR. Retained organic solutes, patient characteristics and all-cause and cardiovascular

mortality in hemodialysis: results from the retained organic solutes and clinical outcomes (ROSCO) investigators. BMC Nephrol. 2013;14:134.

180. Shafi T, Sirich TL, Meyer TW, Hostetter TH, Plummer NS, Hwang S, Melamed ML, Banerjee T, Coresh J, Powe NR. Results of the HEMO study suggest that p-cresol sulfate and indoxyl sulfate are not associated with cardiovascular outcomes. Kidney Int. 2017;92(6): 1484–92.

181. Ross R. Atherosclerosis: an inflammatory disease. N Engl J Med. 1999;340(2):115–26.

182. Kari JA, Doland AE, Vallance DT, Bruckdorfer KR, Leone A, Mullen MJ, Bunce T, Dorado B, Deanfield JE, Rees L. Physiology and biochemistry of endothelial function in children with chronic kidney failure. Kidney Int. 1997;52(2):468–72.

183. Lee TY, Noria S, Lee J, Gotlieb AI. Endothelial integrity and repair. Adv Exp Med Biol. 2001;498:65–74.

184. Chang MC, Chang HH, Chan CP, Yeung SY, Hsien HC, Lin BR, Yeh CY, Tseng WY, Tseng SK, Jeng JH. p-cresol affects reactive oxygen species generation, cell cycle arrest, cytotoxicity and inflammation/atherosclerosis-related modulators production in endothelial cells and mononuclear cells. PLoS One. 2014;9:e114446.

185. Li L, Li J, Li X, Yuan FH. Protein-bound p-cresol inhibits human umbilical vein endothelial cell proliferation by inducing cell cycle arrest at G(0)/G(1). Am J Transl Res. 2017a;9(4): 2013–23.

186. Cerini C, Dou L, Anfosso F, Sabatier F, Moal V, Glorieux G, De Smet R, Vanholder R, Dignat-George F, Sampol J, Berland Y, Brunet P. P-cresol, a uremic retention solute, alters the endothelial barrier function in vitro. Thromb Haemost. 2004;92(1):140–50.

187. Zhu JZ, Zhang J, Yang K, Du R, Jing YJ, Lu L, Zhang RY. p-cresol, but not p-cresylsulphate, disrupts endothelial progenitor cell function in vitro. Nephrol Dial Transplant. 2012;27(12): 4323–30.

188. Ying Y, Yang K, Liu Y, Chen QJ, Shen WF, Lu L, Zhang RY. A uremic solute, p-cresol, inhibits the proliferation of endothelial progenitor cells via the p38 pathway. Circ J. 2011;75 (9):2252–9.

189. Faure V, Dou L, Sabatier F, Cerini C, Sampol J, Berland Y, Brunet P, Dignat-George F. Elevation of circulating endothelial microparticles in patients with chronic renal failure. J Thromb Haemost. 2006;4(3):566–73.

190. Deng F, Wang S, Zhang L. Endothelial microparticles act as novel diagnostic and therapeutic biomarkers of circulatory hypoxia-related diseases: a literature review. J Cell Mol Med. 2017;21(9):1698–710.

191. Ridger VC, Boulanger CM, Angelillo-Scherrer A, Badion L, Blanc-Brude O, Bochaton-Piallat ML, Boilard E, Buzas EI, Caporali A, Dignat-George F, Evans PC, Lacroix R, Lutgens E, Ketelhuth DFJ, Nieuwland R, Toti F, Tunon J, Weber C, Hoefer IE. Microvesicles in vascular homeostasis and diseases. Position paper of the European Society of Cardiology (ESC) working group on atherosclerosis and vascular biology. Thromb Haemost. 2017;117(7): 1296–316.

192. Guerrero F, Carmona A, Obrero T, Jimenez MJ, Soriano S, Moreno JA, Martin-Malo A, Aljama P. Role of endothelial microvesicles released by p-cresol on endothelial dysfunction. Sci Rep. 2020;10(1):10657.

193. Amabile N, Guérin AP, Leroyer A, Mallat Z, Nguyen C, Boddaert J, London GM, Tedgui A, Boulanger CM. Circulating endothelial microparticles are associated with vascular dysfunction in patients with end-stage renal failure. J Am Soc Nephrol. 2005;16(11):3381–8.

194. Garcia-Jerez A, Luengo A, Carracedo J, Ramirez-Chamond R, Rodriguez-Puyol M, Calleros L. Effects of uremia on endothelial cell damage is mediated by the integrin linked kinase pathway. J Physiol. 2015;593(3):601–18.

195. Cho HJ, Youn SW, Cheon SI, Kim TY, Hur J, Zhang SY, Lee SP, Park KW, Lee MM, Choi YS, Park YB, Kim HS. Regulation of endothelial cell and endothelial progenitor cell survival and vasculogenesis by integrin-linked kinase. Arterioscler Thromb Vasc Biol. 2005;25(6): 1154–60.

196. Friedrich EB, Liu E, Sinha S, Cook S, Milstone DS, MacRae CA, Mariotti M, Kuhlencordt PJ, Force T, Rosenzweig A, St-Arnaud R, Dedhar S, Gerszten RE. Integrin-linked kinase regulates endothelial cell survival and vascular development. Mol Cell Biol. 2004;24(18): 8134–44.

197. Kaneko Y, Kitazato K, Basaki Y. Integrin-linked kinase regulates vascular morphogenesis induced by vascular endothelial growth factor. J Cell Sci. 2004;117(3):407–15.

198. Guo Y, Pu WT. Cardiomyocyte maturation: new phase in development. Circ Res. 2020;126 (8):1086–106.

199. Peng YS, Ding HC, Lin YT, Syu JP, Chen Y, Wang SM. Uremic toxin p-cresol induces disassembly of gap junctions of cardiomyocytes. Toxicology. 2012;302(1):11–7.

200. Peng YS, Lin YT, Wang SD, Hung KY, Chen Y, Wang SM. p-cresol induces disruption of cardiomyocyte adherens junctions. Toxicology. 2013;306:176–84.

201. Lano G, Burtey S, Sallee M. Indoxyl sulfate, a uremic endotheliotoxin. Toxins (Basel). 2020;12(4):229.

202. Belghasem M, Roth D, Richards S, Napolene MA, Walker J, Yin W, Arinze N, Lyle C, Spencer C, Francis JM, Thompson C, Andry C, Whelan SA, Lee N, Ravid K, Chitalia VC. Metabolites in a mouse cancer model enhance venous thrombogenicity through the aryl hydrocarbon receptor-tissue factor axis. Blood. 2019;134(26):2399–413.

203. Karbowska M, Kaminski TW, Marcinczyk N, Misztal T, Rusak T, Smyk L, Pawlak D. The uremic toxin indoxyl sulfate accelerates thrombotic response after vascular injury in animal models. Toxins (Basel). 2017;9(7):229.

204. Karbowska M, Kaminski TW, Znorko B, Domaniewski T, Misztal T, Rusak T, Pryczynicz A, Guzinska-Ustymowicz K, Pawlak K, Pawlak D. Indoxyl sulfate promotes arterial thrombosis in rat model via increased levels of complex TF/VII, PAI-1, platelet activation as well as decreased content of SIRT1 and SIRT3. Front Physiol. 2018;9:1623.

205. Dou L, Jourde-Chiche N, Faure V, Cerini C, Berland Y, Dignat-George F, Brunet P. The uremic solute indoxyl sulfate induces oxidative stress in endothelial cells. J Thromb Haemost. 2007;5(6):1302–8.

206. Stinghen AE, Chillon JM, Massy ZA, Boullier A. Differential effects of indoxyl sulfate and inorganic phosphate in a murine cerebral endothelial cell line (bEnd.3). Toxins (Basel). 2014;6 (6):1742–60.

207. Tumur Z, Shimizu H, Enomoto A, Miyazaki H, Niwa T. Indoxyl sulfate upregulates expression of ICAM-1 and MCP-1 by oxidative stress-induced NF-kappaB activation. Am J Nephrol. 2010;31(5):435–41.

208. Yu M, Kim YJ, Kang DH. Indoxyl sulfate-induced endothelial dysfunction in patients with chronic kidney disease via an induction of oxidative stress. Clin J Am Soc Nephrol. 2011;6(1): 30–9.

209. Kaminski TW, Pawlak K, Karbowska M, Mysliwiec M, Pawlak D. Indoxyl sulfate: the uremic toxin linking hemostatic system disturbances with the prevalence of cardiovascular disease in patients with chronic kidney disease. BMC Nephrol. 2017;18(1):35

210. Hung SC, Kuo KL, Wu CC, Tarng DC. Indoxyl sulfate: a novel cardiovascular risk factor in chronic kidney disease. J Am Heart Assoc. 2017;6(2):e005022.

211. Lucas S, Omata Y, Hofmann J, Böttcher M, Iljazovic A, Sarter K, Albrecht O, Schulz O, Krishnacoumar B, Krönke G, Herrmann M, Mougiakakos D, Strowig T, Schett G, Zaiss MM. Short-chain fatty acids regulate systemic bone mass and protect from pathological bone loss. Nat Commun. 2018;9(1):55.

212. Blais A, Rochefort GY, Moreau M, Calvez J, Wu X, Matsumoto H, Blachier F. Monosodium glutamate supplementation improves bone status in mice under moderate protein restriction. JBMR Plus. 2019;3(10):e10224.

213. Lee WC, Guntur AR, Long F, Rosen CJ. Energy metabolism of the osteoblast: implications for osteoporosis. Endocr Rev. 2017;38(3):255–66.

214. Udagawa N, Koide M, Nakamura M, Nakamichi Y, Yamashita T, Uehara S, Kobayashi Y, Furuya Y, Yasuda H, Fukuda C, Tsuda E. Osteoblast differentiation by RANKL and OPG signaling pathways. J Bone Miner Metab. 2021;39(1):19–26.
215. Chalvon-Demersay T, Blachier F, Tomé D, Blais A. Animal models for the study of the relationship between diet and obesity: a focus on dietary protein and estrogen deficiency. Front Nutr. 2017;4:5.
216. Karlamangla AS, Burnett-Bowie SM, Crandall CJ. Bone health during the menopause transition and beyond. Obstet Gynecol Clin N Am. 2018;45(4):695–708.
217. Yan J, Herzog JW, Tsang K, Brennan CA, Bower MA, Garrett WS, Sartor BR, Aliprantis AO, Charles JF. Gut microbiota induce IGF-1 and promote bone formation and growth. Proc Natl Acad Sci U S A. 2016;113(47):E7554–63.
218. Fulzele K, Clemens TL. Novel functions for insulin in bone. Bone. 2012;50(2):452–6.
219. Yakar S, Courtland HW, Clemmons D. IGF-1 and bone: new discoveries from mouse models. J Bone Miner Res. 2010;25(12):2543–52.
220. Yakar S, Rosen CJ, Beamer WG, Ackert-Bicknell CL, Wu Y, Liu JL, Ooi GT, Setser J, Frystyk J, Boisclair YR, LeRoith D. Circulating levels of IGF-1 directly regulate bone growth and density. J Clin Invest. 2002;110(6):771–81.
221. Zhou T, Wang M, Ma H, Li X, Heianza Y, Qi L. Dietary fiber, genetic variations of gut microbiota-derived short-chain fatty acids, and bone health in UK biobank. J Clin Endocrinol Metab. 2021;106(1):201–10.
222. Diaz Heijtz R, Wang S, Anuar F, Qian Y, Björkholm B, Samuelsson A, Hibberd ML, Forssberg H, Petterson S. Normal gut microbiota modulates brain development and behavior. Proc Natl Acad Sci U S A. 2011;108(7):3047–52.
223. Ogbonnaya ES, Clarke G, Shanahan F, Dinan TG, Cryan JF, O'Leary OF. Adult hippocampal neurogenesis is regulated by the microbiome. Biol Psychiatry. 2015;78(4):e7–9.
224. Fung TC, Olson CA, Hsiao EY. Interactions between the microbiota, immune and nervous system in health and disease. Nat Neurosci. 2017;20(2):145–55.
225. De Palma G, Lynch MD, Lu J, Dang VT, Deng Y, Jury J, Umeh G, Miranda PM, Pigrau Pastor M, Sidani S, Pinto Sanchez MI, Philip V, McLean PG, Hagelsieb MG, Surette MG, Bergonzelli GE, Verdu EF, Britz-McKibbin P, Neufeld JD, Collins SM, Bercik P. Transplantation of fecal microbiota from patients with irritable bowel syndrome alters gut function and behavior in recipient mice. Sci Transl Med. 2017;9(379):eaaf6397.
226. Foster JA, McVey Neufeld KA. Gut-brain axis: how the microbiome influences anxiety and depression. Trends Neurosci. 2013;36(5):305–12.
227. Kelly JR, Borre Y, O'Brien C, Patterson E, El Aidy S, Deane J, Kennedy PJ, Beers S, Scott K, Moloney G, Hoban AE, Scott L, Fitzgerald P, Ross P, Stanton C, Clarke G, Cryan JF, Dinan TG. Transferring the blues: depression-associated gut microbiota induces neurobehavioural changes in the rat. J Psychiatry. 2016;82:109–18.
228. Stilling RM, Dinan TG, Cryan JF. Microbial genes, brain and behavior -epigenetic regulation of the gut-brain axis. Genes Brain Behav. 2014;13(1):69–86.
229. Zeng L, Zeng B, Wang H, Li B, Huo R, Zheng P, Zhang X, Du X, Liu M, Fang Z, Xu X, Zhou C, Chen J, Li W, Guo J, Wei H, Xie P. Microbiota modulates behavior and protein kinase C mediated cAMP response element-binding protein signaling. Sci Rep. 2016;6:29998.
230. Srandwitz P. Neurotransmitter modulation by the gut microbiota. Brain Res. 2018;1693 (B):128–33.
231. Rhee SH, Pothoulakis C, Mayer EA. Principles and clinical implications of the brain-gut-enteric microbiota axis. Nat Rev Gastroenterol Hepatol. 2009;6(5):306–14.
232. Sperandio V, Torres AG, Jarvis B, Nataro JP, Kaper JB. Bacteria-host communication: the language of hormones. Proc Natl Acad Sci U S A. 2003;100(15):8951–6.
233. Hughes DT, Sperandio V. Inter-kingdom signalling: communication between bacteria and their hosts. Nat Rev Microbiol. 2008;6(2):111–20.

234. Cogan TA, Thomas AO, Rees LEN, Taylor AH, Jepson MA, Williams PH, Ketley J, Humphrey TJ. Norepinephrine increases the pathogenic potential of Campylobacter jejuni. Gut. 2007;56(8):1060–5.
235. Lyte M. Microbial endocrinology in the microbiome-gut-brain axis: how bacterial production and utilization of neurochemicals influence behavior. PLoS Pathog. 2013;9(11):e1003726.
236. Strandwitz P, Kim KH, Terekhova D, Liu JK, Sharma A, Levering J, McDonald D, Dietrich D, Ramadhar TR, Lekbua A, Mroue N, Liston C, Stewart EJ, Dubin MJ, Zengler K, Knight R, Gilbert JA, Clardy J, Lewis K. GABA-modulating bacteria of the human gut microbiota. Nat Microbiol. 2019;4(3):396–403.
237. Lyte M. The role of microbial endocrinology in infectious disease. J Endocrinol. 1993;137(3): 343–5.
238. Cryan JF, O'Riordan KJ, Cowan CSM, Sandhu KV, Bastiaanssen TFS, Boehme M, Codagnone MG, Cussotto S, Fulling C, Goluveba AV, Guzzetta KE, Jaggar M, Long-Smith CM, Lyte JM, Martin JA, Molinero-Perez A, Moloney G, Morelli E, Morillas E, O'Connor R, Cruz-Pereira JS, Peterson VL, Rea K, Ritz NL, Sherwin E, Spichak S, Teichman EM, van de Wouw M, Ventura-Silva AP, Wallace-Fitzsimons SE, Hyland N, Clarke G, Dinan TG. The microbiota-gut-brain-axis. Physiol Rev. 2019;99(4):1877–2013.
239. Forsythe P, Kunze WA. Voices from within: gut microbes and the CNS. Cell Mol Life Sci. 2013;70(1):55–69.
240. Ornelas A, Dowdell AS, Scott Lee J, Colgan SP. Microbial metabolite regulation of epithelial cell-cell interactions and barrier function. Cell. 2022;11(6):944.
241. Obermeier B, Daneman R, Ransohoff RM. Development, maintenance and disruption of the blood-brain barrier. Nat Med. 2013;19(12):1584–96.
242. Pardridge WM. The blood-brain barrier: bottleneck in brain drug development. NeuroRx. 2005;2(1):3–14.
243. Najjar S, Pearlman DM, Devinsky O, Najjar A, Zagzag D. Neurovascular unit dysfunction with blood-brain barrier hyperpermeability contributes to major depressive disorder: a review of clinical and experimental evidence. J Neuroinflammation. 2013;10:142.
244. Cecchelli R, Berezowski V, Lundquist S, Culot M, Renftel M, Dehouck MP, Fenart L. Modelling of the blood-brain barrier in drug discovery and development. Nat Rev Drug Discov. 2007;6(8):650–61.
245. Hashimoto Y, Campbell M. Tight junction modulation at the blood-brain barrier: current and future perspectives. Biochim Biophys Acta. 2020;1862(9):183298.
246. Hawkins RA, O'Kane RL, Simpson IA, Vina JR. Structure of the blood-brain barrier and its role in the transport of amino acids. J Nutr. 2006;136(1S):218S–26S.
247. Ohtsuki S, Terasaki T. Contribution of carrier-mediated transport systems to the blood-brain barrier as a supporting and protecting interface for the brain; importance for CNS drug discovery and development. Pharm Res. 2007;24(9):1745–58.
248. Patching SG. Glucose transporters at the blood-brain barrier: function, regulation and gateways for drug delivery. Mol Neurobiol. 2017;54(2):1046–77.
249. Banks WA. Characteristics of compounds that cross the blood-brain barrier. BMC Neurol. 2009;9(S1):S3.
250. Lochhead JJ, Yang J, Ronaldson PT, Davis TP. Structure, function, and regulation of the blood-brain barrier tight junction in central nervous system disorders. Front Physiol. 2020;11: 914.
251. Lockwood AH, Finn RD, Campbell JA, Richman TB. Factors that affect the uptake of ammonia by the brain: the blood-brain pH gradient. Brain Res. 1980;181(2):259–66.
252. Skowronska M, Albrecht J. Alterations of blood brain barrier function in hyperammonemia: an overview. Neurotox Res. 2012;21(2):236–44.
253. Kiecker C. The origins of circumventricular organs. J Anat. 2018;232(4):540–53.
254. Guerra MM, Gonzalez C, Caprile T, Jara M, Vio K, Munoz RI, Rodriguez S, Rodriguez EM. Understanding how the subcommissural organ and other periventricular secretory

structures contribute via the cerebrospinal fluid to neurogenesis. Front Cell Neurosci. 2015;9: 480.
255. Joly JS, Osorio J, Alunni A, Auger H, Kano S, Rétaux S. Windows of the brain: towards a developmental biology of circumventricular and other neurohemal organs. Semin Cell Dev Biol. 2007;18(4):512–24.
256. Kaur C, Ling EA. The circumventricular organs. Histol Histopathol. 2017;32(9):879–92.
257. Price CJ, Hoyda TD, Ferguson AV. The area postrema: a brain monitor and integrator of systemic autonomic state. Neuroscientist. 2008;14(2):182–94.
258. Miller AD, Leslie RA. The area postrema and vomiting. Front Neuroendocrinol. 1994;15(4): 301–20.
259. Mayer EA, Tillisch K, Gupta A. Gut/brain axis and the microbiota. J Clin Invest. 2015;125(3): 926–38.
260. Uzbay T. Germ-free animal experiments in the gut microbiota studies. Curr Opin Pharmacol. 2019;49:6–10.
261. Sudo N, Chida Y, Aiba Y, Sonoda J, Oyama N, Yu XN, Kubo C, Koga Y. Postnatal microbial colonization programs the hypothalamic-pituitary-adrenal system for stress response in mice. J Physiol. 2004;558(1):263–75.
262. Swartz TD, Duca FA, de Wouters T, Sakar Y, Covasa M. Up-regulation of intestinal type 1 taste receptor 3 and sodium glucose luminal trasporter-1 expression and increased sucrose intake in mice lacking gut microbiota. Br J Nutr. 2012;107(5):621–30.
263. Crumeyrolle-Arias M, Jaglin M, Bruneau A, Vancassel S, Cardona A, Daugé V, Naudon L, Rabot S. Absence of the gut microbiota enhances anxiety-like behavior and neuroendocrine response to acute stress in rats. Psychoneuroendocrinology. 2014;42:207–17.
264. Desbonnet L, Clarke G, Shanahan F, Dinan TG, Cryan JF. Microbiota is essential for social development in the mouse. Mol Psychiatry. 2014;19(2):146–8.
265. Hsiao EY, McBride SW, Hsien S, Sharon G, Hyde ER, McCue T, Codelli JA, Chow J, Reisman SE, Petrosino JF, Patterson PH, Mazmanian SK. Microbiota modulate behavioral and physiological abnormalities associated with neurodevelopmental disorders. Cell. 2013;155(7): 1451–63.
266. Neufeld KM, Kang N, Bienenstock J, Foster JA. Reduced anxiety-like behavior and central neurochemical change in germ-free mice. Neurogastroenterol Motil. 2011;23(3):255–264, e119.
267. Ngo DH, Vo TS. An updated review on pharmaceutical properties of gamma-aminobutyric acid. Molecules. 2019;24(15):2678.
268. Wagner S, Castel M, Gainer H, Yarom Y. GABA in the mammalian suprachiamastic nucleus and its role in diurnal rhythmicity. Nature. 1997;387(6633):598–603.
269. Benson C, Mifflin K, Kerr B, Jesudasan SJ, Dursun S, Baker G. Biogenic amines and the amino acids GABA and glutamate: relationships with pain and depression. Mod Trends Pharmacopsychiatry. 2015;30:67–79.
270. Nuss P. Anxiety disorders and GABA neurotransmission: a disturbance of modulation. Neuropsychiatr Dis Treat. 2015;11:165–75.
271. Hyland NP, Cryan JF. A gut feeling about GABA: focus on GABA(B) receptors. Front Pharmacol. 2010;1:124.
272. Karatzas KA, Brennan O, Heavin S, Morrissey J, O'Bryne CP. Intracellular accumulation of high levels of gamma-aminobutyrate by listeria monocytogenes 10403S in response to low pH: uncoupling of gamma-aminobutyrate synthesis from efflux in a chemically defined medium. Appl Environ Microbiol. 2010;76(11):3529–37.
273. Nacher A, Polache A, Moll-Navarro MJ, Pla-Delfina JM, Merino M. Intestinal absorption pathway of gamma-aminobutyric acid in small intestine. Biopharm Drug Dispos. 1994;15(5): 359–71.
274. van Berlo CL, de Jonge HR, van den Bogaard AE, van Eijk HM, Janssen MA, Soeters PB. Gamma-aminobutyric acid production in small and large intestine of normal and

germ-free Wistar rats. Influence of food intake and intestinal flora. Gastroenterology. 1987;93 (3):472–9.

275. Van Gelder NM, Elliott KA. Disposition of gamma-aminobutyric acid administered to mammals. J Neurochem. 1958;3(2):139–43.

276. Todd N, Zhang Y, Power C, Becerra L, Borsook D, Livingstone M, McDannold N. Modulation of brain function by targeted delivery of GABA through the disrupted blood-brain barrier. NeuroImage. 2019;189:267–75.

277. Aston-Jones G, Waterhouse B. Locus coeruleus: from global projection system to adaptive regulation of behavior. Brain Res. 2016;1645:75–8.

278. Saboory E, Ghasemi M, Mehranfard N. Norepinephrine, neurodevelopment and bahavior. Neurochem Int. 2020;135:104706.

279. Holland N, Robbins TW, Rowe JB. The role of noradrenaline in cognition and cognitive disorders. Brain. 2021;144(8):2243–56.

280. MacKenzie ET, McCullough J, O'Kean M, Pickard JD, Harper AM. Cerebral circulation and norepinephrine: relevance of the blood-brain barrier. Am J Phys. 1976;231(2):483–8.

281. Chiueh CC, Sun CL, Kopin IJ, Fredericks WR, Rapoport SI. Entry of (3H)norepinephrine, (125I)albumin and Evans blue from blood into brain following unilateral osmotic opening of the blood-brain barrier. Brain Res. 1978;145(2):291–301.

282. Klein MO, Battagello DS, Cardoso AR, Hauser DN, Bittencourt JC, Correa RG. Dopamine: functions, signaling, and association with neurological diseases. Cell Mol Neurobiol. 2019;39 (1):31–59.

283. Berke JD. What dopamine mean? Nat Neurosci. 2018;21(6):787–93.

284. Claus H, Decker H. Bacterial tyrosinases. Syst Appl Microbiol. 2006;29(1):3–14.

285. Qu M, Lin Q, Huang L, Fu Y, Wang L, He S, Fu Y, Yang S, Zhang Z, Zhang L, Sun X. Dopamine-loaded blood exosomes targeted to brain for better treatment of Parkinson's disease. J Control Release. 2018;287:156–66.

286. McCutcheon RA, Abi-Dargham A, Howes OD. Schizophrenia, dopamine and the striatum: from biology to symptoms. Trends Neurosci. 2019;42(3):205–20.

287. Baronio D, Gonchoroski T, Castro K, Zanatta G, Gottfried C, Riesgo R. Histaminergic system in brain disorders: lessons from the translational approach and future perspectives. Ann General Psychiatry. 2014;13(1):34.

288. Church MK. Allergy, histamine and antihistamines. Handb Exp Pharmacol. 2017;241:321–31.

289. Maintz L, Novak N. Histamine and histamine intolerance. Am J Clin Nutr. 2007;85(5): 1185–96.

290. Nomura H, Shimizume R, Ikegaya Y. Histamine: a key neuromodulator of memory consolidation and retrieval. Curr Top Behav Neurosci. 2022;59:329–53.

291. Neuhuber W, Wörl J. Monoamines in the enteric nervous system. Histochem Cell Biol. 2018;150(6):703–9.

292. Colombo FM, Cattaneo P, Confalonieri E, Bernardi C. Histamine food poisonings: a systematic review and meta-analysis. Crit Rev Food Sci Nutr. 2018;58(7):1131–51.

293. Domingos-Lopes MFP, Stanton C, Ross RP, Silva CCG. Histamine and cholesterol lowering abilities of lactic acid bacteria isolated from artisanal Pico cheese. J Appl Microbiol. 2020;29 (6):1428–40.

294. Alstadhaug KB. Histamine in migraine and brain. Headache. 2014;54(2):246–559.

295. Oleskin AV, Shenderov BA, Rogovsky VS. Role of neurochemicals in the interactions between the microbiota and the immune and the nervous system of the host organism. Probiotics Antimicrob Proteins. 2017;9(3):215–34.

296. El-Merahbi R, Löffler M, Mayer A, Sumara G. The roles of peripheral serotonin in metabolic homeostasis. FEBS Lett. 2015;589(15):1728–34.

297. O'Mahony SM, Clarke G, Borre YE, Dinan TG, Cryan JF. Serotonin, tryptophan metabolism and the brain-gut-microbiome axis. Bahav Brain Res. 2015;277:32–48.

298. Gershon MD. 5-hydroxytryptamin (serotonin) in the gastrointestinal tract. Curr Opin Endocrinol Diabetes Obes. 2013;20(1):14–21.

299. Bulat M, Supek Z. The penetration of 5-hydroxytryptamine through the blood-brain barrier. J Neurochem. 1967;14(3):265–71.
300. Winkler T, Sharma HS, Stalberg E, Olsson Y, Dey PK. Impairment of blood-brain barrier function by serotonine induces desynchronization of spontaneous cerebral cortical activity: experimental observations in the anaesthetized rat. Neuroscience. 1995;68(4):1097–104.
301. Raybould HE, Cooke HJ, Christofi FL. Sensory mechanisms: transmitters, modulators and reflexes. Neurogastroenterol Motil. 2004;16(S1):60–3.
302. O'Hara JR, Ho W, Linden DR, Mawe GM, Sharkey KA. Enteroendocrine cells and 5-HT availability are altered in mucosa of Guinea pigs with TNBS ileitis. Am J Phys. 2004;287(5): G998–G1007.
303. Gao K, Mu CL, Farzi A, Zhu WY. Tryptophan metabolism: a link between the gut microbiota and brain. Adv Nutr. 2020;11(3):709–23.
304. Williams BB, Van Benschoten AH, Cimermancic P, Donia MS, Zimmermann M, Taketani M, Ishihara A, Kashyap PC, Fraser JS, Fischbach MA. Discovery and characterization of gut microbiota decarboxylases that can produce the neurotransmitter tryptamine. Cell Host Microbe. 2014;16(4):495–503.
305. Jones RS. Tryptamine: a neuromodulator or neurotransmitter in mammalian brain? Prog Neurobiol. 1982;19(1–2):117–39.
306. Vitale AA, Pomilio AB, Canellas CO, Vitale MG, Putz EM, Ciprian-Ollivier JJ. In vivo long-term kinetics of radiolabeled n,n-dimethyltryptamine and tryptamine. J Nucl Med. 2011;52 (6):970–7.
307. Davila AM, Blachier F, Gotteland M, Andriamihaja M, Benetti PH, Sanz Y, Tomé D. Intestinal luminal nitrogen metabolism: role of the gut microbiota and consequences for the host. Pharmacol Res. 2013;68(1):95–107.
308. Oliphant K, Allen-Vercoe E. Macronutrient metabolism by the human gut microbiome: major fermentation by-products and their impact on host health. Microbiome. 2019;7(1):91.
309. Eklou-Lawson M, Bernard F, Neveux N, Chaumontet C, Bos C, Davila-Gay AM, Tomé D, Cynober L, Blachier F. Colonic luminal ammonia and portal blood L-glutamine and L-arginine concentrations: a possible link between colon mucosa and liver ureagenesis. Amino Acids. 2009;37(4):751–60.
310. Ashy AA, Salleh M, Ardawi M. Glucose, glutamine, and ketone-body metabolism in human enterocytes. Metabolism. 1988;37(6):602–9.
311. Walker V. Ammonia metabolism and hyperammonemic disorders. Av Clin Chem. 2014;67: 73–150.
312. Hazell AS, Butterworth RF. Hepatic encephalopathy: an update of pathophysiologic mechanisms. Proc Soc Exp Biol Med. 1999;222(2):99–112.
313. Frieg B, Görg B, Gohlke H, Häussinger D. Glutamine synthetase as a central element in hepatic glutamine and ammonia metabolism: novel aspects. Biol Chem. 2021;402(9): 1063–72.
314. Fiati Kenston SS, Song X, Li Z, Zhao J. Mechanistic insights, diagnosis, and treatment of ammonia-induced hepatic encephalopathy. J Gastroenterol Hepatol. 2019;34(1):31–9.
315. Lee GH. Hepatic encephalopathy in acute-on-chronic liver failure. Hepatol Int. 2015;9(4): 520–6.
316. Butterworth RF. Pathogenesis of hepatic encephalopathy and brain edema in acute liver failure. J Clin Exp Hepatol. 2015;5(S1):S96–S103.
317. Atterbury CE, Maddrey WC, Conn HO. Neomycin-sorbitol and lactulose in the treatment of acute portal-systemic encephalopathy. A controlled, double-blind clinical trial. Am J Dig Dis. 1978;23(5):398–406.
318. Conn HO, Leevy CM, Vlahcevic ZR, Rodgers JB, Maddrey WC, Seeff L, Levy LL. Comparison of lactulose and neomycin in the treatment of chronic portal-systemic encephalopathy. A double blind controlled trial. Gastroenterology. 1977;72(4P1):573–83.
319. Als-Nielsen B, Gluud LL, Gluud C. Non-absorbable disaccharides for hepatic encephalopathy: systematic review of randomized trials. BMJ. 2004;328(7447):1046.

320. Cordoba J, Lopez-Hellin J, Planas M, Sabin P, Sanpedro F, Castro F, Esteban R, Guardia J. Normal protein diet for episodic hepatic encephalopathy: results of a randomized study. J Hepatol. 2004;41(1):38–43.

321. Rose C, Michalak A, Rao KV, Quack G, Kircheis G, Butterworth RF. L-ornithine L-aspartate lowers plasma and cerebrospinal fluid ammonia and prevents brain edema in rats with acute liver failure. Hepatology. 1999;30(3):636–40.

322. Bai M, Yang Z, Qi X, Fan D, Han G. L-ornithine-L-aspartate for hepatic encephalopathy in patients with cirrhosis: a meta-analysis of randomized controlled trials. J Gastroenterol Hepatol. 2013;28(5):783–92.

323. Kircheis G, Nilius R, Held C, Berndt H, Buchner M, Görtelmeyer R, Hendricks R, Krüger B, Meister H, Otto HJ, Rink C, Rösch W, Stauch S. Therapeutic efficacy of L-ornithine-L-aspartate infusions in patients with cirrhosis and hepatic encephalopathy: results of a placebo-controlled, double-blind study. Hepatology. 1997;25(6):1351–60.

324. Shawcross D, Jalan R. Dispelling myths in the treatment of hepatic encephalopathy. Lancet. 2005;365(9457):431–3.

325. Sharma SR, Gonda X, Tarazi FI. Autism spectrum disorder: classification, diagnosis and therapy. Pharmacol Ther. 2018;190:91–104.

326. Varghese M, Keshay N, Jacot-Descombes S, Warda T, Wicinski B, Dickstein DL, Harony-Nicolas H, De Rubeis H, Drapeau E, Buxbaum JD, Hof PR. Autism spectrum disorder: neuropathology and animal models. Acta Neuropathol. 2017;134(4):537–66.

327. Masi A, DeMayo MM, Glozier N, Guastella AJ. An overview of autism spectrum disorder, heterogeneity and treatment options. Neurosci Bull. 2017;33(2):183–93.

328. Guo H, Wang T, Wu H, Long M, Coe BP, Li H, Xun G, Ou J, Chen B, Duan G, Bai T, Zhao N, Shen Y, Li Y, Wang Y, Zhang Y, Baker C, Liu Y, Pang N, Huang L, Han L, Jia X, Liu C, Ni H, Yang X, Xia L, Chen J, Shen L, Li Y, Zhao R, Zhao W, Peng J, Pan Q, Long Z, Su W, Tan J, Du X, Ke X, Yao M, Hu Z, Zou X, Zhao J, Bernier RA, Eichler EE, Xia K. Inherited and multiple de novo mutations in autism/developmental delay risk genes suggest a multifactorial model. Mol Autism. 2018;9:64.

329. Hallmayer J, Cleveland S, Torres A, Phillips J, Cohen B, Torigoe T, Miller J, Fedele A, Collins J, Smith K, Lotspeich L, Croen LA, Ozonoff S, Lajonchere C, Grether JK, Risch N. Genetic heritability and shared environmental factors among twin pairs with autism. Arch Gen Psychiatry. 2011;68(11):1095–102.

330. Kim JY, Son MJ, Son CY, Radua J, Eisenhut M, Gressier F, Koyanagi A, Carvalho AF, Stubbs B, Solmi M, Rais TB, Lee KH, Kronbichler A, Dragioti E, Shin JI, Fusar-Poli P. Environmental risk factors and biomarkers for autism spectrum disorder: an umbrella review of the evidence. Lancet Psychiatry. 2019;6:590–600.

331. Hertz-Picciotto I, Schmidt RJ, Krakowiak P. Understanding environmental contributions to autism: causal concepts and the state of science. Autism Res. 2018;11:554–86.

332. Wang L, Christophersen CT, Sorich MJ, Gerber JP, Angley MT, Conlon MA. Elevated fecal short-chain fatty acid and ammonia concentrations in children with autism spectrum disorder. Dig Dis Sci. 2012;57(8):2096–102.

333. Al-Owen M, Kaya N, Al-Shamrani H, Al-Bakheet A, Qari A, Al-Muaigl S, Ghaziuddin M. Autism spectrum disorder in a child with propionic acidemia. JIMD Rep. 2013;7:63–6.

334. Altieri L, Neri C, Sacco R, Curatolo P, Benvenuto A, Muratori F, Santocchi E, Bravaccio C, Lenti C, Saccani M, Rigardetto R, Gandione M, Urbani A, Persico AM. Urinary p-cresol is elevated in small children with severe autism spectrum disorder. Biomarkers. 2011;16(3):252–60.

335. De Angelis M, Piccolo M, Vannini L, Siragusa S, De Giacomo A, Serrazzanetti DI, Cristofori F, Guerzoni ME, Gobbetti M, Francavilla R. Fecal microbiota and metabolome of children with autism and pervasive developmental disorder not otherwise specified. PLoS One. 2013;8(10):e76993.

336. Gabriele S, Sacco R, Altieri L, Neri C, Urbani A, Bravaccio C, Riccio MP, Lovene MR, Bombace F, De Magistris L, Persico AM. Slow intestinal transit contributes to elevate urinary p-cresol level in Italian autistic children. Autism Res. 2016;9(7):752–9.
337. Gabriele S, Sacco R, Cerullo S, Neri C, Urbani A, Tripi G, Malvy J, Barthelemy C, Bonnet-Brihault F, Persico AM. Urinary p-cresol is elevated in young French children with autism spectrum disorder: a replication study. Biomarkers. 2014;19(6):463–70.
338. Kang DW, Ilhan ZE, Isern NG, Hoyt DW, Howsmon DP, Shaffer M, Lozupone CA, Hahn J, Adams JB, Krajmalnik-Brown R. Differences in fecal microbial metabolites and microbiota of children with autism spectrum disorders. Anaerobe. 2018;49:121–31.
339. Macfabe DF. Short-chain fatty acid fermentation products of the gut microbiome: implications in autism spectrum disorders. Microb Ecol Health Dis. 2012;23:1.
340. Shaw W. Increased urinary excretion of a 3-(3-hydroxyphenyl)-3-hydroxypropionic acid (HPHPA), an abnormal phenylalanine metabolite of Clostridia spp. In the gastrointestinal tract, in urine samples from patients with autism and schizophrenia. Nutr Neurosci. 2010;13 (3):135–43.
341. Sanctuary MR, Kain JN, Augkustsiri K, German JB. Dietary consideration in autism spectrum disorders: the potential role of protein digestion and microbial putrefaction in the gut-brain axis. Front Nutr. 2018;5:40.
342. Bermudez-Martin P, Becker JA, Caramello N, Fernandez SP, Costa-Campos R, Canaguier J, Barbosa S, Martinez-Gili L, Myridakis A, Dumas ME, Bruneau A, Cherbuy C, Langella P, Callebert J, Launay JM, Chabry J, Barik J, Le Merrer J, Glaichenhaus N, Davidovic L. The microbial metabolite p-cresol induces autistic-like behaviors in mice by remodeling the gut microbiota. Microbiome. 2021;9(1):157.
343. Careaga M, Schwartzer J, Ashwood P. Inflammatory profiles in the BTBR mouse: how relevant are they to autism spectrum disorders? Brain Behav Immun. 2015;43:11–6.
344. Silverman JL, Yang M, Lord C, Crawley JN. Behavioural phenotyping assays for mouse model of autism. Nat Rev Neurosci. 2010;11(7):490–502.
345. Pascucci T, Colamartino M, Fiori E, Sacco R, Coviello A, Ventura R, Puglisi-Allegras S, Turriziani L, Persico AM. p-cresol alters brain dopamine metabolism and exacerbates autism-like behaviors in the BTBR mouse. Brain Sci. 2020;10(4):233.
346. Persico AM, Napolioni V. Urinary p-cresol in autism spectrum disorder. Neurotoxicol Teratol. 2013;36:82–90.
347. Buie T, Campbell DB, Fuchs GJ 3rd, Furuta GT, Levy J, Vandewater J, Whitaker AH, Atkins D, Bauman DL, Beaudet AL, Carr EG, Gershon ED, Hyman SL, Jirapinyo P, Jyonouchi H, Kooros K, Kushak R, Levitt P, Levy SE, Lewis JD, Murray KF, Natowicz MR, Sabra A, Wershil BK, Weston SC, Zeltzer L, Winter H. Evaluation, diagnosis, and treatment of gastrointestinal disorders in individuals with ASDs: a consensus report. Pediatrics. 2010;125(S1):S1–S18.
348. McElhanon BO, McCracken C, Karpen S, Sharp WG. Gastrointestinal symptoms in autism spectrum disorders: a mata-analysis. Pediatrics. 2014;133(5):872–83.
349. D'Eufemia P, Celli M, Finocchiaro R, Pacifico L, Viozzi L, Zaccagnini M, Cardi E, Giardini O. Abnormal intestinal permeability in children with autism. Acta Pediatr. 1996;85(9):1076–9.
350. Calderon-Guzman D, Hernandez-Islas JL, Espitia Vasquez IR, Barragan-Mejia G, Hernandez-Garcia E, Del Angel DS, Juarez-Olguin H. Effects of toluene and cresols on Na+,K+-ATPase and serotonin in rat brain. Regul Toxicol Pharmacol. 2005;41(1):1–5.
351. Goodhart PJ, DeWolf WE Jr, Kruse LI. Mechanisms-based inactivation of dopamine beta-hydroxylase by p-cresol and related alkylphenols. Biochemistry. 1987;26(9):2576–83.
352. Tevzadze G, Zhuravliova E, Barbakadze T, Shanshiashvili L, Dzneladze D, Nanobashvili Z, Lordkipanidze T, Mikeladze D. Gut neurotoxin p-cresol induces differential expression of GLUN2B and GLUN2A subunits of the NMDA receptor in the hippocampus and nucleus accumbens in healthy and audiogenic seizure-prone rats. AIMS Neurosci. 2020;7(1):30–42.
353. Yamamoto H, Hagino Y, Kasai S, Ikeda K. Specific roles of NMDA receptor subunits in mental disorders. Curr Mol Med. 2015;15:193–205.

354. Zhou Q, Sheng M. NMDA receptors in nervous system diseases. Neuropharmacology. 2013;74:69–75.
355. Gacias M, Gaspari S, Santos PMG, Tamburini S, Andrade M, Zhang F, Shen N, Tolstikov V, Kiebish MA, Dupree JL, Zachariou V, Clemente JC, Casaccia P. Microbiota-driven transcriptional changes in prefrontal cortex override differences in social behavior. elife. 2016;5: e13442.
356. Bergles DE, Richardson WD. Oligodendrocyte development and plasticity. Cold Spring Harb Perspect Biol. 2015;8(2):e020453.
357. Tennoune N, Andriamihaja M, Blachier F. Production of indole and indole-related compounds by the intestinal microbiota and consequences for the host: the good, the bad, and the ugly. Microorganisms. 2022;10(5):930.
358. Jaglin M, Rhimi M, Philippe C, Pons N, Bruneau A, Goustard B, Dauge V, Maguin E, Naudon L, Rabot S. Indole, a signaling molecule produced by the gut microbiota, negatively impacts emotional behaviors in rats. Front Neurosci. 2018;12:216.
359. Mir HD, Milman A, Monnoye M, Douard V, Philippe C, Aubert A, Castanon N, Vancassel S, Guerineau NC, Naudon L, Rabot S. The gut microbiota metabolite indole increases emotional responses and adrenal medulla activity in chronically stressed male mice. Psychoneuroendocrinology. 2020;119:104750.
360. Riggio O, Mannaioni G, Ridola L, Angeloni S, Merli M, Carla V, Salvatori FM, Moroni F. Peripheral and splanchnic indole and oxindole levels in cirrhotic patients: a study on the pathophysiology of hepatic encephalopathy. Am J Gastroenterol. 2010;105(6):1374–81.
361. Philippe C, Szabo de Edelenyi F, Naudon L, Druesne-Pecollo N, Hercberg S, Kesse-Guyot E, Latino-Martel P, Galan P, Rabot S. Relation between mood and the host-microbiome co-metabolite 3-indoxylsulfate: results from the observational prospective NutriNet-Sante study. Microorganisms. 2021;9(4):716.
362. Rothhammer V, Mascanfroni ID, Bunse L, Takenaka MC, Kenison JE, Mayo L, Chao CC, Patel B, Yan R, Blain M, Alvarez JI, Kébir H, Anandasabapathy N, Izquierdo G, Jung S, Obholzer N, Pochet N, Clish CB, Prinz M, Prat A, Antel J, Quintana FJ. Type I interferons and microbial metabolites of tryptophan modulate astrocyte activity and central nervous system inflammation via the aryl hydrocarbon receptor. Nat Med. 2016;22(6):586–97.
363. Chyan YJ, Poeggeler B, Omar RA, Chain DG, Frangione B, Ghiso J, Pappolla MA. Potent neuroprotective properties against the Alzheimer beta-amyloid by an endogenous melatonin-related indole structure, indole-3-propionic acid. J Biol Chem. 1999;274(31):21937–42.
364. Medvedev A, Buneeva O, Glover V. Biological targets for isatin and its analogues: implications for therapy. Biologics. 2007;1(2):151–62.
365. Watkins P, Clow A, Glover V, Halket J, Przyborowska A, Sandler M. Isatin, regional distribution in rat brain and tissues. Neurochem Int. 1990;17(2):321–3.
366. Glover V, Halket JM, Watkins PJ, Clow A, Goodwin BL, Sandler M. Isatin: identity with the purified endogenous monoamine oxidase inhibitor tribulin. J Neurochem. 1988;51(2):656–9.
367. Medvedev A, Igosheva N, Crumeyrolle-Arias M, Glover V. Isatin: role in stress and anxiety. Stress. 2005;8(3):175–83.
368. Battacharya SK, Chakrabarti A, Sandler M, Glover V. Anxiolytic activity of intraventricularly administered atrial natriuretic peptide in the rat. Neuropsychopharmacology. 1996;15(2): 199–206.
369. Abel EL. Behavioral effects of isatin on open field activity and immobility in the forced swim test in rats. Physiol Behav. 1995;57(3):611–3.
370. Zhou Y, Zhao ZQ, Xie JX. Effects of isatin on rotational behavior and DA levels in caudate putamen in Parkinsonian rats. Brain Res. 2001;917(1):127–32.
371. Hamaue N, Minami N, Terado M, Hirafuji M, Endo T, Machida M, Hiroshige T, Ogata A, Tashiro K, Saito H, Parvez SH. Comparative study of the effects of isatin, an endogenous MAO-inhibitor, and selegiline on bradykinesia and dopamine levels in a rat model of Parkinson's disease induced by the Japanese encephalitis virus. Neurotoxicology. 2004;25 (1–2):205–13.

372. Yuwiler A. The effect of isatin (tribulin) on metabolism of indoles in the rat brain and pineal: in vitro and in vivo studies. Neurochem Res. 1990;15(1):95–100.
373. Kolla NJ, Bortolato M. The role of monoamine oxidase a in the neurobiology of aggressive, antisocial, and violent bahavior: a tale of mice and men. Prog Neurobiol. 2020;194:101875.
374. Tan YY, Jenner P, Chen SD. Monoamine oxidase-B inhibitors for the treatment of Parkinson's disease: past, present, and future. J Parkinsons Dis. 2022;12(2):477–93.
375. Carpenedo R, Mannaioni G, Moroni F. Oxindole, a sedative tryptophan metabolite, accumulates in blood and brain of rats with acute hepatic failure. J Neurochem. 1998;70(5): 1998–2003.
376. Dong F, Hao F, Murray IA, Smith PB, Koo I, Tindall AM, Kris-Etherton PM, Gowda K, Amin SG, Patterson AD, Perdew GH. Intestinal microbiota-derived tryptophan metabolites are predictive of Ah receptor activity. Gut Microbes. 2020;12(1):1–24.
377. Mannaioni G, Carpenedo R, Pugliese AM, Corradetti R, Moroni F. Electrophysiological studies on oxindole, a neurodepressant tryptophan metabolite. Br J Pharmacol. 1998;125(8): 1751–60.
378. Wang G, Korfmacher WA. Development of a biomarker assay for 3-indoxyl sulfate in mouse plasma and brain by liquid chromatography/tandem mass spectrometry. Rapid Commun Mass Spectrom. 2009;23(13):2061–9.
379. Chu C, Murdock MH, Jing D, Won TH, Chung H, Kressel AM, Tsaava T, Addorisio ME, Putzel GG, Zhou L, Bessman NJ, Yang R, Moriyama S, Parkhurst CN, Li A, Meyer HC, Teng F, Chavan SS, Tracey KJ, Regev A, Schroeder FC, Lee FS, Liston C, Artis D. The microbiota regulate neuronal function and fear extinction learning. Nature. 2019;574(7779): 543–8.
380. Chen JJ, Zhou CJ, Zheng P, Cheng K, Wang HY, Li J, Zeng L, Xie P. Differential urinary metabolites related with the severity of major depressive disorder. Behav Brain Res. 2017a;332:280–7.
381. Brydges CR, Fiehn O, Mayberg HS, Schreiber H, Dehkordi SM, Bhattacharyya S, Cha J, Choi KS, Craighead WE, Krishnan RR, Rush AJ, Dunlop BW, Kaddurah-Daouk R, Mood Disorders Precision Medicine Consortium. Indoxyl sulfate, a gut microbiome-derived uremic toxin, is associated with psychic anxiety and its functional magnetic resonance imaging-based neurologic signature? Sci Rep. 2021;11(1):21011.
382. Alvarez JI, Dodelet-Devillers A, Kebir H, Ifergan I, Fabre PJ, Terouz S, Sabbagh M, Wosik K, Bourbonnière L, Bernard M, van Horssen J, de Vries HE, Charron F, Prat A. The hedgehog pathway promotes blood-brain barrier integrity and CNS immune quiescence. Science. 2011;334(6063):1727–31.
383. Karbowska M, Hermanowicz JM, Tankiewicz-Kwedlo A, Kalaska B, Kaminski TW, Nosek K, Wisniewska RJ, Pawlak D. Neurobehavioral effects of uremic toxin-indoxyl sulfate in the rat model. Sci Rep. 2020;10(1):9483.
384. Adesso S, Magnus T, Cuzzocrea S, Campolo M, Rissiek B, Paciello O, Autore G, Pinto A, Marzocco S. Indoxyl sulfate affects glial function increasing oxidative stress and neuroinflammation in chronic kidney disease: interactions between astrocytes and microglia. Front Pharmacol. 2017;8:370.
385. Lin YT, Wu PH, Tsai YC, Hsu YL, Wang HY, Kuo MC, Kuo PL, Hwang SJ. Indoxyl sulfate induces apoptosis through oxidative stress and mitogen-activated protein kinase signaling pathway inhibition in human astrocytes. J Clin Med. 2019;8(2):191.

Lessons to Be Learned from Clinical and Experimental Research on the Intestinal Microbiota Metabolic Activity for Health Benefit and Perspectives

6

Abstract

Although there is no doubt that numerous metabolites produced by the bacteria from alimentary compounds in the intestinal fluid are active, either positively or negatively, on the colon epithelium in terms of renewal, barrier function, energy metabolism, and physiology, we need additional information on the concentrations of bacterial metabolites in their active form in the vicinity of colonic epithelium and on the effects of mixtures of these compounds on this structure. Many active bacterial metabolites have been shown to be implicated in microbial communication, serving notably as signaling molecules implicated in bacterial physiology and growth. If bacteria can provide compounds such as some vitamins and amino acids to the host, the exchange of substrates is largely orientated toward supply by the host of substrates from alimentary origin to the intestinal microbiota for their metabolic and physiological needs. Several intermediary and metabolic end products produced by the intestinal bacteria can cross the intestinal epithelium, with a part of them being utilized and transformed in the epithelial cells during their transcellular journey. The compounds released in the portal blood can then be further transformed in the liver, giving rise to co-metabolites. Although several among bacterial metabolites and co-metabolites have been shown to be beneficial for some host tissues in different physiological and pathophysiological situations, the same compounds may prove to be deleterious on other tissues, leading to consider effects of these compound according to the host tissues and to the physiological or pathophysiological context. Additional studies are urgently needed concerning the metabolic and physiological relationships between mammals and their microbiota. These studies should include mechanistic studies regarding for instance the processes implicated in the transport of active bacterial metabolites from the luminal fluid to the bloodstream, the enzymatic processes that allow synthesis of co-metabolites in mammalian tissues, as well as entry of these compounds within the target tissues. Dietary intervention in volunteers in different situations is also crucially needed to better consider the microbial-host metabolic communication in both preventive and curative perspective.

As presented in this book the results of published experimental studies clearly indicate that numerous metabolites produced from compounds present in food by the intestinal microbiota and which are found in stools and in the luminal fluid of the large intestine are biologically active on the intestinal mucosa and within the intestinal mucosa on the intestinal epithelium. The effects recorded are related to the metabolism of epithelial cells and on the physiological functions of the different phenotypes present within the intestinal epithelium. Also, the studies of several bacterial metabolites have demonstrated effects on biological processes that are involved in the overall renewal of the epithelial layer and on the selective barrier function of this epithelium. Although it is obvious that the metabolism and physiological functions carried out by the intestinal mucosa in the different anatomical parts of the intestine as well as risks of inflammatory and neoplastic diseases in the intestine are not dependent exclusively on the luminal environment, this parameter appears of prime importance to be considered [1–6].

6.1 Modulating the Intestinal Microbiota Activity for Health Benefit. The Interest of Metabolomics Analysis of Biological Fluids

However, the experimental works performed are characterized by several limitations that need to be taken into consideration for future research. Firstly, the concentrations of the bacterial metabolites tested are not necessarily within the range of concentrations that are present in the vicinity of the intestinal epithelial cells. Since this parameter is not easy to measure, the bacterial concentrations used in experiments often refer to the concentration measured in feces which can be different when compared with the concentrations in the different segments of the intestine. This is an important aspect to be considered, as the effects of several bacterial metabolites on the intestinal mucosa, either beneficial or deleterious, depend on the concentrations tested. Secondly, the intestinal luminal fluid contains a complex mixture of bacterial metabolites that can exert additive, synergistic, or opposite effects on the metabolism and functions of cells present within the intestinal mucosa. In fact, in most experimental works, the bacterial metabolites are tested individually, thus making difficult extrapolation from experimental works to "real life situation."

With these limitations in mind, schematically, it is feasible to classify the different bacterial metabolites as beneficial and deleterious, keeping in mind that some bacterial metabolites have been shown to be beneficial at one given concentration, and deleterious at higher concentrations. Bacterial metabolites can be considered as beneficial toward the intestinal mucosa if they contribute to the maintenance of epithelial homeostasis, of selective barrier function, of normal endocrine activity, and if they maintain an appropriate immune response. Inversely, bacterial metabolites can be considered as deleterious when they exert at excessive concentrations adverse effects on the coordinated renewal of the epithelium, on the barrier function, on the normal physiological functions of the intestine, and if they participate in an inappropriate activation of the immune system. Some limited information indicates that for some bacterial metabolites with deleterious effects on the intestinal epithelium, detoxifying metabolism can avoid excessive accumulation of these compounds inside the intestinal epithelial cells, thus avoiding adverse effects on various cellular targets. The determination for deleterious bacterial metabolites of threshold concentration values above which risk of deleterious effects on intestinal mucosa is increased represents an important information to be determined, as well as the adaptive capacity of the intestinal epithelial cells toward increased concentrations of these compounds in the luminal fluid.

However, at this step of discussion, it appears important to also consider the available data from a strictly "microbial point of view." This point is related to the fact that for several bacterial metabolites, it has been shown that these compounds are implicated in communication between microbes of the same or from different species, either commensal (with no pathogenicity) or conversely pathogenic. Thus, it is plausible that intestinal bacteria produce several metabolites primarily to exchange signals between them and other microbial species, thus regulating their respective growth and biological activities. In addition, to maintain their growth, the bacteria within the intestinal microbiota are dependent on different substrates provided by the host, notably those that are available after food consumption. In the large intestine, where the population of bacteria is abundant, the intestinal microbiota uses dietary substrates that have not been digested (or not fully digested) by the host. While metabolizing these substrates, they produce a myriad of bacterial metabolites that are often metabolic end-products, and among which, numerous ones have been shown to interfere with the host's intestine metabolism and function. Then, it can be hypothesized that the production of bacterial metabolites from food ingested by the host is a way for bacteria to communicate between them, but also, importantly, this production is involved in communication between intestinal bacteria and their hosts.

It should not be ignored however that the communication between bacteria and the host is not unidirectional since a limited amount of nutrients and micronutrients (for instance indispensable amino acids and certain vitamins) can be provided by the bacteria to the host. However, from the available data, although it has not been quantified precisely, it appears that the supply of nutrients is mainly done from host to bacteria rather than from bacteria to host.

The communication between the intestinal microbiota and its host is obviously not restricted to the intestinal ecosystem. Indeed, several bacterial metabolites are absorbed from the luminal fluid to the bloodstream and are firstly recovered in the portal blood. Some of these bacterial metabolites are heavily or partly metabolized during the transepithelial journey through the absorptive intestinal epithelial cells, serving as energy substrates and/or as signaling molecules, or being converted by the host to co-metabolites. These co-metabolites are then, for some of them, released in the portal circulation. Thus, it appears that bacterial metabolites can be modified in the host's cells, pointing out metabolic interactions between the intestinal microbiota and its lodging host. Of major importance, both beneficial and deleterious effects of bacterial metabolites and co-metabolites on host's peripheral tissues have been reported within the gut–liver, gut–endocrine pancreas, gut–kidney, gut–cardiovascular system, gut–bone, and gut–brain axis in different physiological and pathophysiological situations.

The situation is complicated by the fact that some bacterial metabolites or co-metabolites are beneficial for some tissues regarding the physiological functions or recovery of normal functions, while being detrimental to the functions of other tissues and cells. As detailed in this book, the case of indole and indole-related compounds generated from intestinal bacterial activity is an illustrative example of such complexity [7]. Indeed, on one hand, indole and several indole-related compounds produced by the intestinal microbiota exert overall beneficial effects on the intestinal mucosa in different situations, notably in inflammatory situations, while on the other hand the indole-derived compound indoxyl sulfate, a co-metabolite produced by the liver, is well known to be one of the uremic toxins that can seriously aggravate chronic kidney disease when present in excess, notably through deleterious effects on kidney cells.

Then any intervention, either from dietary, microbial, and/or pharmacological origin, which aims at modulating for instance indole and indole-related compounds in the biological fluids, should crucially consider the dual effects of these compounds, either beneficial or deleterious, according to the host tissues considered, and to the physiological or pathophysiological context. For instance, given the deleterious effect of indoxyl sulfate on kidney cells in chronic kidney disease patients, it can be envisaged in that situation to either decrease the production of indole by the colonic microbiota, and/or to decrease indoxyl sulfate production by the liver, and/or to increase its removal from blood by an appropriate dialysis system.

Emerging data indicate that not only the composition of aliments consumed but also the structure of food, represents an element that has an impact on the composition and metabolic activity of the intestinal microbiota. For instance, in experimental works, the lipo-protein emulsion structure in the food, either liquid or gelled, that are chemically identical, but differs at the macro- and microstructure levels, display different effects after 3 weeks on the composition and metabolic activity of the intestinal microbiota [8, 9], thus asking for additional works on the effects of food processing and technology on the production of bacterial metabolites by the intestinal microbiota both in the small and large intestine luminal fluid.

Metabolic analysis of biological fluids may help in estimating the overall metabolic activity of the intestinal microbiota in different situations. The biological fluids under consideration may be blood, fecal, and/or urine samples. Among them, analysis of bacterial metabolites and co-metabolites in morning spot urine offers several advantages in addition to being less invasive and more practical than the utilization of blood or fecal samples. Firstly, it gives an estimation of the metabolites that have been produced overnight by the intestinal microbiota, absorbed through the intestinal epithelium, partly metabolized by the host, and finally excreted in the urine. Since numerous compounds among the bacterial metabolites are not produced by the host, analysis of these compounds (and related co-metabolites) should give a rather specific view of different aspects of the microbiota metabolic activity. Then in a given situation, taking into account the state of the art regarding the effects of the bacterial metabolites and co-metabolites on the splanchnic area organs (intestine and liver), and peripheral tissues, it is conceivable to use the metabolomic analysis of urine as a basis for defining dietary, pharmaceutical or biochemical interventions aiming at modulating the metabolic activity of the intestinal microbiota for the benefit of patients in both a preventive and curative perspective.

For instance, in case of chronic kidney disease, personalized measurement in the urine of bacterial uremic toxins would possibly lead to selective dietary recommendations with the objective to reduce such production. Using such strategy would however need to consider the fact that several bacterial metabolites have been shown to be implicated in communication between bacteria, including communication with some pathogens. Then, the efficacy of such future strategy would depend on the further accumulation of new data regarding the implication of bacterial metabolites and related co-metabolites on both communication in the microbial world and on communication between the microbial world and the host tissues in different pathological situations. As developed in this book, these pathological situations include intestinal mucosa inflammation, colorectal cancer, liver inflammation, chronic kidney disease, disease related to cardiovascular system dysfunction, bone loss, endocrine dysfunction, and certain brain dysfunctions. This is not science fiction as the still limited available experimental and the few clinical data clearly indicate a potential possibility to diminish the risk of several pathological processes by modulating the intestinal microbiota metabolic activity.

6.2   The Needs for Mechanistic Studies for Enlightening the Black Box Between Bacterial Metabolites and Co-Metabolites Production and Effects on Host Tissues

Obviously, new studies are mandatory to enlighten the black box between bacterial metabolites and co-metabolites and effects on the different host tissues. For instance, although it is known that some bacterial metabolites are absorbed and can enter specific regions of the brain, very few data are available on the physiological concentrations that reach the central nervous system, and on the putative transport mechanisms that are involved in such blood-to-brain transfer. Even the systems which allow the transfer of bacterial metabolites from the intestinal luminal fluid to the portal blood remain largely unidentified, as well as possible competition between different bacterial metabolites that enter intestinal epithelial cells through common transporters. In addition, the characteristics of the host enzymatic systems that allow to convert the bacterial metabolites into corresponding co-metabolites, as well as the enzymatic detoxifying systems (and the way by which they are induced) are still little documented.

Regarding the experimental models that will help to better understand the biological effects of bacterial metabolites and derived co-metabolites on the different tissues and cells in the body, special attention is worth to be paid to the utilization of organoids. The intestinal organoids can be prepared from human biopsies and from biopsies obtained from animals and maintained in culture or kept in deep cold for additional experiments. Notably, the use of colonic organoids should help a lot for determining the mechanisms by which the numerous bacterial metabolites, tested individually and in combination, act on the different phenotypes among intestinal epithelial cells [10]. This model of colonic organoids represents in addition an opportunity to reduce the number of animals used in pre-clinical experiments. However, even if the objective of limiting the number of animals for experiments must always be kept in mind by scientists, there are some situations where it will be still necessary to perform experiments with animal models to better understand the impact of bacterial metabolites and related co-metabolites on host tissues in different pathophysiological contexts.

Such information will then help in envisaging dietary and pharmacological interventions aiming at diminishing in different situations the production of deleterious bacterial metabolites and co-metabolites, while increasing the production of beneficial ones.

6.3   The Need for Dietary Intervention Studies in Volunteers

As said above, and in different parts of this book, only few dietary interventions have been made in volunteers to test the potential efficacy of modulation of the bacterial metabolic activity in different situations. In a healthy situation, the results of such dietary intervention could pave the way for future clinical studies aiming at defining, within a preventive strategy perspective, optimal personalized nutrition allowing to increase the beneficial over deleterious bacterial metabolite ratio considering the individual metabolic capacities and relative risk of different pathological processes. The same reasoning can be made regarding patients affected by diseases for which a component related to an inappropriate metabolic activity of the intestinal microbiota is involved in disease development.

Considering the heterogeneity of the intestinal microbiota composition and metabolic activity between individuals, such intervention trials would benefit from randomized and double-blind trials with a parallel design in which each participant will be its own control, thus allowing comparison of the intestinal microbiota individually before and after intervention.

Finally, the progress in the definition of optimal bacterial metabolite and co-metabolites profiles in biological fluids of healthy individuals, in a prevention perspective, and in patients in given pre-pathological and pathological situations in a curative perspective, and the definition of modified dietary conditions that will prove to be beneficial will depend on a constant flow of novel data originating from both experimental and clinical studies.

Key Points

- Numerous bacterial metabolites derived from alimentary compound are active on the intestinal epithelium in terms of renewal, barrier function, energy metabolism and physiology, but the effects of the mixture of compounds remain little studied.
- Many bacterial metabolites intervene in first place in microbial communication, as well as in bacterial metabolism and physiology.
- The exchange of substrates between hosts and their intestinal microbiota is largely orientated toward supply by the host to the intestinal microbiota for metabolic and physiological needs.
- Many bacterial metabolites can cross the intestinal epithelium and be transferred to peripheral tissues for further metabolism, giving rise to co-metabolites.
- In specific situations, several of these co-metabolites have been shown to be beneficial for some tissues, while being deleterious for others.
- Mechanistic studies are required to better enlighten the black box between the intestinal microbiota and host tissues.
- Dietary intervention studies are crucially needed to take into account the microbial-host metabolic interactions in both preventive and curative perspectives.

References

1. Blachier F, Andriamihaja M, Larraufie P, Ahn E, Lan A, Kim E. Production of hydrogen sulfide by the intestinal microbiota and epithelial cells and consequences for the colon and rectum mucosa. Am J Phys. 2021;320(2):G125–35.
2. Blachier F, Beaumont M, Andriamihaja M, Davila AM, Lan A, Grauso M, Armand L, Benamouzig R, Tomé D. Changes in the luminal environment of the colonic epithelial cells and physiopathological consequences. Am J Pathol. 2017;187(3):476–86.
3. Cheng Y, Ling Z, Li L. The intestinal microbiota and colorectal cancer. Front Immunol. 2020;11:615056.
4. Gasaly N, de Vos P, Hermoso MA. Impact of bacterial metabolites on gut barrier function and host immunity: a focus on bacterial metabolism and its relevance for intestinal inflammation. Front Immunol. 2021;12:658354.
5. Louis P, Hold GL, Flint HJ. The gut microbiota, bacterial metabolites and colorectal cancer. Nat Rev Microbiol. 2014;12(10):661–72.
6. Song M, Chan AT, Sun J. Influence of gut microbiome, diet, and environment on risk of colorectal cancer. Gastroenterology. 2020;158(2):322–40.
7. Tennoune N, Andriamihaja M, Blachier F. Production of indole and indole-related compounds by the intestinal microbiota and consequences for the host: the good, the bad, and the ugly. Microorganisms. 2022;10(5):930.
8. Beaumont M, Jaoui D, Douard V, Mat D, Koeth F, Goustard B, Mayeur C, Mondot S, Hovaghimian A, Le Feunteun S, Chaumontet C, Davila AM, Tomé D, Souchon I, Michon C, Fromentin G, Blachier F, Leclerc M. Structure of protein emulsion in food impacts intestinal microbiota, caecal luminal content composition and distal intestine characteristics in rats. Mol Nutr Food Res. 2017;61(10)
9. Oberli M, Douard V, Beaumont M, Jaoui D, Devime F, Laurent S, Chaumontet C, Mat F, Le Feunteun S, Michon C, Davila AM, Fromentin G, Tomé D, Souchon I, Leclerc M, Gaudichon C, Blachier F. Lipo-protein emulsion structure in the diet affects protein digestion

kinetics, intestinal mucosa parameters and microbiota composition. Mol Nutr Food Res. 2018;62(2)
10. Puschhof J, Pleguezuelos-Manzano C, Martinez-Silgado A, Akkerman N, Saftien A, Boot C, de Waal A, Beumer J, Dutta D, Heo I, Clevers H. Intestinal organoid cocultures with microbes. Nat Protoc. 2021;16(10):4633–49.

Printed in the United States
by Baker & Taylor Publisher Services